From Ethnomycology to Fungal Biotechnology

Exploiting Fungi from Natural Resources for Novel Products

From Ethnomycology to Fungal Biotechnology

Exploiting Fungi from Natural Resources for Novel Products

Edited by

Jagjit Singh
Oscar Faber Heritage Conservation
St. Albans, United Kingdom

and

K. R. Aneja
Kurukshetra University
Kurukshetra, India

Kluwer Academic / Plenum Publishers
New York, Boston, Dordrecht, London, Moscow

Library of Congress Cataloging in Publication Data

From ethnomycology to fungal biotechnology: exploiting fungi from natural resources for novel products / edited by Jagjit Singh and K. R. Aneja.
    p.   cm.
    Proceedings of the International Conference "From Ethnomycology to Fungal Biotechnology: Exploiting Fungi from Natural Resources for Novel Products," held December 15–16, 1997, in Simla, India—Verso t.p.
    Includes bibliographical references and index.
    ISBN 0-306-46059-9
    1. Mycology—Congresses. 2. Fungi—Biotechnology—Congresses. 3. Mycorrhizal fungi—Congresses. 4. Fungi as biological pest control agents—Congresses. 5. Fungi—Therapeutic use—Congresses. 6. Phytopathogenic fungi—Congresses. I. Singh, Jagjit, 1912–   . II. Aneja, K. R. III. International Conference on Ethnomycology to Fungal Biotechnology: Exploiting Fungi from Natural Resources for Novel Products (1997: Simla, India)
QK600.3.F76   1999   98-48208
579.5—dc21   CIP

Proceedings of the International Conference "From Ethnomycology to Fungal Biotechnology: Exploiting Fungi from Natural Resources for Novel Products," held December 15 – 16, 1997, in Simla, India

ISBN 0-306-46059-9

© 1999 Kluwer Academic / Plenum Publishers, New York
233 Spring Street, New York, N.Y. 10013

10 9 8 7 6 5 4 3 2 1

A C.I.P. record for this book is available from the Library of Congress.

To my love for mycology and mycologists

— JS

To my wife and children

— KRA

PREFACE

Fungi play a major role in the sustainability of the biosphere, and mycorrhizal fungi are essential for the growth of many of our woods and forests. The applications of fungi in agriculture, industry and biotechnology remain of paramount importance, as does their use as a source of drugs and to help clean up our environment.

This volume contains key papers from the conference 'From Ethnomycology to Fungal Biotechnology: Exploiting Fungi from Natural Resources for Novel Products'. This was the first international scientific conference covering the transfer of traditional remedies and processes in ethnomycology to modern fungal biotechnology. The conference was held at Simla, Himachal Pradesh, India from 15 to 16 December 1997.

The key subject areas addressed in the conference were the issues of exploring and exploiting fungal diversity for novel leads to new antibiotics, enzymes, medicines and a range of other leads for wood preservation, biological control, agricultural biotechnology and the uses of fungi in the food industry. The conference programme included key-note presentations followed by poster sessions and general discussion.

The book is broadly based, covering five main areas: Ethnomycology, Fungal Biotechnology, Biological Control, Mycorrhizal Fungi and Fungal Pests.

There is no doubt that in the past fungi have played a key role in ethnomycological remedies and that in the future they will continue to attract the interest of a wide range of disciplines ranging from environmental conservation, agriculture and the food industry to wood preservation and aerobiological studies.

We hope that this volume will serve as an inspiration and an aid to students, researchers and professionals involved in the fields of ethnomycology, fungal biotechnology, biological control and mycorrhizal fungi studies.

We are grateful to all the speakers and delegates who contributed to the conference and Professor Lakhanpal and Dr Lal Singh for the organisation of the domestic arrangements.

Dr Jagjit Singh and Dr K.R. Aneja

Acknowledgements

We would like to express our sincerest appreciation to Ms Sarah Case for her invaluable and painstaking assistance in the proof-reading, editing and preparation of the manuscripts.

We would also wish to thank our sponsors which include: the Karl Meyer Foundation, Switzerland; Dr John Palfreyman of University of Abertay Dundee, Scotland; and Jørgen Bech-Andersen of Hussvamp Laboratoriet, Denmark.

CONTENTS

ETHNOMYCOLOGY

FUNGAL BIOTECHNOLOGY

BIOLOGICAL CONTROL

MYCORRHIZAL FUNGI

FUNGAL PESTS

FROM ETHNOMYCOLOGY TO FUNGAL BIOTECHNOLOGY: A HISTORICAL PERSPECTIVE

Myank U. Charaya[1] and R.S. Mehrotra[2]

[1]Microbiology Laboratory
Department of Botany
M.M. Postgraduate College
Modinagar-201 204
India

[2]Retd. Professor
Department of Botany
Kurukshetra University
Kurukshetra-136 119
India

INTRODUCTION

Some 1.5 million species of fungi are believed to exist all over the world, of which only about 72,000 (less than 5%) have been described so far (Hawksworth, 1991; Hawksworth et al., 1995). These display a wide variety of morphological forms ranging from the microscopic unicellular yeasts to multicellular macroscopic mushrooms. The vegetative structure of a vast majority of the fungi consists of thin-walled, transparent, branched or unbranched hyphae. In a number of simple fungi (especially yeasts and chytrids), the vegetative structure consists of a single, microscopic cell, spherical, ellipsoidal, tubular or irregular in shape. However, the uniqueness of the fungi lies in (i) their ability to produce a surprisingly large variety of enzymes (conferring upon them the ability to colonise and degrade a huge variety of substrates); and (ii) their potential to synthesise an amazing variety of metabolites with the biological activity. Also, the hyphae present a large surface area through which the fungi can interchange substances with the environment - absorbing essential materials required for growth and development from the environment, and excreting the waste products.

The Role of the Fungi in our Society

Technologies based on the degradative or synthetic activities of the fungi have become an integral part of the human society and, hence, of our commercial set up as well. Current commercial products of the fungi include amino acids, antibiotics, alcoholic beverages (including distilled alcohol), fuel (ethanol, biogas), biopesticides, mycoherbicides, bread, cheeses, fermented foods, food (mushroom, etc.), single-celled protein, flavours, food colourants, preservatives, soy sauce, vitamins, organic acids and mycelial paper. Bioremediation, ensilage, biotransformation and many such processes involve the utilisation of the fungi. In the emerging 'age of biotechnology', the fungi are expected to provide a wider range of useful products and processes for human welfare under the banner of what is called 'fungal biotechnology'. Already, the applications of fungal activities dominate present-day biotechnology (Moss, 1990).

The Origins of Fungal Biotechnology

According to the Office of Technology Assessment of the United States Congress, biotechnology comprises "any technique that uses living organisms, or substances from these organisms, to make or modify a product, to improve plants or animals, or to develop microorganisms for specific use" (see Balasabramaniam, 1996; Subramanian, 1992).

If the term biotechnology is used in a wider context, it clearly implies that fungal biotechnology is not an exclusively modern practice; rather its roots may be traced back thousands of years - "ever since the first toast was proposed over a shell full of wine and the first loaf of leavened bread was baked" (Alexopoulos, 1962). In fact, some of the fungi have been utilised by humans for a variety of purposes since ancient times - as food, for preparing fermented drinks and leavened bread. Also, they have been put to medical uses and have been employed in rituals. Such practices were not confined to any particular region of the world, although different ethnic groups utilised different fungi in their own ways. Even today, primitive tribes and ethnic peoples utilise plants as well as fungi in their own ways. The study of fungi in folklore and rituals, from prehistoric times to the present day, comprises the science of 'ethnomycology'. The late Dr. R.G. Wasson of New York, his wife V.P. Wasson and R. Heim have made significant contributions to the science of ethnomycology (Wasson and Wasson, 1957; Heim et al., 1967).

Emergence of the "Science" of Fungal Biotechnology from the "Art" of the Biotechnologist

Quite obviously, the utilisation of fungi by societies in ancient times was the result of empirical experiences stretching through generations. Thus, what was practiced by them may be regarded as the 'art' of the biotechnologist. The ripening of cheese, cultivation of ergot for medicine, using fungal preparations as prophylactics or cures for many disorders, cultivation of edible mushrooms and truffles are some of the classical examples of this art. The present day 'science' of biotechnology is an offshoot of the art of the biotechnologist (Subramanian, 1992). The emergence and blossoming of this offshoot viz. the 'science' of fungal biotechnology has not been a simple, straightforward affair. It has taken centuries to overcome the impediments - centuries of careful and critical observations, invention of new tools and techniques, experimentation with great patience, and logical unbiased analysis of results obtained. Growth in our knowledge of the structure, physiology, biochemistry and genetics of the fungi coupled with our capability to manipulate them has led to the emergence of modern fungal biotechnology.

The Impact of Microscopy

Invention of the microscope in the seventeenth century was fundamental for the development of mycology. The first illustrations of microfungi - the sporangia of *Mucor* and the teleutospores of *Phragmidium mucronatum* - were provided by Robert Hooke (1665). In 1680, Leeuwenhoek also examined some fungi, his most significant mycological observation being that of *Saccharomyces cerevisiae* in fermented beer (see Chapman, 1931). In the years, decades and centuries that followed, the number of fungi observed and described increased exponentially. New species of fungi are being described currently at a rate of about 800 per year (Hawksworth et al., 1995).

Development of Techniques for Culturing the Fungi

Another technical advance that contributed greatly to our understanding of the fungi and the fungal processes was the development of pure culture techniques (Table 1). The history of mushroom cultivation bears testimony to this fact. The first efforts at culturing mushrooms are believed to have taken place in France during the lifetime of Louis XIV (1683-1715) (Hayes and Nair, 1975). The 'art' was soon practiced in England. Vivid descriptions of mushroom culture practices prevailing in those times had been provided by Abercrombie: the descriptions clearly show that mushroom growing was at that time an uncertain business, obviously due to non-availability of 'pure' spawn (Abercrombie, 1817). In 1890, however, Constantin developed pure culture of spawn (Constantin and Matruchot, 1894), and in 1905 Duggar was able to obtain pure spawn by culturing a piece of tissue from a mushroom cap.

These developments provided a much needed impetus to the mushroom industry - and not only to the mushroom industry but also to all the industries based on microbes including the fungi. It was only by using pure culture techniques that the microorganisms responsible for many infections, fermentations, nitrogen fixation etc. were isolated and identified: pure cultures form the backbone of fungal biotechnology.

Germ Theory of Fermentation

A number of practices which we now classify as the applications of fungal biotechnology have been in use since man's earliest days. Fermentation is one such process. Our ancestors, since the dawn of civilisation, have been producing wine, beer and bread by using fermentation, although of course they did not know that certain organisms were behind these fermentations. Though Leeuwenhoek had, in 1680, observed the budding cells of yeast in beer (Chapman, 1931), the involvement of living cells in alcoholic fermentation was only recognised in the year 1837, by three scientists independently (Cagniard-Latour, 1837; Kutzing, 1837; Schwann, 1837). However, it was Louis Pasteur who was eventually able to convince the scientific world that a living organism (yeast) is required for the chemical change that transforms sugar into alcohol, and that different kinds of organisms were responsible for different kinds of fermentations (Pasteur, 1860). The germ theory of fermentation was thus established and it became the basis for important industrial developments. Today, a number of industries involving fungal biotechnology may be referred to as fermentations (anaerobic, i.e. true, fermentations as well as oxidative fermentations).

Early Attempts at Industrial Exploitation of Fermentation

Strategies for the industrial exploitation of fungal fermentation evolved as a result of the studies of mould biochemistry of which Carl Wehmer (1858-1935) may be regarded as the founder; he was the first to demonstrate that the fungi were capable of producing organic molecules such as oxalic acid, citric acid and fumaric acid from sugars (Wehmer, 1891, 1893). He worked out a process for manufacturing citric acid from a strain of *Citromyces* (now classified as a species of *Penicillium*). A factory attempted to use the process for manufacturing citric acid in 1893 but the project was abandoned ten years later; some of the reasons being long fermentation times, high costs and contamination. Perhaps Wehmer was ahead of his time. An American, J.N. Currie (1917), found that at pH 2.5-3.5, *Aspergillus niger* produced citric acid from sugar in significant quantities. He worked out a suitable medium for citric acid production from sugar using *A. niger* and by 1923 the process was adopted commercially. Even today, virtually all commercial citric acid is manufactured by fermenting sugar using *A. niger*.

Table 1. Some early technical advances that facilitated the cultivation of fungi.

Scientist/ Organisation	Contribution with Reference
A. Micheli (1718)	Cultured fungi by inoculating the spores of *Aspergillus, Mucor, Tuber, Polyporus, Geaster, Lycoperdon*, etc.; Micheli (1729)
C. Vittadini (1842)	First to solidify media (with gelatin); Vittadini (1942)
O. Brefeld (1872)	(i) Extensive use of gelatin for solidifying media (ii) Dilution plate method for single spore culture; Brefeld (1872)
P. van Tieghem and G. le Monnier (1873)	Invented van Tieghem's cell for the microscopical examination of hanging drop cultures; van Tieghem and Le Monnier (1873)
W. Roberts (1874)	Use of cottonwool plugs to seal glass vessels containing sterile infusions; Roberts (1874)
J. Tyndall (1877)	Discovered "discontinous sterilisation" (tyndallisation); Tyndall (1881)
R. Koch (1883)	Devised pour plate method; Ainsworth (1976)
F. Hesse (1883)	Introduction of agar agar as solidifying agent; Carpenter (1967)
Parisian engineering firm Wiesnegg (1884)	Commercial availability of autoclave; Bulloch (1938)
R.J. Petri (1887)	Invented Petri dish; Petri (1887)

Discovery of Penicillin

The studies in mould biochemistry led to the recognition of a number of other fungal products, and industries based on these came into existence. All this happened at quite a rapid pace during the second half of the nineteenth century and the first half of the present century (Table 2). The discovery and development of Penicillin in particular provided great thrust to research activities concerning fungi, especially from an industrial point of view. Extensive details of these are given by Smith and Barry (1975), Bu'Lock et al. (1982), Smith et al. (1983), Elander and Lowe (1991), Letham (1992) and Vaidya (1995). Most of the world's Penicillin today is derived from *Penicillium chrysogenum* (not from *Penicillium notatum*). Although hundreds of fungi have been found to show antibiotic activity, very few have found wider application - cephalosporins (from *Cephalosporium acremonium*) and griseofulvin (from *Penicillium patulum*) being manufactured in large quantities.

Table 2. Early developments in the establishment of fungi-based industries.

Product	Reference	
Gallic acid	P. van Tieghem (1867)	Described the production of gallic acid from tannin by *Aspergillus niger*.
	A. Calmette (1902)	Patented the process of production of gallic acid using *A. niger* and put it on commercial basis.
Gluconic acid	M. Molliard (1922)	Discovered gluconic acid in the culture filtrates of *Sterigmatocystis nigra* (= *A. niger*).
	J.N. Currie, J.H. Kane and A. Finlay (1933)	Got patent for a fermentation process for the production of gluconic acid.
Itaconic acid	K. Kinoshita (1929)	First to characterise itaconic acid as a metabolic product of *Aspergillus itaconicus*.
	C.T. Calam, A.E. Oxford and H. Raistrick (1939)	Demonstrated that *Aspergillus terreus* produces itaconic acid.
	J.H. Kane, A.C. Finlay and P.F. Amann (1945)	Got patent for the production of itaconic acid by fermentation process.
Glycerol	L. Pasteur (1858)	Observed that small amounts of glycerol were formed during alcoholic fermentation of sugar by yeast (*S. cerevisiae*).
	W. Connstein and K. Ludecke (1919)	Published details of the German process which yielded high amounts of glycerol.
Enzymes	J. Takamine (1894) (See Takamine, 1914)	Took out patent for production of enzyme mixtures, mainly diastase.
	E.H. Le Mense, I. Gorman, J.M. van Lanen, A.F. Langlykke (1947)	Demonstrated the production of amylases from *Aspergillus niger*.
Antibiotics	A. Fleming (1929)	Discovered bacterial inhibitory properties of a metabolic product of *Penicillium notatum* and called it Penicillin.
	E.B. Chain, H.W. Florey, A.D. Gardner, N.G. Heatley, M.A. Jennings, J. Orrewing and A.G. Sanders (1940)	Isolated Penicillin in pure form.

Advent of Enzyme Technology

Another dimension to the fungal technology was added by the advent of enzyme technology. Edward Buchner (1897), a German chemist, showed that fermentation can occur in the presence of not only yeast cells, but in the presence of yeast extracts also. This was a turning point in the fungal technology because it led to the realisation that the participation of the entire organism or cell is not absolutely necessary for carrying out a given bioprocess. As more and more living processes (lactic acid fermentation, alcoholic fermentation, respiration etc.) were shown to be the results of sequential action of enzymes, the attempts at isolation, purification, production and commercial utilisation of enzymes began gathering momentum. The foundations for the deliberate use of fungal enzymes had already been laid down by Takamine in 1894 with the production of amylase preparation 'Taka diastase' from *Aspergillus oryzae* grown on wheat bran (Takamine, 1914). A number of fungal enzymes are now produced on an industrial scale which include glucose aero-hydrogenase, proteases, pectinases, amylases, lipolases, cellulases etc. (Lambert, 1983; Wainwright, 1992).

The fungal enzymes have carved out a niche of their own in various industrial processes like bread-making, malting, whey processing, sucrose conversion, starch conversion, fruit processing and cheese making. These are also used as supplements for pancreatic lipase, and for producing soap, lactose-free foods, soft-centred confectionery and so on. The development of enzyme immobilisation techniques helped in overcoming the limitations in the use of enzymes (limited availability, instability, high costs etc.). The first commercial application of immobilised enzyme technology was developed by Tanabe Seiyaku Co. of Japan in 1969 using immobilised L-amino acylase from *Aspergillus oryzae* (see Trevan, 1987).

Improvement of Useful Fungi

Quite often, the organisms isolated from the natural environment produce a desired metabolite in limited amounts, and the fungi are no exception. Industrial uses of fungi, on the other hand, demand high productivity. By optimising the culture conditions, it is possible to get higher yields from a given strain. However, the productivity is ultimately controlled by the genome and therefore scientists are constantly attempting to obtain fungal strains with better productivity. An important example of this is provided by penicillin from *Penicillium chrysogenum*, a better source. The initial attempts at procuring better resource fungi involved selection - testing various available strains, selecting the best and thus capitalising upon natural genetic variation. For example, Mehrotra and Krishna (1966) determined the best producer of amylase among 875 strains of *Aspergillus niger*, 14 strains of *A. oryzae* and 8 strains of *A. flavus.* Some other examples of using selection only for strain improvement are provided by the work of Keay (1971) on *Aspergillus* and *Rhizopus* for proteases, and that of Fukomoto and Schichiji (1971) on *Coriolus versicolor* for laccase.

Inducing Mutations in Fungi for Improving Yields

Mutations contribute to the genetic variability of organisms and each gene has its own characteristic mutational behaviour. Some genes undergo mutations more frequently than others, even in the same organism. Quite often, the frequency of natural mutation of a desired gene is very low resulting in much lower levels of variation. Muller (1927) demonstrated that mutations could be induced by an external agent. Subsequent discoveries of various physical and chemical mutagens made it possible to use a combination of mutation, screening and selection coupled with modification of the parasexual cycle for strain improvement (Queener and Lively, 1986) and these have yielded good results. For example, the original strain of *Penicillium chrysogenum* produced 100 units per ml of penicillin; by comparison, the recently produced strains by mutation, screening and selection have a titre of 50,000 units per ml. In fact, in the absence of sexuality, the parasexual cycle provides the basis for 'breeding' a number of industrially important fungi.

However, one of the major hurdles in the utilisation of the parasexual cycle for strain improvement has been the esstablishment of the heterokaryon. The problem was overcome with the development of the technique of protoplast fusion.

Improvement through Protoplast Fusion

The protoplast fusion initiates the parasexual cycle leading to diploidisation, mitotic recombination and segregation. Efforts began to establish protoplast fusion in a wide variety of industrial fungi. A number of these were successful; for example, several recombinant strains of *Cephalosporium acremonium* with increased cephalosporin formation, good sporulation and high growth rate were isolated after protoplast fusion (Hamlyn and Ball, 1979;

6

Ferenczy, 1981). Similarly, Chang et al. (1982) fused the protoplasts of two strains of *Penicillium chrysogenum* for the development of strains which could synthesise penicillin V and did not have undesirable morphology. An analysis of what can be achieved using this technique has been presented by Peberdy (1989).

Impact of Molecular Biology

The rise of molecular biology in recent decades has opened up unlimited possibilities of commercial exploitation of the fungi and the fungal genome (Upshall, 1986). The discovery of plasmids in *Saccharomyces cerevisiae* in 1970 proved to be a shot in the arm (Broach, 1982) since it facilitated transformation in yeasts; and the latter has virtually become the '*E. coli* of the eukaryotes' - the 2μ plasmid forming the basis for a number of specially engineered, sophisticated vectors.

Efficient transformation systems have already been reported in *Acremonium chrysogenum, Aspergillus nidulans, Aspergillus niger, Aspergillus awamori, Aspergillus terreus, Mucor circinalloides, Neurospora crassa, Penicillium chrysogenum, Podospora anserina* (Elander and Lowe, 1991). To facilitate gene isolation, and to build an ordered library as well as a database for cloned genes, cosmid libraries of *Aspergillus nidulans* and *Neurospora crassa* are now available (Turner, 1993).

An overview of the rapid progress achieved in recent years in the field of fungal biotechnology with the help of genetic engineering may be obtained from the following account:

(a) In *Penicillium chrysogenum* and *Cephalosporium acremonium* (the commercial ß-tactam antibiotic producers), transformations have been used extensively for (i) identifying cloned genes of antibiotic biosynthetic pathways, and (ii) improving antibiotic yield by inserting additional copies of such genes. Skatrud et al. (1989) were able to improve the yield of *C. acremonium* by inserting additional copies of the expandase/hydrxylase enzyme.

(b) Gutierrez et al. (1991) have been able to construct hybrid pathways for the synthesis of ß-lactam antibiotics by transferring genes for particular steps from *Penicillium chrysogenum* into *Cephalosporium acremonium*.

(c) The genes encoding a number of useful fungal enzymes have been isolated and attempts have been made to improve the yields by transformation (Penttila et al., 1991). Huge-Jensen et al. (1989) have been able to increase the yield of lipases by getting the gene of fungal lipase from *Humicola* expressed in *Aspergillus oryzae* using the α-amylase promoter of the latter.

(d) The fungi have now been established as potent host organisms for the production of heterologous proteins (Kalman et al., 1990). A genetically engineered yeast developed at Delta Biotechnology, U.K. can make haemoglobin (*The Good Yeast Guide,* National Centre for Biotechnology Education, Reading University, U.K., 1991). *Aspergillus awamori* has been developed as a host for the expression of bovine chymosin (Ward, 1990).   Given the remarkable permissiveness of the fungi for expressing foreign genes, their ability to grow on low cost media, and inexpensive harvesting of the mycelia, these hold great promise for the production of a number of useful proteins - mammalian or non-mammalian - including chymosin, tissue plasinogen activator and interferon (Covert and Cullen, 1992).

FROM ETHNOMYCOLOGY TO FUNGAL BIOTECHNOLOGY

The foregoing account presents a picture of what have been the major turning points in the field of fungal biotechnology. No attempt has been made to discuss the various fields of application of the fungal biotechnology which include mycoherbicides and mycopesticides, biological control systems, bioremediation, biological bleaching of kraft paper pulp, food fermentations, biofertilisers, waste management (including that of crop residues), biotransformation and so on. A number of these have been dealt with in other chapters.

The highly advanced form of fungal biotechnology we have at our command today has evolved over a long period of time from the 'art of the biotechnologist'. All the modern high technologies for utilising the fungi, the fungal products and their genomes began with their use by different human societies solely on the basis of their empirical experiences. Their scientific exploitation at the organism level, subcellular level and finally at genome level constitute different steps in the evolution of fungal biotechnology.

As discussed earlier, many primitive tribes and ethnic peoples in many parts of the world still use a number of fungi in different ways unknown to the developed world (Subramanian, 1992). Many benefits may be derived from the valuable information that rests with them. It must be our endeavour to document this information and subject it to scientific validation and utilisation in order to obtain a variety of useful products.

REFERENCES

Abercrombie, J., 1817, *Abercrombie's Practical Gardener, or Improved System of Horticulture*, 2nd ed. (revised by J. Mean), Cadell and Davies, London.

Ainsworth, G.C., 1976, *Introduction to the History of Mycology*, Cambridge University Press, Cambridge.

Alexopoulos, C.J., 1962, *Introductory Mycology*, 2nd ed., John Wiley & Sons, New York.

Balasubramanian, D., 1986, From cell biology to biotechnology, in: *Concepts in Biotechnology* (D. Balasubramanian, C.F.A. Bryce, K. Dharmalingam, J. Green, K. Jayaraman, eds.), pp. 1-5, Universities Press (India) Limited, Hyderbad, India.

Brefeld, O., 1872, *Untersuchungen aus dem Gesammtgebiete der Mykologie*, Heft I, Leipzig, Munster, Berlin.

Broach, J.R., 1982, The yeast plasmid 2μ circle, *Cell* 28: 203-204.

Buchner, E., 1897, Alkoholische Gährung ohne Hefzellen, *Ber. Deutsch. Chem. Ges.* 30: 117-124, 1110-1113.

Bulloch, W., 1938, *The History of Bacteriology*, Oxford University Press, Oxford.

Bu'Lock, J.D., Nisbet, L.J. and Winstanley, D.J., 1982, *Bioactive Microbial Products: Search and Discovery*, Academic Press, London.

Cagniard-Latour, C., 1837, Mémoire sur la fermentation vineuse, *C.R. Acad. Sci. Press* 4: 905-906.

Calam, C.T., Oxford, A.E. and Raistrick, H., 1939, Studies in the biochemistry of microorganisms. LXII. Itaconic acid, a metabolic product of a strain of *Aspergillus terreus* Thom., *Biochem. J.* 33: 1488-1495.

Calmette, A., 1902, *Ver fahren zur Umwandlung von Tannin in Gallussäure*, German Patent 129, 164.

Carpenter, P.L., 1967, *Microbiology*, W.B. Saunders Company, Philadelphia, London.

Chain, E.B., Florey, H.W., Gardner, A.D., Heatley, N.G., Jennings M.A., Orrewing, J. and Sanders, A.G., 1940, Penicillin as a chemotherapeutic agent, *Lancet* 2: 226-228.

Chang, L.T., Teraska, D.T. and Elander, R.P., 1982, Protoplast fusion in industrial fungi, *Dev. Ind. Microbiol.* 23: 21-29.

Chapman, A.C., 1931, The yeast cell: What did Leeuwenhoeck see? *J. Inst. Brewing* 37: 433-436.

Connstein, W. and Lüdecke, K., 1919, Über Glycerin-Gewinnung durch Gärung, *Ber. Deutsch. Chem. Ges.* 52: 1385-1391.

Constantin, J. and Matrochot, L., 1894, Culture d'un champignon lignicole, *C.R. Acad. Paris* 119: 752-753.

Covert, S.F. and Cullen, D.J., 1992, Heterologous protein expression in filamentous fungi, in: *Frontiers in Industrial Mycology* (G.F. Letham, ed.), pp. 66-77, Chapman & Hall, New York.

Currie, J.N., 1917, The citric acid fermentation of *Aspergillus nider*, *J. Biol. Chem.* 31: 15-37.

Currie, J.N., Kane, J.H. and Finlay, A., 1933, *Process for Producing Gluconic Acid by Fungi*, United States Patent 1, 893, 819.

Duggar, B.M., 1905, The principles of mushroom growing and mushroom spawn making, *Bulletin of the United States Bureau of Plant Industries*, No. 55.

Elander, R.P. and Lowe, D.A., 1991, Fungal biotechnology: an overview, in: *Handbook of Applied Mycology* Vol. IV (D.K. Arora, R.P. Elander and K.G. Mukerji, eds.), pp.1-34, Marcel Dekker Inc., New York.

Ferenczy, L., 1981, Microbial protoplast fusion, in: *Genetics as a Tool in Microbiology* (S.W. Glover and D.A. Hopwood, eds.), pp. 1-34, Cambridge University Press, Cambridge.

Fleming, A., 1929, On the antibacterial action of cultures of a *Penicillium*, with special reference to their use on the isolation of *B. influenzae*, *Brit. J. Exp. Path.* 10: 226-236.

Fukumoto, J. and Schichiji, S., 1971, Industrial production of enzymes, in: *Biochemical and Industrial Aspects* of Fermentation (K. Sakaguchi, T. Uemura and S. Kinoshita, eds.), pp. 91-117, Kodansha Lt., Tokyo.

Gutierrez, S., Diez, B., Alvarez, E., Barredo, J.L. and Martin, J.F., 1991, Expression of the *pen DE* gene of *Penicillium chrysogenum* encoding isopenicillin N acyltransferase in *Cephalosporium acremonium* - production of benzylpenicillin, *Mol. Gen. Genet.* 255: 56-64.

Hamlyn, P.F. and Ball, C., 1979, Recombination studies with *Cephalosporium acremonium*, in: *Genetics of Industrial Microorganisms* (O.K. Sebek and A.I. Laskin, eds.), pp. 185-191, American Society for Microbiology, Washington D.C.

Hawksworth, D.L., 1991, The fungal dimension of biodiversity: magnitude, significance and conservation, *Mycol. Res.* 95: 641-655.

Hawksworth, D.L., Kirk, P.M., Sutton, B.C. and Pegler, D.N. (eds.), 1995, *Ainsworth and Bisby's Dictionary of Fungi*, 8th ed., CAB International, International Mycological Institute, Egham, U.K.

Hayes, W.A. and Nair, N.G., 1975, The cultivation of *Agaricus bisporus* and other edible mushrooms, in: *The Filamentous Fungi Vol. I: Industrial Mycology* (J.E. Smith and D.R. Berry, eds.), pp. 212-248, Edward Arnold, London.

Heim, R., Cailleux, R., Wasson, R.G. and Thévenard, P., 1967, *Nouvelle Investigations sur les Champignons Hallucinogénes*, Paris (Mus. Nat. Hist. Natur.).

Hooke, R., 1665, *Micrographia*, London (Reprinted in 1961, Dover Publications, New York).

Huge-Jensen, B., Adreasen, F., Christensen, T., Christensen, M., Thim, L. and Boel, E., 1989, *Rhizomucor meihei* triglyceride lipase is processed and secreted from transformed *Aspergillus oryzae*, *Lipids* 24: 781-785.

Kalman, M., Cserpan, I. and Bajsar, G., 1990, Synthesis of a gene for human serine albumin and its expression in *Saccharomyces cerevisiae*, *Nucl. Acid Res.* 18: 6075-6081.

Kane, J.H., Finlay, A.C. and Amann, P.F., 1945, *Production of Itaconic Acid*, U.S. Patent 2, 385, 283.

Keay, L., 1971, Microbial proteases, *Process Biochemistry* 6: 17-21.

Kinoshita, K., 1929, Formation of itaconic acid and mannite by a new filamentous fungus, *J. Chem. Soc. Japan* 50: 583-593.

Kützing, F., 1837, Microscopische Untersuchungen über die Hefe und Esigmutter nebst mehreren andern dazu gehorigen vegetabilischen, *J. Prakt. Chemie* 11: 385-409.

Lambert, P.W., 1983, Industrial enzyme production, in: *The Filamentous Fungi, Vol IV, Fungal Technology* (J.E. Smith, D.R. Berry and B. Kristiansen, eds.), pp. 210-237, Edward Arnold, London, U.K.

Le Mense, E.H., Gorman, I., van Lanen, J.M and Langlykke, A.F., 1947, The production of mold amylases in submerged culture, *J. Bact.* 54: 149-159.

Letham, G.F. (ed.), 1992, *Frontiers in Industrial Mycology,* Chapman & Hall, New York, London.

Mehrotra, B.S. and Krishna, N., 1966, Amylase from Indian strains of moulds, in: *Annual National Academy of Sciences of India* 1966: 86.

Micheli, P.A., 1729, *Nova Plantarum Genera Juxta Tournefortii Methodum Disposita,* Florence.

Molliard, M., 1922, Sur une nouvelle fermentation acide produite par le *Sterigmatocystis nigra, Compt. rend.* 174: 881-883.

Moss, M.O., 1990, Fungal biotechnology, *National Centre for School Biotechnology (U.K.) Newsletter* 8: 1-2.

Muller, H.J., 1927, Artificial transmutation of the gene, *Science* 66: 84-87.

Pasteur, L., 1858, Production constante de glycerine dnas la fermentation, *Compt. rend.* 46: 857.

Pasteur, L., 1860, Memoire sur la fermentatioin alcoolique, *Ann. Chimie Phys.*ser. 3, 58: 323-426.

Peberdy, J.F., 1989, Fungi without coats - protoplasts as tools for mycological research, *Mycol. Res.* 93: 1-20.

Penttila, M., Teeri, T.T., Nevalainen, H. and Knowles, J.K.C., 1991, The molecular biology of *Trichoderma reesei* and its application to biotechnology, in: *Applied Molecular Genetics of Fungi* (J.F. Peberdy, C.E. Caten, J.E. Ogden, J.W. Bennett, eds.), pp. 85-102, Cambridge University Press, Cambridge.

Petri, R.J., 1887, Eine klaine Modification des Koch'schen Plattenver fahrens, *Zbl. Bakt.* 1: 279-280.

Queener, S.W. and Lively, D.H., 1986, Screening and selection for strain improvement, in: *Manual of Industrial Microbiology and Biotechnology* (A.L. Demain and N.A. Solomons, eds.), pp. 155-169, American Society for Microbiology, Washington.

Roberts, W., 1874, Studies on biogenesis, *Phil Trans.*169: 457-477.

Schwann, T., 1837, Vorläufige Mittheilung betreffend Versuche über die Weingährung und Fäulniss, *Ann. Physik. Chemie* 41: 184.

Skatrud, P.L., Tietz, A.J., Ingolia, T.D., Cantwell, C.A., Fisher, D.L., Chapman, J.L. and Queener, S.W., 1989, Use of recombinant DNA to improve production of cephalosporin C by *Cephalosporium acremonium, Biotechnology* 7: 477-485.

Smith, J.E., 1981, *Biotechnology,* 1st ed., Edward Arnold (Publishers) Ltd., London.

Smith, J.E., Berry, D.R. and Kristiansen, B.C. (eds.), 1983, *The Filamentous Fungi, Vol. IV, Fungal Technology,* Edward Arnold, London.

Subramanian, C.V., 1992, Tropical mycology and biotechnology, *Current Science* 63: 167-172.

Takamine, J., 1914, Enzymes pf *Aspergillus oryzae* and the application of its amyloclastic enzyme to the fermentation industry, *Chem. News* 110: 215-218.

Trevan, M.D., 1987, Enzyme applications, in: *Biotechnology: The Biological Principles* (M.D. Trevan, S. Boffey,

K.H. Goulding, P. Stanbury, eds.), pp. 178-228, Tata McGraw Hill Publishing Co. Ltd., New Delhi.

Turner, G., 1993, Genetic engineering of filamentous fungi, in: *Genetic Engineering of Microorganisms* (A. Puhler, ed.), pp. 157-171, VCH, Veinheim, West Germany.

Tyndall, J., 1881, *Essays on the floating matter of the air in relation to putrefaction and infection,* London.

Upshall, A., 1986, Filamentous fungi in biotechnology, *Biotechniques* 4: 158-166.

Vaidya, J.G., 1995, *Biology of the Fungi,* Satyajeet Prakashan, Pune, India.

van Tieghem, P., 1867, Sur la fermentation gallique, *Comptes Rendus Hebdomadaires des Seances de l'Academie des Sciences* 65: 1091-1094.

van Tieghem, P. and Le Monnier, G., 1873, Recherches sur lea Mucorinees, *Ann. Sci. Nat. Paris* ser.6, 17: 261-399.

Vattadini, C., 1842, *Monographia Lycoperdineorum,* Turin.

Wainwright, M., 1992, *An Introduction to Fungal Biotechnology,* John Wiley & Sons, Chichester, U.K.

Ward, M., 1990, Chymosin production in *Aspergillus,* in: *Molecular Indstrial Mycology: Systems and Applications in Filamentous Fungi* (S. Leong and R. Berka, eds.), pp. 83-105, Marcel Dekker, New York.

Wasson, V.P. and Wasson, R.G., 1957, *Mushrooms, Russia and History, Vols 1 and 2,* Pantheon, New York.

Wehmer, C., 1891, Entstehung und physiologische Bedeutung der Oxalsäare im Stoffwechsel einiger Pilze, *Bot. Z.* 49: 233-638.

Wehmer, C., 1893, *Beiträge zur Kenntnis einheimischer Pilze,* Hannover and Jena, Hansche Buchhandlung.

ETHNOMYCOLOGY AND FOLK REMEDIES: FACT AND FICTION

Jagjit Singh

Oscar Faber Heritage Conservation
Marlborough House
Upper Marlborough Road
St Albans
Herts AL1 3UT
U.K.

INTRODUCTION

The History of Medicine

"I have an earache..."

2000 BC:	"Here, eat this root."
AD 1000:	"That root is heathen. Here, say this prayer."
AD 1850:	"That prayer is superstition. Here, drink this potion."
AD 1940:	"That potion is snake oil. Here, swallow this pill."
AD 1985:	"That pill is ineffective. Here, take this antibiotic."
AD 2000:	"That antibiotic is unnatural. Here, eat this root."

(New Scientist, 6 September 1997)

A comprehensive account on the study of fungi in folklore and rituals, from prehistoric times to the present day, was given by Wasson (1968). The study of fungi in folklore, fiction and rituals from prehistoric times to the modern era is called the science of ethnomycology (see also Charaya and Mehrotra in this volume).

Throughout the history of mankind, fungi have been regarded with fear and fascination; sometimes revered, sometimes hated, but always considered mysterious. They have been a source of food since times of antiquity and there are many recipes for cooking fungi in a book written by one Caelius Apicus in the third Century A.D. (probably the oldest cookery book written in Europe). However, the artificial cultivation of mushrooms for food does not appear to have been practised until the 17th Century. The physician and poet Nicander, born about 150 A.D., wrote (Wasson, 1968):

"The rank in smell and those of livid show,
All that at roots of oak and olive grow
Touch not! But those upon the fig tree's rind
Securely pluck, a safe and savory kind."

The physician Galen expressed his view about fungi as follows: "Fungi after being eaten in large quantities yield cold, clammy, noxious juices as their nourishing qualities; the Boleti are the most harmless and after them the Amanitae, as for the rest it is far safer to have nothing to do with them" (Findlay, 1982). Moreover, Dioscorides, the celebrated Greek writer on medicine, stated that even the good kinds "if partaken of too freely are injurious being indigestible causing stricture and cholera," and he advised an emetic being taken after meals where they had been eaten (Findlay, 1982).

The word "fungus" may be derived from "funus", a corpse, and "ago", I make. The best authenticated and ingenious case of fungal poisoning is that of the emperor Claudius who succeeded Caligula in A.D.41. Emperor Claudius' 4th wife Agrippina was determined that her son from a former marriage should succeed as emperor instead of the Emperor's son Britannicus. She prepared a dish composed of Amanita of the Caesars steeped in juice extracted from the deadly *Amanita phalloides*. Claudius died from fungal poisoning and Nero succeeded him to the throne (Findlay, 1982).

The fungus *Fomes officinalis* was thought by Dioscorides (first century A.D.) to be a powerful drug which could relieve almost all complaints. He wrote, "Its properties are styptic and heat-producing, efficacious against colic and sores, fractured limbs and bruises from falls...It is given in liver complaints, asthma, jaundice, dysentery, kidney diseases and cases of hysteria...In cases of phthisis it is administered in raisin wine, in affections of the spleen with honey and vinegar. By persons troubled with pains in the stomach and by those who suffer from acrid eructations the root is chewed and swallowed without any liquid."

For many centuries fungi were regarded as the result of decomposition not the cause. But it was due to the work of C.H. Pearsoon (1775-1835) and E. Fries (1794-1878) that a new era in our knowledge of fungi began (Findlay, 1982). The sudden appearance of so-called 'fairy rings' or the luminosity of certain wood-rotting fungi provided the early herbalists, naturalists and poets with fascinating material with which to write interesting poems and fiction.

The fungus *Fomes fomentarius* has been used as tinder and its medicinal use in India was introduced by the Portuguese in Goa (Dymock et al., 1890).

The Mexican Indians seem to regard hallucinogenic plants (and mushrooms) as mediators with God, not as a god themselves. However, the Nahuma Aztecs called the mushrooms *teo-nanacatl*, meaning 'God's flesh'.

In Vedic times Soma was drunk by priests only (Wasson, 1968). Some of their hymns are of so exalted, even delirious, a tenor that the modern leader was led to exclaim: "This surely was composed under the influence of a divine inebriant." It takes little perception to sense the difference in tone between awe-inspired hymns to Soma and the rowdy drinking songs of the West prompted by alcohol. "In a word, my belief is that Soma is the divine mushroom of Immortality, and that in the early days of our culture, before we made use of reading and writing, when the Rgveda was being composed, the prestige of this miraculous mushroom ran by word of mouth, far and wide throughout Eurasia, well behind the regions where it grew and was worshipped."

The identity of Soma is *Amanita muscaria* (Fr. ex L.) Quel., in English, Fly Agaric. The Fly Agaric has been the sacred element in the Shamanic rites of many tribes of Northern Siberia. Alcohol was introduced by Russians in the 16th and 17th Centuries, but Fly Agaric had been their precious possession long before then.

Mushroom intoxication had a quite different effect from alcoholic drunkenness, since the former put the Kamchatka natives into a peaceful and gentle mood (Erman, 1828-1830).

According to Von Maydell (1861-1871) "...the mushroom produces only a feeling of great comfort, together with outward signs of happiness, satisfaction and well-being. Thus far the use of Fly Agaric has not been found to lead to any harmful results, such as impaired health or reduced mental powers".

Erman (1828-1830) states "There is no doubt.......about a marvellous increase in physical strength", which the man praised as still another effect of the mushroom intoxication. "In harvesting hay," he said "I can do the work of three men from morning to nightfall without any trouble if I have eaten a mushroom."

Fungi in Folk Medicine

The ergotism epidemic at the beginning of the eleventh century affected poor people in many parts of Europe. The disease was also known as the 'Sacred Fire' (*ignis sacer*) because a sensation of being burnt was felt by those who contracted it. St. Anthony was believed to have power over fire and thus in the 12th Century, those who blasphemed against him were thought liable to meet their deaths through burning.

Ergot was also used as a drug as early as 1582, when it was discovered that preparations of ergot caused contractions of the muscles of the uterus, thus hastening the birth of a child. A whole range of alkaloids have been isolated from the ergot and they contain as a basic constituent lysergic acid, from which LSD25, an extremely potent drug can be prepared.

Wainwright (1989) quotes the following examples of mould therapy which predated Fleming's original discovery in 1928. A letter of *The News Chronicle*, for example, quotes from the *Heritage of the West* by E.W. Martin: "In many Cornish and Devonshire farmhouses the Good Friday bun was allowed to hang suspended from the grimy beams of the kitchen ceiling and there were a number of superstitions attended to it. Dominant among them perhaps was the tradition that the mouldy portions removed from time to time and mixed with water were suitable as curative agents for any complaints or disease and this pseudo-remedy was employed to treat both human beings and cattle" (Townend, 1944).

A similar account was given by Dr. A.E.Cliffe, a biochemist from Montreal, Canada: "It was during a visit through central Europe in 1908 that I came across the fact that almost every farmhouse followed the practice of keeping a mouldy loaf on one of the beams in the kitchen. When I asked the reason for this I was told that this was an old custom and that when any member of the family received an injury such as a cut or bruise, a thin slice from the outside of the loaf was cut off, mixed into a paste with water and applied to the would with a bandage. It was assumed that no infection would result from such a cut" (Lechevalier & Solotorovsky, 1965).

In a similar vein, the following quote appeared in *The Daily Express* in 1943: "Mrs Eva Wood of Bungay, Suffolk, is a little scornful of the wonder drug that has been discovered from mould called penicillin. Her great-grandmother used to collect all the new copper pennies she could, old copper kettles, smear them with lard and leave them in a damp place. When the mould had formed she would scrape it off into little boxes and everyone for miles around came to her for the remedy for what ailed them."

A Mr H. C. Watkinson, Headmaster of Mexborough Grammar School, Yorkshire, recalled a twentieth-century use of moulds to treat infections as follows: "During my undergraduate days at Cambridge within the period 1911-1913 I called at the Botany Laboratory to do a little extra work. We were studying fungi and the activities of the class were centred at that time on *Penicillium glaucum*. It was the custom in the class to be provided with a growth of the fungus which had previously been growth on old scraps of shoeleather. Only a portion of the growth was used and on the occasion of my calling, the old laboratory attendant was collecting the stuff left on the students' benches. I was somewhat curious and asked why he was so carefully scraping off the fungus into a jar. He told me that he used it as a salve which had been used in his family for a very long time. It was used for what he called gatherings. I presumed by this he meant septic wounds" (Townend, 1944).

The discovery of two sets of fungal fruiting bodies (Basidiomes) strewn on leather straps of pre-historic man in a Tyrolean glacier in 1990 provided invaluable biological and anthropological information (Chapela and Lizon, 1993). The dried perforated fruiting bodies were still held after 5200 years by a knot tied at the end of the strap. This extraordinary ethnomycological find leads to two hypotheses: (i) Fungal fruiting bodies may be used as a first aid kit on a string - medicinal spiritual hypothesis; and (ii) The brack fungi may be used as tinder. Thus in the light of our current anthropological, historical and myco-pharmaceutical knowledge, the glacier man may have carried fungus for use as tinder or for medicinal purposes (Dymock et al., 1890).

Table 1. Some examples of ethno-medically important fungi (from J.G. Vaidya & A.S. Rabba 1993).

Name of Fungus	Folk Medicines
Field mushroom (*Agaricus campestris* Linn.: Fr.)	Nutritious tonic, used againist tuberculosis, anti-inflammation and sinusitis.
God's Bread or Little Man's Bread (*Mylitta lepidescens* Horan)	Treatment of renal ailments, diuretic used by South Indians.
Giant Puffball (*Langermannia gigantea* [Pers] Rostk.)	Used against external swellings as an embrocation, against haemorrhoids and as a remedy for convulsions.
Ergot of Rye (*Claviceps purpurea* (Fr.) Tul.) Ergot of Bajra (*C. microcephala* (Wallr.) Til.) Grain Smut of Wheat (*Ustilago tritici* (Pers) Rostrup) Smut of Maize (*Ustilago maydis* (DC) Corda)	} Traditionally used for many of the disorders } encountered during pregnancy for uterine and } minor arterial contractions. }
Yeasts e.g. Magic Roti (*Torula & Saccharomyces* spp.)	Against fever, blood disorders, dysentery, diarrhoea and wound dressing.
Larch quinine fungus (*Laricifomes officinalis* (Vaill.:Fr.) Kotl. & Pouzar	Against haemorrhoids, water motions, vomiting, etc.
Chaga (Tchaga) fungus (*Inonotus obliquus* (Pers.:Fr.) Pilat)	Befungin - trade name, against chronic gastritis and ulcers, and early growth of tumours.
Snuff fungus (*Daedaleopsis flavida* (Lev.) Roy & Mitra)	Treatment of jaundice.
Umbarache Kan - fruiting body grows on *Ficus religiosa* Linn.	Against kidney disorders.
Phansomba - produces mango-shaped fruiting bodies on Phanas tree (*Artocarpus heterophyllus* Rotb.	Gums, diarrhoea and dysentery.
Tinder fungus (*Fomes fomentarius*)	Cauterisation of burnt tissues.
Shiitake mushroom (Japan) (*Lentinula edodes*)	Lowers cholesterol content of blood.

Novel Leads to New Medicines

Fungi are widely used for the production of commercially important products and are under constant investigation for other potentially useful products (see Figure 1). Fungal biodiversity has been estimated at some 1.5 million species and less than 0.1 million have been characterised to date. This clearly shows that there is unrivalled chemical diversity which could give us novel leads to molecules for the generation of new medicines. The top twenty selling prescription medicines (see Table 2) are of fungal origin.

Table 2. Top twenty selling pharmaceuticals, worldwide, in 1995 (from Langley, 1997).

Compound	Mode of Action	Sales $m
Ranitidine	H2-antagonist	3914
Omeprazole	proton pump inhibitor	2539
Enalapril	ACE inhibitor	2220
Nifedipine	calcium antagonist	2191
Pravastatin	hypolipidaemic	1980
Fluoxetine	antidepressant	1790
Simvastatin	hypolipidaemic	1615
Lovastatin	hypolipidaemic	1345
Acyclovir	antiviral	1313
Ciprofloxacin	antibiotic	1279
Amoxicillin +clavulanate	antibiotic	1176
Captopril	ACE inhibitor	1110
Cyclosporin	immunosuppressant	993
Diclofenac	NSAI	938
Ceftriaxone	antibiotic	909
Famotidine	H2-antagonist	850
Methyltestosterone	hormone	840
Salbutamol	b-2 agonist	825
Fluconazole	antifungal	800
lisinopril	ACE inhibitor	772

Primary Metabolites

Fungi are very versatile and can use a range of difference sources of nutrients which are assimilated into the primary metabolic pathways at different points. There are a number of commercially important primary metabolites, for example citric acid, ethanol, enzymes, amino acids and vitamins. Primary metabolites are formed during the active growth of the fungus. The fungus will take from the natural environment those nutrients which it can utilise as an energy source to produce materials such as proteins, lipids and nucleic acids for continued growth and biomass production (Isaac, 1997).

Secondary Metabolites

Some commercially important secondary products, including antibiotics (e.g. penicillin from *Penicillium chrysogenum*, cephalosporin from *Cephalosporium acremonium*, griseofulvin from *Penicillium griseofulvum*) and alkaloids (*Claviceps* spp.), are derived from filamentous fungi and used medicinally (see Table 2 and Figure 2).

Secondary metabolites are not essential for fungal growth but are produced naturally by many fungi. Many of the compounds produced have antifungal and antibacterial activity (e.g. antibiotics, mycotoxins) and may therefore impart a competitive advantage, acting as weapons for survival. The compounds have antimicrobial activity and producer organisms may well be sensitive to these. Most have mechanisms to prevent their own demise from the effects of the compounds they produce. In most cases the products are formed after active growth and by that time the mycelium is able to detoxify the compound or prevent entry of the antibiotic through the cell wall by a change in the permeability of the plasma membrane (Isaac, 1997).

Figure 1. The diversity of fungi used for the production of commercially important products.

Biodiversity

Biological diversity, or biodiversity, is the total variability of life on Earth. Biodiversity is not simply the total number of species; it encompasses the complexity, richness and abundance of nature at all levels from the genes carried by lcoal populations to the layout of communities and ecosystems across the landscape.

Human activities in the search for novel leads to compounds for pharmaceutical and other commercial use are leading to the destruction of biodiversity at an unprecedented rate. Because of this, some ecologists and conservationists have been heartened as the economic benefits of biodiversity have started to become apparent in recent decades and the direct economic value of the natural world has been recognised as a source of a vast range of products from antibiotics and novel medicines to Brazil nuts, salmon, spices, mushrooms, mahogany, oils and ecotourist dollars.

Some recent biodiversity initiatives have been set up to manage biodiversity and its sustainability, for example:
- Global Biodiversity Strategy
- Global Biodiversity: status of the Earth's living resources
- Caring for the Earth: a strategy for sustainable living
- Global Mean Biodiversity Strategy: a strategy for building conservation into decision-making
- From Genes to Ecosystem: a research agenda for biodiversity

Strategies for Biodiversity Conservation and Sustainable Use

Conservation and preservation are the maintenance of some or all components of biological diversity and also includes sustainable use of the components. 'Sustainable use' means the use of components of biological diversity in a way and at a rate that does not lead to the long term decline of biological diversity, thereby maintaining its potential to meet the needs and aspirations of present and future generations (Heywood, 1995).

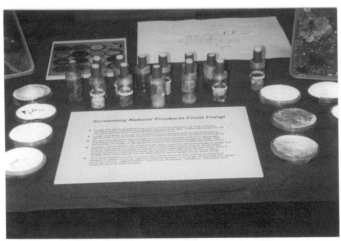

Figure 2. Screening natural products from fungi.

The global policy-makers tend to behave as though the survival of most non-human organisms is an amenity, one that future generations of humans can live without. However, everything that scientists are learning about the Earth's life processes argues against this view (Baskin, 1997). In the exploration and exploitation of biodiversity, therefore, it is fundamental to investigate the functional role of biodiversity at all levels, from genes to species, communities, ecosystems and landscapes.

REFERENCES

Baskin, Y., 1997, *The Work of Nature: How the Diversity of Life Sustains Us*, Island Press, USA.

Chapela, I. H. and Lizon, P., 1993, *Mycologist* 7(3): 122.

Dymock, W., Warden, C.J.H. and Hooper, D., (1890), *Pharmocographia Indica - A History of the Principal Drugs of Vegetable Origin, Part III*, Education Society Press, Byculla, Bombay, pp. 629-635.

Erman, A., 1833-1848, *A Journey Around the World through Northern Asia and Both Oceans in 1828, 1829 and 1830*, Berlin 1833-1848, p. 223.

Findlay, W.P.K., 1982, *Fungi, Folklore, Fictioni and Fact,* The Richmond Publishing Company Ltd., U.K.

Heywood, V.H., 1995, *Global Biodiversity Assessment*, UNEP, Cambridge University Press, UK.

Isaac, S., 1997, Mycology answers, *The Mycologist* II(4): 182-183.

Langley, D., 1997, Exploiting the fungi: novel leads to new medicines, *The Mycologist* 11(4): 165-167.

Lechevalier, H. A. and Solotorovsky, M., 1965, *Three Centuries of Microbiology*, McGraw Hill, New York, p. 461.

Townend, B. R., 1944, Penicillin in folk medicine, *Notes and Queries* 186: 158-159.

Vaidya, J.G. and Rabba, A.S., 1993, Fungi in folk medicine, *The Mycologist* 7(3): 131-133.

Von Maydell, 1861-1871, *Journeys and Investigations in the Jakutskaia Oblast of Eastern Siberia in 1861-1871.*

Wainwright, M., 1989, Moulds in folk medicine, *Folklore* 100(ii): 162-166.

Letter to *Daily Express*, 19 Nov 1943.

Wasson, R.G., 1968, *Soma, Divine Mushroom of Immortality*, Harcourt Brace Jovanovich, New York.

SOME ETHNOMYCOLOGICAL STUDIES FROM MADHYA PRADESH, INDIA

N.S.K. Harsh[1], B.K. Rai[2] and V.K. Soni[1]

[1]Forest Pathology Division
Tropical Forest Research Institute
Jabalpur-482 021
Madhyar Pradesh
India

[2]Plant Pathology Department
Jawaharlal Nehru Krishi Vishwa Vidyalaya
Jabalpur
Madhyar Pradesh
India

INTRODUCTION

Madhya Pradesh in central India lies between 17°48'N and 26°52'N latitude and 70°2'E and 84°24'E longitude and abounds in biodiversity of flora and fauna due to its rich forest cover. It has a geographical area of 443,446 sq. km, of which 30.5% is under forest (Anon., 1995). It is home to a number of primitive tribes, namely Gond, Bhil, Baigas, Saharia, Kol, Abujhamarias, Bharias, Hill Korwa and Kamar (Tiwari, 1994). The ethanobotanical treasure occurring in the forests of Madhya Pradesh provides sustenance and livelihood to the tribal communities and rural poor which constitute nearly 23% of the population of the state. The higher fungi in particular have an important place in the diet and medicines of the tribes, about which little is known to the so-called civilized world. The present study therefore attempts to provide an insight into the ethnomycology of this region.

Information about the edible fungi and those sold in markets was gathered from the tribal peoples as well as from the weekly local markets (bazaars) of Balaghat, Jabalpur, Jagdalpur, Mandla, Rajnandgaon and Shahdol districts of Madhya Pradesh. Details about medicinal fungi were collected through inquiries and personal approaches with the assistance of knowledgeable tribal medicine men. Fruit bodies of the edible and medicinal fungi were collected from their natural habitats and brought to the laboratory for detailed taxonomic studies and identification was done with the aid of standard monographs (Bakshi, 1971; Bondartsev, 1953; Coker and Couch, 1969; Cunningham, 1942; Ryvarden and Johansen, 1980).

From Ethnomycology to Fungal Biotechnology
Edited by Singh and Aneja, Plenum Press, New York, 1999

OBSERVATIONS

Table 1 lists the edible fungi used by different tribes of Madhya Pradesh along with the method of preparation.

An account of the fungi used in folk medicine is presented in Table 2.

Methods of Collection, Drying and Storage

There are no systematic methods of collection for these fungi. Fruit bodies are collected haphazardly wherever they are encountered in the forest areas. However, the tribals are well acquainted with the habitat and period of occurrence of the edible and medicinal fungi from personal experience (Harsh et al., 1993a,b,c). Whole families including men, women and children come out to hunt these useful fungi during the rainy season (June-September). However, the tribal women are particularly active as they have the responsibility of preparing food for the family while gathering other food, fodder and fuel items for day to day needs (Harsh et al., 1996). The collection is taken in cloth bags and bamboo baskets to homes and markets. Most species have a very short shelf life and cannot be stored for more than one day, but a few of them like *Termitomyces heimii* can be dried under the sun or over flames for longer storage.

For drying during cloudy weather, specially designed bamboo mats are used (Figure 1). Bamboo sticks 65 in number, each 40 cm long and 4 mm in cross section are horizontally placed (2 sticks per cm) and tied with a thin wire to form a mat. Four bamboo strips of 35 cm length and 1 cm breadth are placed at the ends of the mat and two strips of 51 cm length and 1 cm breadth are placed diagonally and tied with a thin wire. This mat is hung over the *'chulha'* (fireplace) some 40-50 cm above the flames. The collected mushrooms are kept on this mat during cloudy days. The dried mushrooms are used during off-season, but are not seen to be sold in the markets.

Marketing

Some of the edible fungi were also seen to be sold in the local weekly markets of the tribal areas (these markets are held on a fixed day during each week where tribals sell their collections from the forests and in turn buy items like salt, oil, soap, clothes, combs and other utility goods). A detailed study about their role in the tribal economy was made (Harsh et al., 1993b). Table 3 gives an account of estimated quantities of some edible fungi being sold in weekly markets during a period of approximately three rainy months. About 2.5 tonnes of *Termitomyces heimii* alone were estimated to be sold in 15 local markets during the season (Table 4) fetching a price of about Rs 25000 (Rs 40 = $1). However, if these edible mushrooms are compared with the cultivated edible button mushroom *Agaricus bisporus* (sold at Rs 70-80 per kg fresh or Rs 30-40 per 250 g canned) and with the wild growing *Morchella* spp. (sold at Rs 1500 per kg fresh and Rs 3000 per kg dried) which are marketed in the urban areas of India and the latter exported too, it is evident that the fungi sold in the tribal markets are sold very cheaply, although in nutrition and taste they are no inferior. *Termitomyces heimii* in particular has a resemblance to *Morchella* spp., as it also grows wild in abundance, can be dried and packed easily and is rich in nutrition and taste, but the sale value is very low in comparison to the latter.

The sale of these fungi provides sustenance and livelihood to the tribal poor, particularly in lean periods when other non-wood forest products like myrobalans, leaves, seeds, flowers and fruits are not available. Sharma et al. (1997) estimated that the sale of edible fungi contributes about 2% to the annual income of a tribal family in the Amarkantak plateau of Madhya Pradesh. There is considerable potential for boosting the economy through the sale

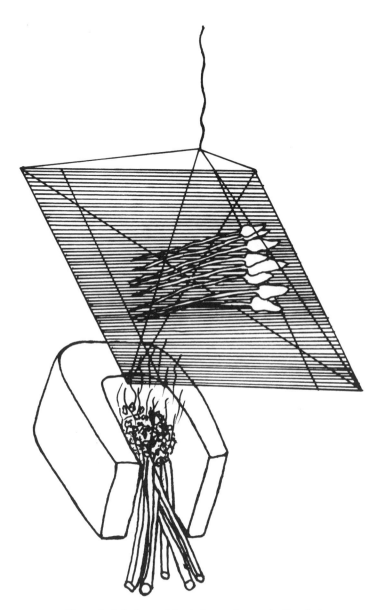

Figure 1. A device used for drying mushrooms by the tribals.

Table 1. Edible fungi used by tribal people of Madhya Pradesh.

Local name	Etymology	Scientific name	Method of cooking	Habitat description	Time of occurrence	Remarks
Sawan putpura (in Balaghat), puttu (in Mandla), putpura (in Shahdol), Band chatti (in Jagdalpur)	Sawan (a Hindu calendar month) =July-Aug; putpura, puttu = erupting out of ground; band = close; chatti = umbrella	*Astraeus hygrometricus* (Pers.) Morg.	Washed and sliced into pieces like potato, fried in mustard oil with spices and salt, eaten with chapaties of corn flour	Found in ectomycorrhizal association with roots of *Shorea robusta* (sal) trees in the forests	July-August	The fungus is eaten all over the sal belt in the forests of Madhya Pradesh. This is sold in the tribal markets also.
Dharati ka phool	Dharati = earth; ka = of; phool = flower	*Calvatia cyathiformis* (Bosc.) Morg.	Fried in mustard oil and eaten with chapaties of wheat	In open grasslands, pasture lands and forest floor	June-July	Baiga and Bharia tribes eat this fungus.
Phutpura	Phutpura = erupting out of ground	*Geastrum fimbriatum* (Fr.) Fischer	Prepared like egg curry with spices and eaten with rice, also eaten as boiled eggs	In open sandy soil, as well as on forest floor	July-August	Baiga and Bharia tribes use this fungus.
Ghundi	Ghundi = nipple shaped, button	*Lycoperdon pusillum* Pers.	Surface washed, scrubbed and cooked as a vegetable with potatoes	In gardens, fields, grasslands and freshly worked earth with sandy soil	June-September	It fruits soon after start of rains.
Asharh phutpura (in Balaghat), Asharhi band chatti (in Shahdol), puttu or phutpura (in Jagdalpur)	Asharh (a Hindu calendar month) = June-July	*Mycenastrum corium* (Guersent) Desvaux	Cooked as egg curry and eaten with rice	On sal forest floor	June-August	This is sold in the local village markets.
Bhat pihiri	Bhat = cooked rice; pihiri = mushroom	*Podabrella microcarpa* (Berk. & Br.) Singer	Cooked in mustard oil with fried onions and considered a delicacy	Grows gregariously over termite mounds	July-August	It has a very short shelf life but is sold fresh in the village markets.

Local name	Etymology	Scientific name	Method of cooking	Habitat description	Time of occurrence	Remarks
Pihiri, bhoron pihiri	Bhoron = early morning	*Termitomyces heimii* Natarajan	Cooked in mustard oil/groundnut oil with fried onions and spices, a delicacy	Emerge gregariously from nderground termite fungal combs on open fields, grasslands, forest floor	July-September	Most sought after fungus among tribal and urban people, dominates local markets.
Phuttu, putu	Phuttu, putu = erupting out of ground	*Scleroderma radicans* Lloyd	Washed fruit bodies, boiled and cooked with potatoes in curry, eaten with chapaties made of wheat and corn flour	On sandy soil among grasses	August-September	Not very common

Table 2. List of fungi used in folk medicine by the tribals of Madhya Pradesh.

Local name	Etymology	Scientific name	Disease/disorder in which used	Method of preparation and dose	Habitat	Time of occurrence	Remarks
Sawan putpura, puttu, putpura, band chatti	See Table 1	*Astraeus hygrometricus*	Burns	Spore mass (powder) mixed in mustard oil and applied over the burnt skin	See Table 1	See Table 1	Used by Gond tribes of Madhya Pradesh
Phoosh	Phoosh = dust, powder blowing	*Bovista apedicellata*	Wounds	Spore mass mixed with mustard oil and applied over wound surface	On sandy grasslands	August-September	Used by Baiga tribes
Nir ghanti	Nir = nest; ghanti = bell	*Cyathus limbatus* Tulasne	Eye redness, pain, conjunctivitis	Peridoles crushed with clean water or rose petals extract, filtered through clean cotton and filtrate is used as eye drop	Growing over dead stored bamboo	August	Used by Baiga people
Ghundi	See Table 1	*Lycoperdon pusillum*	Bleeding from deep cuts and wounds	Spore powder mixed with mustard oil and applied over cuts/wounds	See Table 1	See Table 1	Used by Baiga people
Jhirri-pihiri	Jhirri = crack, slit, net; pihiri = mushroom	*Phallus impudicus* L.: Pers.	1. Typhoid (known as motijhara among tribes)	Fruit bodies crushed and suspension with old jaggery and water given to patients 1 tsp. 3 times a day for 4 days.	Leaf litter and old bamboo chips	August	Used by Baigas

Local name	Etymology	Scientific name	Disease/disorder in which used	Method of preparation and dose	Habitat	Time of occurrence	Remarks
			2. Labour pains	Same preparation as above given to ladies to give relief during labour pains.			
Jarh phorh	Jarh = root; phorh = rupture, break, burst	*Ganoderma lucidum* (Leyss.) Karst.	1. Cataract	Fruit bodies dried over flames, powdered and mixed with coconut oil (5 g powder, 25 g oil) as an ointment which is applied around the eye socket twice daily for 45 days.	Grows on the roots (causing root rot in living trees) and stumps of various forest trees	July-September	Used by Baigas and Abujmariahs
			2. Asthma	Fruit bodies dried and ground to powder and mixed with county liquor made from *Madhuca latifolia* flowers, given as 1 tsp. 3 times daily for 2-3 months to cure the disease.			

Local name	Etymology	Scientific name	Disease/disorder in which used	Method of preparation and dose	Habitat	Time of occurrence	Remarks
			3. Hydrocele	Fruit bodies boiled with equal amounts of coconut oil, mustard oil and linseed oil for about an hour, cool filtrate is applied around the scrotum 3 times a day for a month.			
Phoot doodh	Phoot = to gush, erupt, start; doodh = milk	*Xylaria polymorpha* Pers.: Merat.	Activation of lactation after delivery	Dried fruit bodies powdered and mixed with old jaggery to make pea sized tablets, one tablet twice daily with cow milk before meals for 5 days to give effect	Grows on the roots and stumps of *Shorea robusta* trees in the forests	July-September	Used by Baigas and Gonds
Saja pihiri	Saja = *Terminalia tomentosa* tree	*Microporus xanthopus* (Beac.:Fr.) Kunt.	1. Ear pain	Dried fruit bodies ground with water and filtered through cotton, filtrate is used as ear drops	Grows on dead twigs	July-September	Fruit bodies growing on *Terminalia tomentosa* twigs are preferred
			2. Fever and vomiting	Dried fruit bodies ground with gum of *Shorea robusta* tree and given as 1 tsp. with water 3 times daily.			

Local name	Etymology	Scientific name	Disease/disorder in which used	Method of preparation and dose	Habitat	Time of occurrence	Remarks
Kala pihiri	Kala = black; pihiri = mushroom	*Daldinia concentrica* (Bolt.:Fr.) Ces. & de Not	Chronic cough	Dried fruit bodies and old earthen vessel piece powdered separately, mixed in equal amounts in honey, given 1 tsp. twice daily.	Grows on dead wood, bark of trees	July-September	Fruit bodies growing on old *Butea monosperma* trees are particularly used

Table 3. Estimated quantities of some edible fungi sold in the weekly markets of tribal area.

Name of the market	Name of fungi			
	Astraeus hygrometricus	*Mycenastrum corium*	*Podabrella microcarpa*	*Termitomyces heimii*
Bajag (Mandla)	+++	++++	+++	++++
Dindori (Mandla)	++	++	++	+++
Garhasarai (Mandla)	++	++	+++	+++
Gorakhpur (Mandla)	+	+++	++	+++
Karanjia (Mandla)	++	+++	++	+++
Niwas (Mandla)	+	-	+	++
Kundam (Jabalpur)	-	-	+	+++
Amarkantak (Shahdol)	+++	++++	+++	++++

- = not marketed; + = Below 25 kg; ++ = 26-50 kg; +++ = 51-75 kg; +++ = Above 75 kg

Table 4. An account of the sale of *Termitomyces heimii* in some tribal markets of Madhya Pradesh.

Name of the market	Approx. quantities (in kg fresh weight) marketed during the season	Cost / kg (in Rs)		Money earned after sale (in Rs)
		Minimum	Maximum	
Dhanpunji (Jagdalpur)	240	10	12	2640
Jagdalpur	200	8	12	2000
Kurandi (Jagdalpur)	180	8	10	1620
Pandutala (Balaghat)	200	8	10	1800
Bajag-Chanda (Mandla)	300	8	10	2700
Bichia (Mandla)	140	8	10	1260
Ghotas (Mandla)	180	8	10	1620
Karanjia-Rupa (Mandla)	160	8	10	1440
Medha (Mandla)	120	10	12	1320
Motinala (Mandla)	100	8	10	900
Sijhora (Mandla)	140	6	10	1120
Bodla (Rajnandgaon)	200	8	12	2000
Chilpi (Rajnandgaon)	140	6	10	1120
Mangli (Rajnandgaon)	120	8	10	1080
Amarkantak (Shahdol)	120	8	12	1200
Total	**2540**	Average Rs 9.96		**25620**

of these fungi if they are popularised among the masses and sold in an organised and systematic manner along the lines of *Morchella* spp.

Fruit bodies of some of the wood-decaying fungi are collected in bulk by tribal people from the forest areas of Bilaspur, Jagdalpur, Mandla, Rajnandgaon and Shahdol districts of Madhya Pradesh from October to December and dried under the sun. The fungi are purchased from them by middlemen at a cost of Rs 2-3 per kg (dry weight) and in turn are sold at Rs 20-22 per kg to export houses of New Delhi, Tuticorin and Calcutta, who sell them at $2-3 per kg. The purpose for which these fungi are being exported is not fully known although it has been found that they may be used in making decorative materials. There is a preference for fungi with white fruit bodies such as *Lentinus tigrinus* Fr., *Lenzites acuta* Berk., *Lenzites vespacea* (Pers.) Ryv., *Microporus xanthopus* (Beac.:Fr.) Kunt., *Trametes cingulata* Berk., *T. elegans* (Spreng.:Fr.) Fr. and *T. lactinea* Berk.

Medicinal Fungi

Tribal people in remote areas depend upon folk medicines and household remedies to a great extent. The prevalent practices of indigenous herbal medicines including fungi have descended from generation to generation and include cures for both simple and complicated diseases. The information given in Table 2 is confined to the tribals only and that too to a few tribal medicine men. The knowledge is usually passed from one generation to another among the families of tribal medicine men and the danger of its being lost is ever present. There is a need to explore and use this knowledge in medical science. Generally this information is a closely guarded secret and the information given here was only obtained through considerable persuasion and after spending time with the tribals to win their confidence. Our attempt to document this piece of information must therefore be seen in this context.

One of the fungi listed, *Ganoderma lucidum* is being investigated for its antitumour properties in China, Japan and United States of America (Miyajaki and Nishijima, 1981; Lee et al., 1984; Furusawa et al., 1992). *Ganoderma* tea has come from China to the U.K. and is much in demand (Singh, pers. comm.). The fruit bodies of *Ganoderma lucidum* are being bought from India by South Koreans for the extraction of certain compounds which are known to provide youth and vigour to the aged (Mukerji, pers. comm.).

REFERENCES

Anon., 1995, *Forestry Statistics India,* 1995, Indian Council of Forestry Research and Education, Dehra Dun, India.

Bakshi, B.K., 1971, *Indian Polyporaceae on Trees and Timber*, ICAR Publ., New Delhi.

Bondartsev, A.S., 1953, *The Polyporaceae of the European USSR & Caucasia,* Keter Press, Jerusalem.

Coker, W.C. and Couch, J.N., 1969, *The Gasteromycetes of Eastern United States and Canada,* J. Cramer, Lehre.

Cunnigham, G.H., 1942, *The Gasteromycetes of Australia and New Zealand,* Donedin, N.Z.

Furusawa, E., Chou, S.C., Furusawa, S., Hirazumi, A. and Dang, Y., 1992, Antitumour activity of *Ganoderma lucidum,* an edible mushroom, on intraperitoneally implanted lewis lung carcinoma in synergenic mice, *Phytotherapy Research* 6: 300-304.

Harsh, N.S.K., Tiwari, C.K. and Jamaluddin, 1993a, Market potential of wild edible fungi in Madhya Pradesh, *Indian J. Trop. Biodiv.* 1: 93-98.

Harsh, N.S.K., Rai, B.K. and Ayachi, S.S., 1993b, Forest fungi and tribal economy - a case study in Baiga tribe of Madhya Pradesh, *J. Trop. For.* 9: 270-279.

Harsh, N.S.K., Rai, B.K. and Tiwari, D.P., 1993c, Use of *Ganoderma lucidum* in folk medicine, *Indian J Trop. Biodiv.* 1: 324-326.

Harsh, N.S.K., Tiwari, C.K. and Rai, B.K., 1996, Forest fungi in the aid of tribal women of Madhya Pradesh, *Sustainable Forestry* 1: 10-15.

Lee, S.S., Chen, F.D., Chang, S.C., Wei, Y.H., Liu, I., Chen, C.F., Wei, R.D., Chen, K.Y. and Han, P.W., 1984, *In vivo* antitumor effect of crude extracts from the mycelium of *Ganoderma lucidum, J. Chi. Oncol. Soc.*

5: 22-28.

Miyazaki, T. and Nishijima, M., 1981, Studies of fungal polysaccharides-XXVII. Structural examination of a water-soluble, antitumor polysaccharide of *Ganoderma lucidum, Chem. Pharm. Bull.* 29: 3611-3616.

Ryvarden, L. and Johansen, I., 1980, *A Preliniinary Polypore Flora of East Africa,* Fungiflora Oslo, Norway.

Sharma, M.C., Masih, S.K. and Sharma, C.B., 1997, Participation in collection of NTFP and their share in tribal economy, *J. Trop. For.* 13: 220-225.

Tiwari, S.K., 1994, *Encyclopaedia of Indian Tribals, Vol. 1,* Rahul Publishing House, Delhi.

SOME INDIAN SPICES AND THEIR ANTIMICROBIAL PROPERTIES

Daljit S. Arora

Department of Microbiology
Guru Nanak Dev University
Amritsar-143 005
India

INTRODUCTION

The chemotherapy of infectious diseases has proved to be a continuous struggle. A broad spectrum of antimicrobial agents are being developed to combat infections but these agents are rendered ineffective because microorganisms become resistant to these agents. Antimicrobial agents obtained from indigenous plants are quite effective against these resistant strains. Among these plants, spices occupy an important position.

Spices are documented in history as far back as 5000 B.C. with Malabar, the western part of Southern India, considered to have been a spice kingdom in olden times (Marck and Thornton, 1993). The history of Indian spices dates back to the period before the birth of Jesus Christ; for instance, pepper and cloves were described in the stories of the Arabian Nights. In early Egypt, spices were used as aromatic body ointments and pomades. By 600 B.C. onions and garlic had become popular condiments in Persia, and in 559 to 529 B.C., essential oils were produced there from coriander and saffron. During the Middle Ages, several kingdoms of Kerala were interested in the spice trade. Among these kingdoms, Cochin was considered to be the 'Venice of the East' in view of its commanding position in the spice trade.

The term 'spice' denotes the plant or vegetable product or part used for imparting flavour and aroma and for seasoning food. Spices affect biological functions and have been traditionally used for many disorders, for example clove has been used for gum disorders and garlic for eyesight (Ridley, 1993). They are also known for their antimicrobial properties and this chapter deals with some of the spices and their antimicrobial properties.

Black Pepper (*Piper nigrum*)

Black pepper ('kali mirch') is the most important spice of India since it earns a considerable share in the international trade and hence is known as the 'black gold of India' (Pruthi, 1992).

Extracts of *Piper* species have insecticidal (Atal et al., 1967; Asprey and Thornton, 1976;

Escobar, 1972) and medicinal properties (Loder et al., 1969). Black pepper at a concentration of 1% inhibits the growth of *Escherichia coli, Salmonella virchow* (Nair et al., 1996), *Bacillus subtilis* (Ito and Meixu, 1994) and *Clostridium botulinum* (Huhtanen, 1980; Ito and Meixu, 1994). Black pepper is recommended as a remedy for cholera. It is also used externally as a mouthwash in toothache, as an application of inflamed parts and also for pimples and boils.

There are very few reports on *Piper nigrum* but some work has been done on *Piper hispidum*. 4,5-dimethoxy-2,3-(methylenedioxy)-1-allylbenzene, which is a natural isolate of *Piper hispidum*, along with three related compounds synthesized from piperonal have been screened for their antibacterial and antifungal activity. All these compounds inhibit the growth of *Xanthomonas campestris* and *Xanthomonas cartone* at concentrations of 10 to 50 ppm. For *Agrobacterium tumefaciens*, the dose requirement is above 200 ppm and for *Rhizobium japonicum* above 500 ppm (Nair and Burke, 1990). The higher sensitivity of *Xanthomonas* can make pepper a potential control agent for the dreaded canker disease caused by *Xanthomonas* species on citrus trees. Moreover, all the test compounds are active against fungi, namely *Cladosporium herbarum, Helminthosporium carbonum, Alternaria brassicola, Pyrenochaete terrestris* and *Alternaria chrysanthemi* (Naire and Burke, 1990).

Capsicum Species

Aqueous and heated extracts of *Capsicum* species have been found to exhibit varying degrees of inhibition against *Bacillus cereus, Clostridium sporogenes, Clostridium tetani* and *Streptococcus pyogenes*. A survey of the Mayan pharmocopoeia revealed that tissues of *Capsicum* species are included in a number of herbal remedies for a variety of ailments of microbial origin (Cichewicz and Thorpe, 1996).

Caraway (*Carum carvi*)

Caraway ('shia jeera') oil is used to increase the shelf life of food products by assuring protection from microbial spoilage. It contains carvone and p-cymene as the major components along with minor quantities of gamma-terpinene, caryophyllene and carvol (Farag et al., 1989b).

Few reports are available on the inhibitory potency of caraway oil on bacteria and fungi. The antibacterial activity of essential oil of caraway has been tested against Gram-negative bacteria (*Pseudomonas fluorescens, Serratia marcescens* and *E. coli*), Gram-positive bacteria (*Staphylococcus aureus, Micrococcus* spp., *Sarcina* spp. and *Bacillus subtilis*), acid fast bacteria (*Mycobacterium phlei*) and one yeast (*Saccharomyces cerevisiae*). Gram-positive bacteria were more sensitive to the caraway oil than Gram-negative bacteria. *Mycobacterium phlei* and *Saccharomyces cerevisiae* also show higher sensitivity than Gram-negative bacteria (Farag et al., 1989a). Caraway oil also exhibits antifungal activity against *Aspergillus parasiticus*. At a concentration of 0.8 mg/ml, growth and aflatoxin production are completely inhibited (Farag et al., 1989a).

Cardamom (*Ellettaria cardamomum*)

Cardamom ('choti elaichi') is widely consumed as a food spice, employed as a remedy to stomach ache and for making mouthwash and soaps. According to Husain (1996) cardamom, along with pepper, ginger, clove, casia, nutmeg and allspice, contributes to 90% of the total spice trade.

The distillate of n-hexane extract of cardamom seeds have been found to possess active antimicrobial activity. Around ten major volatile compounds, viz. 1,8-cineole, alpha-terpinyl acetate, linalol, linalyl acetate, geraniol, limonene, alpha-terpinene, safrole, methyleuginol and euginol, obtained from the active distillate fraction (Abo Khatawa and Kubo, 1987) have been found to possess variable antimicrobial activity against bacteria, fungi and yeast (Kubo et al., 1991). The different bacteria tested were *Bacillus subtilis, Brevibacterium ammoniagenes, Staphylococcus aureus, Streptococcus mutans, Propionibacterium acnes, Escherichia coli, Pseudomonas aeruginosa, Enterobacter aerogenes* and *Proteus vulgaris*. It has been observed that Gram-positive bacteria are more sensitive than Gram-negative bacteria. A similar observation has been made by Farag et al. (1989b) in a study on the inhibitory effect of some Egyptian spice essential oils.

In fact, the Gram-negative bacteria are sensitive to only a few phytochemicals. *Propionibacterium acnes* is the most sensitive followed by *S. mutans*. *P. acnes* causes acne: the bacterium produces lipase that hydrolyses sebum triglycerides to free fatty acids and causes inflammation and comedones. Cardamom can provide protection against *P. acnes* infection. In fact cardamom is being used as a fragrance in soaps to control acne. Similarly *S. mutans* (causative agent of dental caries) can also be eliminated by providing cardamom oil in mouthcare products.

Antifungal activity in cardamom oil has been reported for *Pityrosporum ovale* and *Trichophyton mentagrophytes* (Kubo et al., 1991). The fungus *P. ovale* occurs on human skin and causes pityriasis versicolor which is a mild, chronic, superficial human dermatomycosis, whereas *T. mentagrophytes* is a parasite of hair. Hence, cardamom oil may prove to be a promising antimicrobial agent in skin and hair care products for controlling dermatomycotic fungi. Its antifungal activity has also been reported against mycotoxigenic moulds such as *Aspergillus flavus, A. parasiticus, A. ochraceus, Penicillium palulum, P. roquefortii* and *P. citrinum* (Badei, 1992).

Cinnamon (*Cinnamomum zeylanicum*)

Cinnamon ('dalchini') is a plant of great economic importance and contains significant amounts of proteins, carbohydrates, vitamins and trace elements. Its antimicrobial activity has been detected against a variety of bacteria and fungi. Many workers have found cinnamon, mustard and clove to be potent spices in checking microbial growth (Anand and Johar, 1957; Pruthi, 1980). At 2% concentration cinnamon is active against the food-borne fungi *Trichoderma harzianum, Alternaria alternata, Fusarium culmorum, Aspergillus versicolor, Cladosporium cladosporoides* and *Penicillium citrinum* (Schmitz et al., 1993). At a concentration of 1%, it delays acid production by *Lactobacillus plantarum* and *Pediococcus cerevisiae* (Zaika and Kissinger, 1979). It is also effective against *Aspergillus niger, Bacillus cereus, Saccharomyces cerevisiae* and *Mycoderma* spp.

Vapours from cinnamon have an antibacterial effect on typhoid and paratyphoid bacteria. Cinnamon oil has antimicrobial activity against *Staphylococcus aureus, E. coli, Salmonella typhosa* and *Bacillus dysentriae* (George and Pandalai, 1958). During the course of its testing, the essential oil of cinnamon showed antibacterial activity against *Vibrio cholerae, Salmonella paratyphi* (human pathogens); *Bacillus anthracis* (pathogen of domesticated animals); and *Xanthomonas malvacearum* (plant pathogen). It was found to possess inhibitory activity only against human pathogens and was ineffective against *Bacillus anthracis* and *Xanthomonas malvacearum* (Nigam, 1982). Cinnamon oil is also a potential inhibitor of *Penicillium expansum* which is a cause of spoilage of apples. Patulin is not detected in apple juice containing 0.3% cinnamon oil after incubation with *P. expansum* for seven days at 25°C (Ryu and Holt, 1993). In further studies cinnamon water extract alone was found to be ineffective against fungal activity but in combination with other antifungal chemicals such as sorbic acid formed an effective antifungal agent (Sebti and Tantaoui, 1994).

Clove (*Eugenia caryophyllous*)

Essential oil of clove ('loung') is used as an analgesic for toothache and when mixed with zinc oxide it forms a temporary filling for deep cavities (Anon, 1979). Eugenol is the active principle of cloves and comprises 90 to 95% of the essential oil of clove (Ridley, 1983). At a concentration of 400μg/ml or even less, eugenol completely inhibits the growth of *Aspergillus flavus* and *A. versicolor* (Hitokoto et al., 1980). At 500 μg/ml clove oil inhibits the growth of *Aeromonas hydrophila* which is a cause of spoilage in non-cured meat (Stecchini et al., 1993). It also checks the growth of *Lactobacillus planterum* but the concentration required is much higher (4 mg/ml) as observed by Zaika and Kissinger (1979). Clove oil has also been found to be an effective inhibitor for *Trichoderma harzianum, Alternaria alternata, Fusarium oxysporum, F. culmorum, F. griseocyanus, Mucor circinelloides, Rhizopus stolonifer, Cladosporium cladosporoides, Penicillium citrinum, Saccharomyces cerevisiae* and *Aspergillus niger* (Schmitz et al., 1994; Meena and Sethi, 1994).

Eugenol is a broad spectrum antibacterial agent and is reported to be a more potent inhibitor for *E. coli* than ampicillin, cephaloridine, cotrimoxazole and gentamycin. Its antibacterial activity against *Staphylococcus aureus* is comparable to the effects of cephalexin, cotrimoxazole, chlorotetracycline, gentamycin and sulphamethezole (Suresh et al., 1992). *Micrococcus* spp., *Bacillus* spp., *Streptococcus* spp. and *Mycoderma pheli* are also significantly inhibited by clove oil (Farag et al., 1989b). Clove oil is practically insoluble in water and must be dispersed in a vehicle before use. Concentrated sugar solution is one such vehicle. Addition of clove oil at 0.4% v/v to a concentrated sugar solution (63% w/v) results in a marked germicidal effect against various bacteria (*Staphylococcus aureus, Klebsiella pneumoniae* and *Clostridium perfringens*) pathogenic for humans and *Candida albicans* (Jorge Briozzo et al., 1989).

Cumin (*Cuminum cyminum*)

Cumin ('jeera') is considered to act as a preservative controlling microbial spoilage in food. This preservative activity is due to the presence of cuminaldehyde, p-pinene and terpinotene. Among the various microorganisms that occur in foods, fungal and yeast cultures are more sensitive to cumin volatile oil and cuminaldehyde than bacteria. Among Gram-negative bacteria, *E. coli* is the most sensitive (Shetty et al., 1994). The essential oil in cumin is active against *Staphylococcus aureus, Salmonella typhi, Shigella dysentriae* and *Vibrio cholerae* (Syed et al., 1986). Moreover, the concentration required for antimicrobial activity is very low (800-1200 ppm). It is fungistatic at lower doses but at higher doses is fungicidal against aflatoxin-producing strains of *Aspergillus flavus* and *Aspergillus parasiticus* (Dubey et al., 1991). In a similar study, Farag et al. (1989a) observed that at 0.2 mg/ml cumin oil produces a highly significant decrease in both mycelium growth and aflatoxin production by *A. parasiticus*. In another study *Mycoderma* sp. is found to be more susceptible to cumin oil than *Aspergillus niger* (Meena and Sethi, 1994). Moreover, cumin oil has insect-repellant properties (Dubey et al., 1991) and cumin seeds possess anticarcinogenic activity (Aruna and Shivaramakrishna, 1992).

Curcuma (*Curcuma amada*)

Curcuma, commonly known as mango ginger, has many medicinal properties. Essential oil of curcuma is active against a wide range of bacteria and fungi, many of which are plant and human pathogens. Curcuma oil exhibits antibacterial activity against *Bacillus subtilis, E. coli, Klebsiella aerogenes, Salmonella paratyphi, Salmonella typhi, Staphylococcus aureus, Erwinia carotovora, Pseudomonas solanacearum, Xanthomonas citri* and *Xanthomonas malvacearum.* Antifungal activity has been reported for *Aspergillus niger, A. flavus, Penicillium lilacinum, P. javanicum, Trichoderma viridae, Curvularia oryzae, Pestalotia lapagericola, Microsporum gypseum* and *Trichophyton mentagrophytes* (Banerjee et al., 1982). For most of the organisms, the minimum dose required for the formation of the inhibition zone was 0.1 ml except for a few cases where it was 0.15 ml (*E. carotova, Pseudomonas solanacearum, Penicillium javanicum* and *Curvularia oryzae*) or even 0.2 ml (*Klebsiella aerogenes, Trichoderma viride* and *Trichophyton mentagrophytes*).

Garlic (*Allium sativum*)

Garlic ('lahsun', 'thom') has been known to mankind as a flavouring and therapeutic agent for thousands of years. Much of the garlic produced in Asia is used in unprocessed form while in developed countries 90% of the garlic crop is processed into 25 different kinds of flakes, salts and granules. Smaller amounts of garlic are processed to produce garlic juice and oil (Dalvi and Salunkhe, 1993). The unique pungent odour of garlic is due to the presence of the chemical diallyl disulphide together with smaller quantities of diallyl trisulphide and diallyl polysulphide and a little of diethyl disulphide. The antimicrobial activity of garlic has been well studied and is attributed to 'allicin'. When garlic is macerated its principal sulphur component 'allin' is broken down by an enzyme known as allinase to a range of flavour components including allicin (Dalvi and Salunkhe, 1993). Allicin exhibits strong antibacterial activity. One mg of allicin is equivalent to 15 oxford units of penicillin (Sreenivasa Murthy et al., 1983).

Several reports on garlic extract are available which show inhibition of growth of bacteria as well as of fungi. Fresh garlic extract shows germicidal activity against *Bacillus cereus* and *B. brevis*. For *Salmonella typhimurium* 4% garlic extract is bacteriostatic while a 5% concentration is bactericidal (El-Khateib and Abd-El-Rehman, 1987). The coagulase activity of *Staphylococcus aureus* is inhibited by 5% garlic extract (Fletcher et al., 1974). At 1% concentration, growth of *Lactobacillus plantarum* is inhibited (Karaioannoglou et al., 1977) and spores of *Clostridium perfringens* do not germinate (Mantis et al., 1979).

Compared with bacteria, the fungi are sensitive to garlic extract at much lower concentrations. *Aspergillus parasiticus* is inhibited at 0.23 to 0.4% concentration as observed by Graham and Graham (1987). A significant antifungal activity of garlic extract has been observed against the plant pathogenic fungi *Colletotrichum gleosporoides* and *Curvularia* sp. (Prodesimo and Ilag, 1976). Ajoene, one of the fractions derived from garlic, inhibits the growth of *Aspergillus niger* and *Candida albicans* at a concentration of less than 20 μg/ml (Yoshida et al., 1987). Garlic extract exerts its anticandidal effect by oxidising the thiol(L-cysteine glutathione 2-mercaptoethanol) groups present in essential proteins causing enzyme inactivation and subsequent inhibition of microbial growth (Ghannosum, 1988). Other workers have studied the biological activity of CHC13 extract of *Allium sativum* in aerosol therapy (Alkiewicz and Lutonski, 1992) and it has been proved that inhalation treatment by means of biological standardised concentration from *Allium sativum* was very effective in elimination of *C. albicans* from the respiratory tract in children.

Other bacteria inhibited by garlic extract include food-poisoning bacteria (Sato et al., 1990), *Proteus, Enterobacter, Klebsiella* (Sharma et al., 1977), *S. typhosa, Shigella paradysentrica* (Dold and Knapp, 1948), acid fast bacteria *Mycobacterium tubercolosis* and *M. leprae* (Suri, 1951). The active principle allicin is quite effective against *M. tuberculosis* H 3TRV and *M. tuberculosis* TRC-1193 which are isoniazid-resistant strains (Ratnakar and Murthy, 1995). Garlic extract retains its antimicrobial activity even on preincubation for 24 hours at 55 to 60°C against *E. coli* and *S. faecalis* (Shashikanth et al., 1981).

Apart from garlic extract, garlic oil also exhibits antimicrobial activity. Toxin production by *Clostridium botulinum* strain A is inhibited by garlic oil (Jacora et al., 1976). The volatile vapours from garlic oil and leaf extract inhibit the growth of *Alternaria* sp., *Curvularia* sp., *Fusarium* sp., *Helminthosporium* sp. (Singh et al., 1979) and *Rhizoctonia solani* (Singh and Singh, 1980). Essential oil of garlic has been reported to be a potent inhibitor of yeast growth at a concentration of 25 ppm (Conner and Beachut, 1984).

Mint (*Mentha* species)

Mentha species possess antispasmodic, chloretic and carminative properties which have been exploited for their medicinal value. Commonly known as 'pudina', there has been very little systematic work done on their antimicrobial properties. However, it has been demonstrated that the differences in chemical composition of essential oils of various species are responsible for their differential antimicrobial activity. Essential oils from two different species, namely *M. arvensis* and *M. longifolia*, exhibit strong antibacterial activity. Thirty-six compounds have been identified in both of these species. Essential oils from both species show strong antibacterial activity against *Bacillus subtilis*. However, the two species differ in activity against *Pseudomonas aeruginosa* which is sensitive only to *M. arvensis* oil (Mimica-Dukic et al., 1993).

According to Mimica-Dukic et al. (1993), *M. arvensis* did not exhibit any antifungal activity which is in contrast to some earlier reports. It could be attributed to differences in chemical composition of essential oil (chemotypes). Although the constituents of essential oils from *Mentha* have been shown to possess relatively higher antibacterial activity in comparison to fungi, *M. longifolia* showed much higher antifungal activity, especially against *Candida albicans* (Mimica-Dukic et al., 1993). This may be due to the high content of thymol which has strong antifungal activity.

Essential oils from other *Mentha* species have also been reported to possess antibacterial activity. *Mentha piperita* shows antibacterial activity against *Salmonella enteritidis* and *Listeria monocytogenes* (Tassou et al., 1995). Essential oil obtained from two mint species, *M. pulegium* and *M. spicata*, were extremely bactericidal at high concentrations whereas lower concentrations caused a dose dependent decrease in bacterial growth rate (Sivropoulau et al., 1995).

CONCLUSIONS

The spices, their extracts, oils and other derivatives are useful in various ways. In general, their volatile components are used for fragrances and as flavour ingredients in foods. They also act as preservatives. Certain indigenous plants with antimicrobial properties including spices and their products have come to the rescue of public health. Further exploration of such plants holds a great potential to meet the requirements of microbiologists to combat the menace of ever developing resistance of microbes to antibiotics. In India, spices grow abundantly and play quite a significant role in the country's economy. Spices have undoubtedly been in use for many years for various ailments but as far as their antimicrobial properties are concerned, significant systematic work is not generally available except for garlic. Some of these spices do have potential antimicrobial properties but these need to be explored fully for various groups of microorganisms, exact component analyses to be made and the underlying mechanisms understood. Thus, there is the potential for spices not only to remain important from a culinary point of view but also to become important as antimicrobial agents.

ACKNOWLEDGEMENTS

The author wishes to thank Miss Harbinder Kaur and Miss Paramjeet Kaur for their help in collection and arranging of literature.

REFERENCES

Abo-Khatawa, N. and Kubo, I., 1987, Chemical composition of the essential oil of cardamom seeds, *Elattaria cardamomum, Proc. Saudi. Biol. Soc.* 10: 297-305.

Alkiewicz, J. and Lutonski, J., 1992, Study on the activity of the $CHCl_3$ extract of garlic on *Candida albicans, Herba Polonica* 38: 79-84.

Anand, J.C. and Johar, D.S., 1957, Effect of condiments on the growth of *Aspergillus* niger in mango pickle, J. *Sci. Ind. Res.* 16: 370-373.

Anon., 1979, *British Pharmaceutical Codex 11th Edn.* London: Pharmaceutical Index.

Aruna, K. and Shivaramakrishna, V.M., 1992, Anticarcinogenic effects of some Indian plant products, *Food Chem. Toxicol.* 30: 953-956.

Asprey, G.F. and Thornton, P., 1976, Medicinal plants of Jamaica, *West Indies Med. J.* 3: 17-20.

Atal, C.K., Dhar, K.C. and Pelter, A., 1967, Isolation and structure determination of (+)-diaeudesmin, the first naturally occurring diaxially substituted 3,7-dioxabicyclo[3,3,30] octane lignan, *J. Chem. Soc. C.* 2228-2231.

Badei, A.Z.M., 1992, Antimycotic effects of cardamom essential oil against mycotoxigenic moulds in relation to its chemical composition, *Chemie Mikrobiologic Technologic Der Lebensmittel* 14: 177-182.

Banerjee, A., Kaul, V.K. and Nigam, S.S., 1982, Antimicrobial activity of the essential oil of *Curcuma* amada Roxb. *Ind. J. Microbiol.* 22: 154-155.

Cichewicz, R.H. and Thorpe, P.H., 1996, The antimicrobial properties of Chillie peppers (*Capsicum* species) and their uses in Mayan medicine, *J. Ethnopharmacology* 52: 61-70.

Conner, D.E. and Beachut, L.R., 1984, Effect of essential oils from plants on the growth of food spoilage yeast, *J. Food Sci.* 49: 429-434.

Dalvi, R.R. and Salunkhe, D., 1993, An overview of medicinal and toxic properties of garlic, *J. Maharashtra Agric. Univ.* 18: 378-381.

Dold, M. and Knapp, A.Z., 1948, *Chem. Abstr.* 47: 94-96.

Dubey, S., Upadhyay, P.D. and Tripathi, S.C., 1991, Fungitoxic and insect repellant efficacy of some spices, *Indian Phytopathol.* 44: 101-105.

El-Khateib, T. and Abd-El-Rehman, H., 1987, Effect of garlic on the growth of *Salmonella typhimurium* in Egyptian fresh sausage and beer burger, *J. Food Protection* 50: 310-311.

Escobar, A.N., 1972, *Flora Toxica de Panama,* Editorial Universitaria, Panama, pp. 239-281.

Farag, R.S., Daw, Z.Y. and Abo-Raya, S.H., 1989a, Influence of some spice essential oils on *Aspergillus parasiticus* growth and production of aflatoxins in a synthetic medium, *J. Food Sci.* 54: 74-76.

Farag, R.S., Daw, Z.Y., Hewedi, F.M. and El-Baroty, G.S.A., 1989b, Antimocrobial activity of some Egyptian spice essential oils, 52: 665-667.

Fletcher, R.D., Parker, D.B. and Hassit, M., 1974, *Folia Microbiologica* 19: 494.

George, M. and Pandalai, K.M., 1958, Antimicrobial activity of few common essential oils, *Proc. Symp. Essential Oils Aromat. Chem.* 158.

Ghannosum, M.A., 1988, Studies on the anticandidal mode of action of *Allium sativum* (garlic), *J. Gen. Microb.* 134: 2917-2924.

Graham, H.D. and Graham, E.J.F., 1987, Inhibition of *Aspergillus* parasiticus growth and toxin production by garlic, *J. Food Safety* 8: 101-108.

Hitokoto, H., Morozumi, S., Wauke, T., Sakai, S. and Kurata, H., 1980, Inhibitory effects of spices on growth and toxin production by toxigenic fungi, *Appl. Environ. Microbiol.* 39: 818-822.

Huhtanen, C.N., 1980, Inhibition of *Clostridium botulinum* by spice extracts and aliphatic alcohols, *J. Food Protection* 43: 195.

Husain, F., 1996, Trends in the international spice trade. International Trade Forum, International Trade Centre UNCTAD/WTO, 2: 14-15, 33-34.

Ito, H. and Meixu, G., 1994, Inhibition of microbial growth by spice extracts and their effect of irradiation, *Food Irradiation Japan* 29: 1-7.

Jacora, C., Notermans, S., Gorin, N. and Kampelmacher, E.H., 1976, Volatile organic sulfur compounds in garlic and onion, *J. Food Prot.* 42: 222.

Jorge Briozzo, Nunez, L., Chirife, J., Herszage, L. and Aquino, M.D. (1989), Antimicrobial activity of clove oil dispersed in a concentrated sugar solution, *J. Appl. Bacteriol.* 66: 69-75.

Karaioannoglou, P.G., Mantis, A.S. and Panetosoos, A.G., 1977, The effect of garlic extracts on lactic acid bacteria, *Lactobacillus plantarum* in culture medium, *Wiss. Technol.* 10: 148.

Kubo, I., Himejima, M. and Muroi, H., 1991, Antimicrobial activity of flavor components of cardamom *Elattaria cardamomum* (Zingiberaceae) seed, *J. Agric. Food Chem.* 39: 1984-1986.

Loder, J.W., Moorhouse, A. and Russel, G.B., 1969, Tumor inhibitory plants: amides of *Piper novae-Hollondiae, Aust. J. Chem.* 22: 1531-1538.

Mantis, A.J., Koidis, P.A., Karaioannoglou, P.G. and Panetosoos, A.G., 1979, *Lebnsm Wiss. in Technol.* 12: 350.

Marck, G.F. and Thornton, J.P., 1993. History of Indian spices, *Ind. J. Med. Res.* 24: 2-4.

Meena, M.R. and Sethi, V., 1994, Antimicrobial activity of essential oils from spices, *J. Food Sci. Technol.* 31: 68-70.

Mimica-Dukic, N., Kite, G., Gasic, O., Stajner, D., Pavkov, R., Jancic, R. and Fellows, L., 1993, Comparative study of volatile constituents and antimicrobial activity of *Mentha* species, *J. Acta Hortic.* 344: 110-115.

Nair, M.G. and Burke, B.A., 1990, Antimicrobial *Piper* metabolite and related compounds, *J. Agric. Food Chem.* 38: 1093-1096.

Nair, K.K.S., Rao, D.N. and Nair, R.B., 1996, Effect of spices and organic acids used in traditional meat products on some bacteria, *Ind. Food Packer* 50: 5-10.

Nigam, S.S., 1982, Antimicrobial activity and essential oils, *Ind. Perfumer* 26: 249-254.

Prodesimo, A.N. and Ilaq, L.L., 1976, *Kalikasan,* 5: 251.

Pruthi, J.S., 1980, *Spices and Condiments: Chemistry, Microbiology and Technology,* Advances in Food Research Suppl. 4, Academic Press, New York, London.

Pruthi, J.S., 1992, *Spices and Condiments,* National Book Trust of India, New Delhi.

Ratnakar, P. and Murthy, P.S., 1995, Purification and mechanism of action of antitubercular principle from garlic, active against isoniazid susceptible and resistant *Mycobacterium tuberculosis* H 37 RV, *Ind. J. Clinical Biochem.* 10: 14-18.

Ridley, H.R., 1983, *Spices,* MacMillan & Co., London.

Ryu, D. and Holt, D.L., 1993, Growth inhibition of *Penicillium expansum* by several commonly used food ingredients, *J. Food Prot.* 56: 862-867.

Sato, A., Terao, M. and Honma, Y.L., 1990, Antimicrobial action of garlic extract on food poisoning bacteria, *J. Food Hygiene Society of Japan* 31: 328-332.

Schmitz, S., Weidenbgner, M. and Kunaz, B., 1993, Herbs and spices as selective inhibitors of mould growth, *Chemie, Mikrobiologie Technologie der Lebensmittel* 15: 175-177.

Sebti, F. and Tantaoui-Elaraki, A., 1994, In vitro inhibition of fungi isolated from 'Pastilla' papers by organic acids and cinnamon, *Lebensmittel-Wissenschaft and Technologie* 27: 370-374.

Sharma, V.D., Sethi, M.S., Kumar, A. and Rarotra, J.R., 1977, *Ind. J. Exp. Biol.* 15: 466.

Shashikanth, K.N., Basappa, S.C. and Sreeinvasa Murthy, V., 1981, Studies on antimicrobial and stimulatory factors of garlic *(Allium sativum), J. Food Sci. Technol.* 18: 44-47.

Shetty, R.S., Singhal, R.S. and Kulkarni, P.R., 1994, Antimicrobial property of cumin, *World J. Microb. Biotech.* 10: 232-233.

Singh, U.P., Pathak, K.K., Khare, M.N. and Singh, R.B., 1979, Effect of leaf extract of garlic on *Fusarium oxysporum* F. sp. *ciceri, Sclerotinia sclerotiorum* and on gram seeds, *Mycologia* 71: 556-564.

Singh, H.B. and Singh, V.P., 1980, *Mycologia,* 72: 1022.

Sivrapoulou, A., Kokkini, S., Lanaras, T. and Arsenakis, M., 1995, Antimicrobial activity of mint essential oils, *J. Agric. Food Chem.* 43: 2384-2388.

Sreenivasa Murthy, V., Shashikanth, K.N. and Basappa, S.C., 1983, Antimicrobial action and therapeutics of garlic, *J. Sci. Ind. Res.* 42: 410-414.

Stecchini, M.L., Sarais, I. and Giavedoni, P., 1993, Effect of essential oils on *Aeromonas hydrophila in a culture medium and in cooked pork, J. Food Protection* 56: 406-409.

Suresh, P., Ingle, V.K. and Vijayalakshi, 1992, Antibacterial activity of eugenol in comparison with other antibiotics, *J. Food Sci. Tech.* 29: 256-257.

Suri, J.C., 1951, *Ind. J. Med. Res.* 39: 441.

Syed, M., Hanif, M., Chaudhary, F.M. and Bhatty, M.K., 1986, Antimocrobial activity of the Umbelliferae family, 1: *Cuminum cyminum, Corianderum sativum, Foeniculum vulgare* and *Bunium persicum* oils, *Pakistan J. Sci. Ind. Res.* 20: 183-188.

Tassou, C.C., Drosinos, E.H. and Nychas, G.J.E., 1995, Effects of essential oil from mint (*Mentha piperita*) on *Salmonella enteriditis* and *Listeria monocytogenes* in model food systems at 4°C and 10°C, *J. Appl. Bacteriol.* 78: 593-600.

Yoshida, S., Kasuga, S., Hayashi, N., Ushiroguchi, T., Matsuura, H. and Nakagawa, S., 1987, Antifungal activity of ajoene derived from garlic, *Appl. Environ. Microb.* 53: 615-617.

Zaika, L.L. and Kissinger, J.C., 1979, Effect of some spices on acid production by starter culture, *J. Food Protection* 42: 572-576.

EXPLORING MUSHROOM DIVERSITY FOR PHARMACEUTICAL UTILITY

S.S. Saini and N.S. Atri

Botany Department
Punjabi University
Patiala-147 002
India

INTRODUCTION

Mushrooms are an important group of neutriceuticals which are used for an immense variety of purposes (Chang and Buswell, 1996). Besides their edibility, mushrooms have long been considered to have medicinal properties. Due to their unique composition, they have played an important part in folk medicines as therapy for a variety of ailments. As part of the diet, they are excellent for sufferers of diabetes, obesity, hyperacidity, hypertension, atherosclerosis, high blood pressure, anaemia and constipation. A large number of mushroom species including *Ganoderma lucidum, Coriolus versicolor, Fomes fometarius, Tremella fuciformis* and *Lentinus edodes* are traditionally used in Chinese folk medicines. Besides the above, other mushrooms, for example *Agrocybe cylindracea, Tricholoma mongolicum, Inonotus obliquus, I. hispidus, Pleurotus ostreatus, Collybia dryophila, C. radicata, C. peronata, Suillus bovinus, Coprinus plicatilis, Hypholoma fasciculare, Leucopaxillus giganteus* and *Pholiotina appendiculata,* are being explored extensively for their pharmaceutical utility.

With the signing of the global biodiversity convention at Rio de Janeiro in 1992 on the conservation of natural resources and our rediscovered interest in traditional systems of medicines based on the utilization of harmless bioactive substances, the exploration of this important group of fungi has become all the more important. Scientists are now paying considerable attention to investigation of the medicinal utility of plants in general but mushrooms in particular. A large number of bioactive substances from mushrooms which are effective against fungi, bacteria and viruses have already been identified.

BIOACTIVE PRODUCTS

There is considerable interest in obtaining new products from natural ecosystems. In fungi, besides enzymes of biotechnological utility and other products including biocontrol agents, the metabolites of pharmaceutical utility are of great interest for counteracting common ailments. Besides antibiotics, a large number of substances known as Host Defence Potentiators (HDPs), Protein bound Polysaccharide or Polysaccharide-Protein Complexes (PSPCs) have been isolated from mushrooms (Subramanian, 1995). Such bioactive mushroom metabolites are believed to have the capability to help in the revitalization of our immune system against a large number of pathogenic and non-pathogenic diseases. They are reported to serve as biological response modifiers with the capability to activate macrophages and T-cells, and to produce cytokines, interleukines and tumour necrosis factors. Some such reported bioactive substances from mushrooms are pleurotin, lepiochlorin, clavicin, sparassol, triterpenes, ganoderols, armillarin, dictyophorin, cylindan, adenosine, etc. In this account, the pharmaceutical applications of various bioactive substances derived from mushrooms are discussed.

Antibacterial Properties

Large numbers of chemical constituents from some mushroom genera (*Clitocybe, Agrocybe, Pleurotus, Psathyrella, Tricholoma, Amanita, Collybia, Coprinus, Agaricus*, etc.) have been reported to possess antibacterial activity. Phenolic and quinoid derivatives from *Agaricus bisporus* have been described to be bactericidal (Vogel et al., 1974). Neubularine from *Clitocybe nebularis* has also been reported to be bacteriostatic (Loefgren et al., 1954). Hervey and Nair (1979) reported lepiochlorin, an antibacterial antibiotic from *Lepiota* cultivated by gardening ant *(Cyphomyrmex costatus)*. Min et al. (1996) reported the antibiotic activities of watery extract of *Amanita pantherina* and *Lycoperdon perlatum* against Gram negative and Gram positive bacteria respectively.

Bianco et al. (1996) tested the mycelial and cultural filtrates from 25 basidiomycetes for their antibiotic activity against different bacteria. Filtrate of *Pholiotina appendiculata*-135A has been reported to be active against *Bacillus cereus, B. subtilis, Staphylococcus aureus, Escherichia coli, Proteus mirabilis, Salmonella typhimurium* and *Candida albicans.* The filtrates of *Albatrellus confluens, Postia styptica* and *Lycoperdon pyriforma* have been reported to inhibit the growth of *Bacillus cereus, B. subtilis, E. coli* and *Salmonella typhimurium.* The extracts of *Clitocybe infundibuliformis, Collybia peronata, Suillus bovinus, Coprinus plicatilis, Cyathus striatus, Hypholoma fasciculare* and *Leucopaxillus giganteus* have been reported to be active against *Bacillus cereus* and *B. subtilis* and that of *Polysticus tomentosus* against *Salmonella typhimurium.* The extracts from *Stereum hirsutum, Phellinus torulosus, Tramates suaveolens, Collybia dryophila (*120A and 120B) and *C. radicata* are reported to be active only against *Bacillus cereus.*

Further investigations on the above fungal taxa could lead to the extraction and identification of some specific antibiotic substances. Antibacterial antibiotic substances have also been reported from *Clitocybe veneriata* (Diatretine - Singer, 1986) and *Agrocybe dura* (Agrocybin - Kavanagh et al., 1950). For more information on the subject reference can be made to Bianco et al. (1995) wherein the authors have provided a review on antibiotics from Basidiomycetes.

Antifungal Properties

Mushrooms which are reported to have a pronounced antifungal active principle include *Lentinus edodes, Coprinus comatus* and *Oudemansiella mucida.* Sparassol (orsellinic acid monomethyl ether) from *Sprassis ramosa* (Falck, 1923) and Cortinellin from *Lentinula edodes* (edible shiitake mushroom) (Herrman, 1962) are important mushroom substances with significant antifungal properties. Martinkova et al. (1995) recorded the antifungal activity of polyketide pigments (Monoascoruberin and Rubropunctatin) from *Monoascus purpureus* which were effective against some species of yeasts and filamentous fungi. Gamble et al. (1995) isolated a new pentacyclic triterpenoid compound (Polytolypin) from *Polytolypa hystricis* exhibiting antifungal activity. Min et al. (1996) reported antifungal activity of the extracts obtained from *Agaricus subrutilescens* and *Amanita virosa*.

Anticancer Properties

Some mushroom substances, for example Lentinan (a polysaccharide from *Lentinus edodes*), Flammulin (a protein from *Flammulina velutipes*), Proteoglucan (a protein with ß-glucan from *Grifola frondosa*), Clavicin (from *Clavatia gigantea*), Retine (α-keto aldehyde from *Agaricus campestris*), Terpenoids (from *Ganoderma lucidum*), Porocin (from *Poria corticola*) etc. are well known to have anti-tumour properties. Pioneers who have contributed significantly towards our knowledge about the anti-tumour properties of the substances include Chedd (1967), Ikekawa et al. (1968), Shibata et al. (1968), Ruelius et al. (1968), Nanba (1993) and King (1993).

Host-mediated anti-tumour polysaccharides from *Ganoderma lucidum*, *G. tsugae* and *G. boninense* are some of the most important immunopotentiators currently in use for the treatment of cancer in Japan and USA. Over 100 terpenoids (Ganoderic acids, Ganoderol, Lucidernic acids and Lucidone etc.) have been reported from the fruit bodies and mycelium of *G. lucidum* (Subramanian, 1995). The anti-tumour effects of *G. lucidum* are reported to be mediated by cytokines released from the activated macrophages and T. lymphocytes (Wang et al. 1997). These cytokines are reported to act synergistically on the inhibition of leukaemic cell growth.

Hyun et al. (1996) reported a neutral protein-bound polysaccharide fraction (Cylindan) from *Agrocybe cylindracea* with marked anti-tumour activity. It is reported to have shown about 70% tumour inhibition against the solid form of sarcoma 180 when a dose of 30 mg/kg/day was intraperitoneally injected into the mice. While conducting immunological studies with Cylindan, Kim et al. (1997) reported a marked life extension effect in mice against the ascite form of sarcoma 180 tumour and lewis lung carcinoma at a dose of 50 mg/kg/day. It is also reported to have restored the decreased immune response of the tumour-bearing mice.

Wang et al. (1996a) isolated two lectins (TML-1 and TML-2) from *Tricholoma mongolicum*. Both these lectins have been reported to possess immunomodulatory and anti-tumour properties with the capability to inhibit the growth of implanted sarcoma 180 cells by 68.84 % and 92.39% respectively. Wang et al. (1996b) reported a polysaccharide-peptide complex with immunoenhancing and anti-tumour activities from the mycelial cultures of *Tricholoma mongolicum*. It inhibited the growth of sarcoma 180 cells implanted in mice. Besides the above mushrooms, *T. matsutake* and *Pholiota nameko* are also receiving the attention of scientists for their anti-tumour properties. Lim et al. (1996) reported an anti-tumour active fraction (PK-1, PK-2, PK-3A and PK-3B) from sclerotium of *Poria cocas* (HH-1). When assayed against sarcoma 180 mouse with PK-1 (100,50 mg/kg), PK-2 (40,20 mg/kg), PK-3A (4,2 mg/kg) and PK-3B (4 mg/kg), the reported inhibition ratio of each fraction was 79.3%, 90.2%, 85.2%, 90.4%, 81.2%, 73.9% and 75.6%, respectively. The life span prolongation effect of each fraction was reported to be much better than that of the control group.

Antiviral Properties

Of the large variety of mushrooms tested against the poliomyelitis virus in mice, some of them, namely *Boletus frostii*, *Clavatia gigantea*, *Cholorophyllum molybdites*, *Lepiota morgani*, *Russula emetica*, *Paneolus subalteatus*, *Armillaria mellea*, *Coprinus micaceus*, *Agaricus campestris* and *A. placomyces* have been reported to possess significant potential (Cochran,

1978). In *Clavatia gigantea,* some high molecular weight derivatives are reported to be effective against poliomyelitis and influenza viruses (Cochran, 1978). Tsunoda et al. (1970) and Sujuki et al. (1976) reported interferon inducing capability in extract of *Lentinus edodes.* In this fungus Eritadenine has been reported to be active against the influenza virus in mice. Kahlos et al. (1996) reported that the black thin external surface of *Inonotus obliquus* strains (A/HINI, A/H$_3$N$_2$, A/Equine 2, B/Yamagata/16/18) grown in birch showed 100% inhibition against human influenza viruses A and B and horse influenza virus A. The antiviral activity of this fungus is thought to be due to Betulin, Lupeol and Mycosterols.

Anti-HIV Properties

The extract of *Grifola frondosa* has been shown to kill the AIDS virus and is reported to be capable of enhancing the activity of helper-T cells. The extract of this fungus is reported to be as effective against HIV as the widely used toxic drug AZT (Nanba, 1993; King, 1993). Lentinan from *Lentinus edodes* also possesses the ability to enhance host resistance to a variety of infections including HIV-1 (Subramanian, 1995). Walder (1995) reported the strong anti-HIV-1 activity of aqueous extracts from *Fomitella supina, Phellinus rhabarbarinus, Trichaptum perrottettii* and *Trametes cubensis.* The active principle is reported to have acted by the mechanism of direct virion inactivation and by inhibition of synchytium formation. The unknown active components of these extracts individually or in combination may have therapeutic relevance. Collins and Tazi (1997) isolated a polysaccharopeptide (PSP) from *Coriolus versicolor* which has potential for use against HIV-1 infection. It acts by inhibition of the interaction between HIV-1 group 120 and immobilised CD$_4$ receptor (IC$_{50}$ =150 μg/ml), recombinant HIV-1 reverse transcriptase (IC$_{50}$ = 125 μg/ml), and glycohydrolase enzyme associated with viral glycosylation. Such properties, coupled with its high solubility in water, heat stability and low cytotoxicity, make it a useful compound for controlling HIV infections.

Hypocholestrolemic and Hypolipidemic Effects

Some of the edible mushrooms, for example *Lentinus edodes, Agaricus bisporus, Pleurotus florida, P. ostreatus* and *Auricularia auricula,* are reported to possess the ability to lower blood cholesterol. Suzuki and Oshima (1976) reported the hypocholestrolemic effects of Shiitake in man. Bhandari et al. (1991) recommended *Pleurotus florida* as the potential source of active ingredients required for sufferers of high blood cholesterol. Hypolipidemic properties of Shiitake are reported to be due to Eritadenine (= Lentysine, Lentinacin), 2(R), 3(R)-dihydroxy-4-(9-adenyl)-butyric acid (Chibata et al., 1969; Rokujo et al., 1970). In *Auricularia polytricha,* an anti-platelet substance (Adenosine) has been reported to have the capability to inhibit platelet aggregation (Markhija and Bailley, 1981). While working with rats, Luthra et al. (1991) reported the interference of the mushroom in the mobilization of lipids. The inclusion of dried *Agaricus bisporus* sporophores at 5% or 10% level in the diet of rats has been reported to have resulted in the accumulation of lipids in the liver with simultaneous decrease in the circulatory lipids, except phospholipids, in plasma. Bobek et al. (1995) reported the antioxidative effect of Oyster mushroom in hypercholestrolemic rats.

Anti-Diabetic Properties

Edible mushrooms are known for their low calorific value (25 to 30 calories/100 g fresh weight) and low carbohydrate content in comparison to other food items (Rai, 1995). For this reason they are considered excellent for diabetic patients. Freeze-dried powder containing mycelium of *Ganoderma lucidum* has been shown to bring down blood sugar levels in experimental diabetic rats. Three hypoglycaemic principles, namely Ganoderans A, B and C, are reported to have been isolated from the fruit bodies of *G. lucidum,* and these have been characterized as peptidoglycans. Of these Ganoderan-B is considered to be the most important as far as antidiabetic properties are concerned (Subramanian, 1995).

Blood Building Properties and Immunity

Mushrooms contain vitamins of the B-complex (Crisan and Sands, 1978). Folic acid, which is a blood building vitamin, is good for persons suffering from anaemia. Ascorbic acid (Vitamin C), present in edible mushrooms, increases resistance in the human body (Crisan and Sands, 1978). Along with these, other vitamins (for example pantothenic acid and niacin) and minerals (calcium, phosphorus, potassium, copper, iron, etc.) add to the vitality and immunity of the body (Chang and Miles, 1989). Qian et al. (1997) reported a protein-bound polysaccharide (PSP) from *Coriolus versicolor* with immunopotentiating effects when administered at 2 g/kg/day to rats. The active principle is reported to have restored the cyclophosphamide-induced immunosuppression such as depressed lymphocyte proliferation.

Hepato-protective Properties

Extracts of *Ganoderma lucidum* have been shown to be hepato-protective. Apart from liver regeneration, beneficial effects in counteracting hepatic necrosis and hepatitis have been reported. Ganodosterone from *G. lucidum* is reported to be a liver protectant with the ability to stimulate liver function. Similarly Ganoderic acids T, S and R from *G. lucidum* and triterpenoids from *G. tsugae* (Lucidone-A, Lucidenol, Ganoderic-B, Ganoderic Acid C_2) are reported to be hepatoprotective (Subramanian, 1995). Park et al. (1997a,b) suggested that polysaccharides from the mycelium of *G. lucidum* serve as promising agents for the inhibition of hepatic cirrhosis. They further suggested that these polysaccharides could serve as promising antifibrotic agents because of their capability to lower the collagen content, serum aspartate transaminase (AST), alanine transaminase (ALT), alkaline phosphatase (ALP) and total bilirubin in the liver.

Anti-coagulant and Anti-thrombic Properties

An anti-platelet substance (Adenosine) from *Auricularia polytricha* is known to inhibit platelet aggregation (atherosclerosis) and prolong bleeding time. The ingestion of this fungus as food is reported to result in reducing the chances of heart attack (Hammerschmidt, 1980). *Auricularia* is said to have been used in folk medicine in Hongkong to thin the blood and reduce clotting problems in post-partum women (Singer and Harris, 1987).

The oral administration of fructo-oligosaccharide mixture from *Lentinus edodes* (SK-204) to rats for ten weeks is reported to have anti-thrombic action (Otsuka et al., 1996) due to the promotion of fibrinolysis and thrombolysis.

Other Therapeutic and Miscellaneous Applications

Besides the above applications there are many other areas in therapeutics and research where mushrooms are finding an increasing role. Coatney et al. (1953) reported terpenoids of *Clitocybe illudens* to be effective against *Plasmodium gallinaceum.* Takemoto (1961) reported insecticidal properties of an amino acid derivative, Tricholomic acid from *Tricholoma muscarum.* Aqueous extracts of *Pleurotus sajor-caju* have been reported to reduce the rates of nephron deterioration in persons suffering from renal failure (Tam et al., 1986). *Grifola frondosa* has been reported to be beneficial for lowering blood pressure, diabetes and constipation. *Fomes officinalis* is another fungus which has been listed as a universal remedy for a variety of ailments. Spores and capillitia of *Lycoperdon* are known to be in use for stopping bleeding from wounds (Hawksworth et al., 1995). Wang et al. (1996c) isolated a hypotensive and vasorelaxing lectin from *Tricholoma mongolicum.* This lactin on administration in rats at a dose of 10 mg/kg body weight resulted in the reduction of arterial blood pressure. The hypotensive activity of lectins is reported to be mediated through vasorelaxation via adenosine A_2 receptors and/or nitric oxide production. *Inonotus hispidus,* which produces styrylpyrones (hispidin) and derivatives of caffeic acid (hispolon) as pigments, has been suggested as a valuable source of new drugs (Pilgrim et al., 1997). Two novel eudesmane type sesquiterpenes, dictyophorines A and B and a known compound teucrenone isolated from *Dictyophora indusiata,* have been reported to promote nerve growth fraction (Kawagishi et al., 1997). Eleven species of bracket mushrooms belonging to *Phellinus (P. badius, P. chinchonensis, P. durrissimus, P. gilvus, P. linteus, P. merrilli, P. pachyphloes, P. pectinatus, P. robiniae, P. senex, P. sublinteus* and two species of *Ganoderma (G. applanatum* and *G. lucidum)* are reported to be in extensive use as Phanasomba or Phanas alombe by the Ayurvedic Vedas in the Pune region of India. A paste prepared out of these is applied to gums for stopping excessive salivation and it has been reported to act as a good styptic (Vaidya, 1991).

MUSHROOM MEDICINES AND PRESCRIPTION

Mushroom bioactive substances are being marketed as medicines by some medical companies, as a consequence of which some mushroom medicines are now available for the treatment of common human ailments. PSK, an anticancer drug (Pai et al., 1990) from *Coriolus versicolor* is reported to be quite popular for cancer treatment in Japan. Soo (1996) prescribed four *Ganoderma lucidum* capsules of 500 mg each three times daily for the treatment of diabetes, three capsules three times daily for rheumatism, six capsules three times daily for cancer, one capsule three times daily for five days for treating hypocholestrolemia and one to two capsules three times daily for general health maintenance, etc. A Hongkong based private company of Australia (Concord Sunchih Concord International Trading) have marketed capsules of *Ganoderma* with recommendations of 1-2 capsules twice daily for general health maintenance and 3 capsules three times daily for therapeutic purposes. Grifron Pro D-fraction, a medicinal product of *Grifola frondosa* has been marketed by Maitake Products, Inc., New Jersey, U.S.A. A 30 ml bottle marketed by the company is reported to contain 900 mg of pure Maitake Pro D-fraction having immunopotentiating properties. For general health maintenance, administration of 5-6 drops of this medicine three times daily and 0.5 mg to 1 mg/kg of body weight for the treatment of suffering persons have been recommended. The recommended prescriptions are reported to be safe for long term use with remote possibilities of any ill effects (Rai, 1997).

FUTURE STRATEGIES

The latest research on the pharmaceutical utility of mushrooms has opened up a new field for mushroom scientists to explore. There is a need to unearth the vast treasure of wild and cultivated mushrooms for their medicinal properties. Extraction and chemical characterisation of medicinally important substances from as yet unexplored mushroom germplasm and their utilization for therapeutic purposes could provide alternative medicines with minimum side effects for the human population. India, being a vast country with varied climatic conditions supporting a rich mushroom flora, so far has a poor record of contributions in this area of research. Except for a few scattered reports little worthwhile research has been done in this direction in India. There is a need to start tapping the mushroom flora for the extraction of medicinally active products with a view to utilizing this rich resource for human welfare.

CONCLUSIONS

Besides being a healthy food, mushrooms are low calorie neutriceuticals best suited for persons suffering from cancer, heart ailments, diabetes, high blood pressure, constipation, renal failure, etc. Due to their unique chemistry and high fibre content mushrooms possess the ability to revitalise immunity and by doing so can increase the life span of persons consuming them. The vast treasure of medicinally important mushrooms in the Indian subcontinent needs exploration for human welfare. There is a requirement to raise awareness of their importance so that the myth about fungi being only pathogenic and harmful is discredited.

Some of these mushrooms could be cultivated and produced under a controlled environment. Their production per unit area is enormous when compared with the production of common food items in our diet. By increasing their production and encouraging the masses to change their dietary habits, the prevailing stress on conventional food items could be reduced. Moreover, mushrooms will provide better nutrition and health, thus lessening our demand for the therapeutic chemicals in use at present for the treatment of different ailments.

Considering their immense utility, it is high time that edible fungi formed a permanent part of the dietary recommendations by medical practitioners for ailing persons.

REFERENCES

Bianco, C., Ausilia, M., Gillone, C. and Cesano, C., 1995, Antibiotics from Basidiomycetes: a review, *Alliona (Turin)* 33 (0): 7-68.

Bianco, C., Ausilia, M. and Giardino, L., 1996, Antibiotic activity in Basidiomycetes X. Antibiotic activity of mycelia and cultural filtrates of 25 new strains, *Alliona (Turin)* 34(0): 39-43.

Bhandari, R., Khanna, P.K., Soni, G.L., Garcha, H.S. and Mittar, D., 1991, Studies on hypocholesterolemic/hypolipidemic action of *Pleurotus florida* in albino rats, in: *Indian Mushrooms* (M.C. Nair, ed.), Kerala Agricultural University, India.

Bobek, P., Ozdin, L. and Kuniak, L., 1995, Antioxidative effect of oyster mushroom (*Pleurotus ostreatus*) in hypercholesterolemic rats, *Pharmazie* 50(6): 441-442.

Chang, S.T. and Buswell, J.A., 1996, Mushroom neutriceuticals, *World J. Microbiol. Biotechnol.* 12(5): 473-476.

Chedd, G., 1967, A chemical to combat cancer, *New Sci.* 34: 324-325.

Chibata, I., Okumura, K., Takeyama, S. and Kotera, K., 1969, Lentinacin: a new hypocholesterolemic substance in *Lentinus edodes, Experientia* 25: 1237-1238.

Coatney, G.R., Cooper, W.C., Eddy, N.B. and Greenberg, J., 1953, Survey of antimalarial agents, *Public Health Monogr.* 9, Fed. Security Agency, Washington D.C.

Cochran, K.W., 1978, Medical effects, in: The *Biology and Cultivation of Edible Mushrooms* (S.T. Chang and W.A. Hayes, eds.), Academic Press Inc., London.

Collins, R.A. and Ng, Tazi Bun, 1997, Polysaccharopeptide from *Coriolus versicolor* has potential for use against human immunodeficiency virus type-1 infection, *Life Sciences* 60(25): 383-387.

Crisan, E.V. and Sands, A., 1978, in: *Biology and Cultivation of Edible Mushrooms* (S.T. Chang and W.A. Hayes, eds.), Academic Press, New York.

Falck, R., 1923, Uber ein krystallisiertes Stoffwechselproduct von *Sprassis ramosa* Shiff., *Chem. Ber.* 56B: 2555-2556.

Gamble, W. R., Gloer, J.B., Scott, J.A. and Malloch, D., 1995, Polytolyin, a new antifungal triterpenoid from the coprophilous fungus *Polytolypa hystricis, J. Natural Products (Lloydia)* 58(12): 1984-1986.

Hammerschmidt, D.E., 1980, Szechwan purpura, *New England J. Medicine* 320(21): 1191-1193.

Hawksworth, D.L., Krick, P.M., Sutton, B.C. and Pegler, D.N., 1995, *Ainsworth & Bisby's Dictionary of the Fungi,* CAB International, Wallingford, Oxon.

Herrman, H., 1962, Cortinellin, eine antibiotisch wirksam Sunstanz aus *Cortinellus shiitake, Naturwissenschaften* 49: 542.

Hervey, A.H. and Nair, 1979, *Mycol* 71: 1064 C.f. Hawksworth, D.L., Krick, P.M., Sutton, B.C. and Pegler, D.N., 1995, *Ainsworth & Bisby's Dictionary of the Fungi,* CAB International, Wallingford, Oxon.

Hyun, J.W., Chae, K.K., Seol, H.P., Jong, M.Y., Mi, J.S., Chary, Y.K., Eung, C.C. and Byong, K.K., 1996, Antitumor components of *Agrocybe cylindracea, Archives of Pharmacal Res. (Seoul)* 19(3): 207-212.

Ikekawa, J., Nakamishi, M., Uehara, N., Chihata, G. and Fukuoka, F., 1968, Antitumor action of some basidiomycetes, especially *Phellinus linteus, Gann.* 59: 155-157.

Kahlos, K., Lesnan, A., Lange, W. and Lindequist, U., 1996, Pleliminary tests of antiviral activity of two *Inonotus obliquus* strains, *Fitoterapia* 67(4): 344-347.

Kavanagh et al., 1950, *Proc. Natn. Acad. Sc. Wash.* 36: 102-106, C.f. Singer, R., 1986, *The Agaricales in Modern Taxonomy,* Sven Koeltz Scientific Books, Germany.

Kawagishi, H., Daisuke Ishiyama, Horonobu Mori, Hideku Sakamoto, Yukio Ishiguro, Shoei Furukawa and Jingxuan Li, 1997, Dictyophorines A and B, two stimulators of NGF-synthesis from the mushroom *Dictyophora indusiata, Phytochemistry* (Oxford) 45(6): 1203-1205.

Kim, Byong Kak, Jin Won Hyun, Jong Myung Yoon and Eung Chil Choi, 1997, Immunological studies on the anti-tumor components of the basidiocarps of *Agrocybe cylindracea, Archives of Pharmacal Research* (Seoul) 20(2): 128-137.

Lim, T.A., 1993, Mushrooms, the ultimate health food but little research in US to prove it, *Mushroom News* 41(2): 26-29.

Loefgren et al., 1954, *Acta. Chem. Scand.* 8: 670-680 cf. Rolf Singer, 1986, *The Agaricales in Modern Taxonomy,* Sven Koeltz Scientific Books, Germany.

Markhija, A.N. and Bailley, J.M., 1981, Identification of the antiplatelet substance in Chinese black tree fungus, *N. Engl. J. Med.* 304: 175.

Martinkova, L., Juzlova, P. and Vesely, D., 1995, Biological activity of polyketide pigments produced by the fungus *Monoascus, J. Appl. Bacteriol.* 79(6): 609-616.

Min, Tae-Jin, Eun-Mi Kim and Sun-Hoo You, 1996, The screening of antifungal and antibacterial activities of extracts from mushrooms in Korea (II), *Korean J. Mycol.* 24(1): 25-37.

Nanba, H., 1993, Maitake mushroom - The Kinf of mushrooms, *Mushroom News* 41(1): 21-25.

Otsuka, Masamichi, Kazumasa Shinozuka, Gochi Hirata and Masaru Kunitomo, 1996, Influences of a Shiitake (*Lentinus edodes*) - fructo-oligosaccharide mixture (SK-204) on experimental pulmonary thrombosis in rats, *Yakugaku Zasshi* 116(2): 169-173.

Pai, S.H., Jong, S.C. and Low, D.W., 1990, Usages of mushrooms, *Bioindustry* 1: 126-131.

Park, Eun-Jeon, Geonil Ko, Jae Baek Kim and Dong Hwan Sohn, 1997a, Antifibrotic effects of a polysaccharide extracted from *Ganoderma lucidum,* glycyrrhizin and pentoxifylline in rats with cirrhosis induced in biliary obstruction, *Biol. and Pharm. Bulletin* 20(4): 417-420.

Park, Eun-Jeon, Geonil Ko, Jae Baek Kim and Dong Hwan Sohn, 1997b, Dose dependent antifibrotic effects of polysaccharide from mycelium of *Ganoderma lucidum* on liver biliary cirrhosis in rats, *Yakhak Hoeji* 41(2): 220-224.

Pilgrim Horst, Ulrike Lindequist and Nasser Abdullah Awad Ali, 1997, *Inonotus hispidus,* a source of new drugs, *Zeitschrift fuer Mykologie* 62(2): 169-194.

Qian, Zhong-Ming, Mei Feng Xu and Pak Lai Tang, 1997, Polysaccharide peptide (psp) restores immunosuppression induced by cyclophosphamade in rats, *American J. Chinese Medicine* 25(1): 27-35.

Rai, R.D., 1997, Medicinal Mushrooms, in: *Advances in Horticulture Vol. 13: Mushrooms* (K.L. Chadha and S.R. Sharma, eds.), Malhotra Publishing House, New Delhi.

Rai, R.D., 1997, Medicinal mushrooms, in: *Advances in Mushroom Biology and Production* (R.D. Rai, B.L. Dhar and R.N. Verma, eds.), National Research Centre Mushrooms, Chambaghat, Solan.

Rokujo, T., Kikuchi, H., Tensho, A., Tsukitane, Y., Takenawa, T., Yoshida, K. and Kamiya, T., 1970, Lentysine: A new hypolipidemic agent from a mushroom, *Life Sci.* 9 Part II: 379-385.

Ruelius, H.W., Jannssen, F.W., Kerwin, R.M., Goodwin, C.W. and Schillings, R.T., 1968, Porocin, an acid protein with antitumour activity from a basidiomycetes - i. Productioin, isolation and purification, *Arch. Biochem. Biophys.* 125: 126-135.

Shibata, S., Nishikawa, Y., Mai, C.F., Fukuoka, F. and Nakanishi, M., 1968, Antitumor studies on some extracts of basidiomycetes, *Gann.* 59: 159-161.

Singer, R., 1986, *The Agaricales in Modern Taxonomy,* Sven Koeltz Scientific Books, Germany.

Soo, T.S., 1996, Effective dosage of the extract of *Ganoderma lucidum* in the treatment of various ailments, in: *Mushroom Biology and Mushroom Products* (D.J. Royse, ed.), Penn. State Univ. Pennsylvania.

Subramanian, C.V., 1995, Mushrooms: Beauty, diversity, relevance, *Curr. Sci.* 69(12): 986-988.

Suzuki, F., Koide, T., Tsunoda, A. and Ishida, N., 1976, Mushroom extract as an interferon inducer-I. Biological and physiochemical properties of spore extracts of *Lentinus edodes, Mushroom Sci.* 9 (Part I): 509-520.

Takemoto, 1961, *Jap. J. Pharm. Chem.* 33: 252 c.f. D.L. Hawksworth, P.M. Krick, B.C. Sutton and D.N. Pegler, 1995, *Ainsworth & Bisby's Dictionary of the Fungi,* CAB International, Wallingford, U.K.

Tam, S.C., Yip, K.P., Fund, K.P. and Chang, S.T., 1986, Hypotensive and renal effect of an extract of the edible mushroom *Pleurotus sajor-caju, Life-Sci.* 38: 1155.

Tsunoda, A., Zuzuki, F., Sato, N., Miyazaki, K. and Ishida, N., 1970, A mushroom extract as an interferon inducer, *Prog. Antimicrob. Anticancer Chemother.* (Proc. 6th Int. Cong. Chemother., 1969) 2: 832-838.

Vaidya, J.G., 1991, Phanas alombe - a true mushroom or Phanasomba, in: *Indian Mushrooms* (M.C. Nair, ed. in chief), Kerala Agricultural University, Vellanikkara 680 654, Kerala.

Vogel, F.S., McGarry, S.J., Kemper, L.A.K. and Graham, D.G., 1974, Bactericidal properties of a class of quinoid compounds related to sporulation in the mushroom *Agaricus bisporus, Am. J. Pathol.* 76: 165-174.

Walder, Raul, Zlatko Kalvatchev, Domingo Garzaro and Miguel Barrios, 1995, Natural products from the Amazonian rainforest of Venezuela as inhibitors of HIV-1 growth, *Acta Cientifica Venezolana* 46(2): 110-114.

Wang, H.X., Liu, W.K., Ng, T.B., Ooi, V.E.C. and Chang, S.T., 1996a, The immunomodulatory and antitumor activities of lectins from mushroom *Tricholoma mongolicum, Immunopharmacology* 31(2-3): 205-211.

Wang, H.X., Ng, T.B., Ooi, V.E.C., Liu, W.K. and Chang, S.T., 1996b, A polysaccharide peptide complex from cultured mycelia of the mushroom *Tricholoma mongolicum* with immunoenhancing and antitumor activities, *Biochemistry and Cell Biology* 74(1): 95-100.

Wang, H.X., Ooi, V.E.C., Ng, T.B., Chiu, K.W. and Chang, S.T., 1996c, Hypotensive and vasorelaxing activities of a lectin from the edible mushroom *Tricholoma mongolicum, Pharmacology and Toxicology* 79(6): 318-323.

Wang, Sheng-Yuan, Ming-Ling Hsu, Hui-Chi Hsu, Cheng-Hwai Tzeng, Shiu-Sheng Lee, Ming-Shi Shiao and Chi-Kuan Ho, 1997, The antitumor effects of *Ganoderma lucidum* is mediated by cytokines released from activated macrophages and T lymphocytes, *Int. J. Cancer* 70(6): 699-705.

FUNGAL BIOTECHNOLOGY FOR THE DEVELOPMENT AND IMPROVEMENT OF INSECTICIDAL FUNGI

J.L. Faull

Biology Department
Birkbeck College
University of London
Malet Street
London WCIE 7HX
U.K.

INTRODUCTION

Insecticidal fungi come from many different genera. The possibility of using them for control of insect pests was recognised long ago, when in the late 19th Century a natural epidemic of Muscardine disease of silk worms was identified as having a fungal origin (Roberts and Hajek, 1992). There were many attempts to use fungal diseases of insects to control insect pests with little success until comparatively recently when a limited number of them were successfully commercialised. However, there remains considerable difficulty in exploiting these species further because of limitations in their biology, notably the rapidity with which they kill their targets in situations where speed of effect is important. My interests, and those of the groups at Birkbeck and IACR Rothamsted, are in the conidial fungi belonging to the genera *Verticillium, Metarrhizium, Beauvaria* and *Tolypocladium*. Our objective is that through collaborative research between the two laboratories we can identify key limits to the use of these fungi as biological control agents (BCAs).

All of these fungi are Deuteromycetes. They are usually generalist pathogens of a wide range of insect species. They are easily cultured in the laboratory on simple media and are also easy to mass-produce commercially. *Verticillium lecanii* is already available as a commercial formulation for use in integrated pest management (IPM) programmes in glasshouses (Lisansky et al. 1991). *Beauvaria bassiana* is available in the US from Troy, and other fungi are produced and used on a wide scale in South America, the ex-Soviet Republics, People's Republic of China and Central European Republics, but registration requirements and high commercial costs have limited European availability. Only eight different fungi are registered as products against insects pests in the European and US markets (WWDA, 1997).

From Ethnomycology to Fungal Biotechnology
Edited by Singh and Aneja, Plenum Press, New York, 1999

EVENTS OF INFECTION

The exact sequence of events that occurs when a pathogenic fungus lands on an insect cuticle varies with the different insect/fungus pairings, but generally follows a sequence of spore attachment to the insect cuticle, followed by germination, cuticle penetration (either from the outside of the insect or from within the insect mouth parts or gut) using enzymes and physical force, and entry and colonisation of the haemocoel (Hajek and St Leger, 1994) (Figure 1).

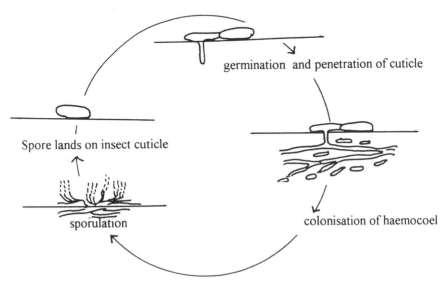

Figure 1. Generalised pattern of insect colonisation by fungi.

During this colonisation phase, the fungus can spread though the haemocoel by the production of mycelium, blastospores or cell wall-less protoplasts. It is at this stage that there may be production of fungal secondary metabolites which act as pathogenicity factors that accelerate the death of the insect (Gillespie and Claydon, 1989; Hajek and St Leger, 1994). The fungi complete their life cycle by extensively colonising the insect and producing large numbers of spores on the outside of the near-empty insect cadaver.

THE ROLE OF THE PATHOGENICITY FACTORS IN INSECT PATHOLOGY

A very wide range of secondary metabolites with very different structures has been reported as being produced by fungi. These include macrocyclic compounds that originate from many different metabolic pathways, and these can behave as antibiotics, immunosuppressants or mycotoxins. The role and influence of secondary metabolites in the interactions between fungi and other organisms has been hotly debated, but their activity as pathogenicity factors in plant pathology has been accepted. The range of compounds produced by insect-pathogenic fungi is also diverse, and again seems to have a significance in pathogenicity, for instance the *in vitro* production of destruxens from *Metarrhizium* spp. has been correlated with toxicosis of mycosed insects (Roberts, 1981; Hajek and St Leger, 1994).

The exact modes of action of these secondary metabolites has yet to be fully elucidated.

Some toxins have an effect on the insect immune system, modifying responses and allowing for more rapid colonisation of the insect haemocoel. Other structures seem more directly toxic to the insect. These include peptide toxins like the destruxens (Figure 2).

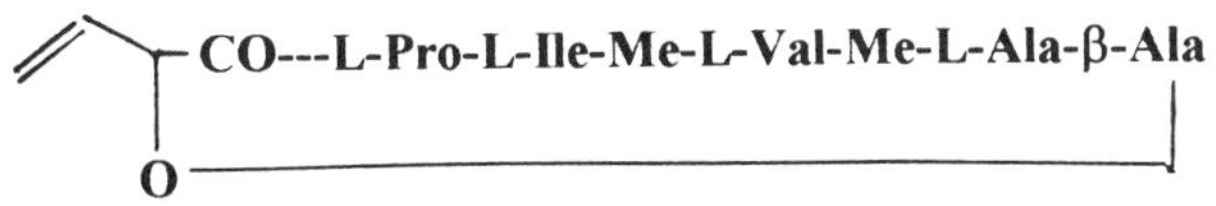

CO---L-Pro-L-Ile-Me-L-Val-Me-L-Ala-β-Ala

O

Key: L=Leucine, Pro=Proline, Ile=Isoleucine, Me=Methionine, Val=Valine, Ala=Alanine, ß-Ala=ß-alanine

Figure 2. The structure of the peptide Destruxen A.

One member of the fungal pathogens of insects worked on extensively by my group is the genus *Tolypocladium*, white-sporing filamentous fungi only recently separated from the Beauvaria group on the basis of differences in conidial apparatus morphology (Bissett, 1983) (Figure 3). An interesting feature of these fungi is that the spores and conidial apparatus of *Tolypocladium* species are hydrophilic and form large spore drops around active phialides. The fungus had been regularly isolated not only from soil but also from aquatic environments, suggesting a rather different specialisation to a habitat that many other of the insect-pathogenic fungi do not happily colonise.

Figure 3. Conidial apparatus of *Beauvaria bassiana* and *Tolypocladium cylindrosporum.*

Tolypocladium species have been reported as naturally attacking many different species of insect including *Aedes sierrensis* (Soares et al., 1989) and *Drosophila* (Bissett, 1983). In the laboratory it has successfully attacked many different species of insect and it has been able successfully to parasitise every type of mosquito and midge offered to it (Goettal, 1987).

The biology of infection is similar to many other entomopathogenic fungi, involving the penetration of the cuticle, followed by haemocoel colonisation and eventually the death of the insect. Two major groups of secondary metabolites from *Tolypocladium* species have been identified in liquid culture filtrate, cyclosporins and efrapeptins. The cyclosporins are cyclic oligopeptides with immunosuppressant activity in animals and insects. These compounds contain eleven mostly methylated amino acids that make the compound hydrophobic. This compound has revolutionised transplantation of organs across histocompatability barriers because it suppresses antibody formation and cell mediated responses in higher animals. Efrapeptins are mixtures of peptide antibiotics containing 15 amino acid residues (Figure 4). They are inhibitors of mitochondrial oxidative phosphorylation and ATPase activity, which they achieve by binding to the soluble (F1) part of the ATPase molecule. Unsurprisingly efrapeptins also inhibit photophosphorylation in the chloroplast by a similar action.

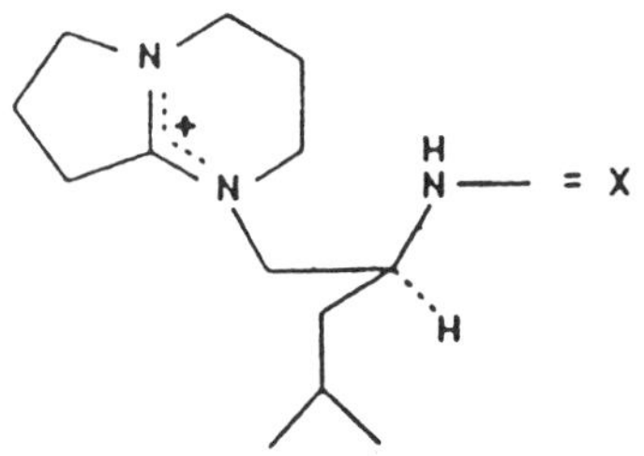

Efrapeptin E Ac-Pip-Aib-Pip-Iva-Aib-Leu-ß-Ala-Gly-Aib-Aib-Pip-Aib-Gly-Leu-Iva-X

Key: Ac=Acetyl, Ala=Alanine, Gly=Glycine, Leu=Leucine, Aib=α aminoisobutyric acid, ß-Ala=ß-Alanine, Iva=Isovaline, Pip=Pipecolic acid.

Figure 4. Structure of the Efrapeptins

The aim of our current work is twofold. Firstly, we need to link some of the effects of infection of insects by *Tolypocladium* species with production of the toxin *in vivo*. Once the link is proven it is our intention to try to enhance efrapeptin production in terms of speed and final titre of toxin for some applications, whilst reducing it in others. Table 1 summarises our strategy for development of *Tolypocladium cylindrosporum*.

Table 1. Strategy for the development of *Tolypocladium cylindrosporum*.

Pathogenicity strategy	
rapin toxin kill	colonisation
suitable for rapid knockdown	suitable for slow knockdown
biotechnological input improves toxin production and speed of kill	biotechnological input represses toxin production - organism suitable for sensitive environments

Rapid Kill Option: Proof of Toxin Involvement in Pathogenicity

Our approach to proving the link between efrapeptin production and insect mortality has been to try and demonstrate that the toxin present in spores is capable of causing insect death. Support for this supposition is provided by our early results from bioassays using third instar larvae of *Aedes albopictus*. Results indicate that it is possible to cause 100% mortality of larvae within 24 hours when using an oral dose of 10^7 blastospores per ml, a far too rapid reaction to be explained by colonisation. We are currently taking these studies further and attempting to show, using isolated fractions of insect mid-gut, whether or not the efrapeptins can be released from spores without degradation by insect gut enzymes to cause rapid knock-down effects when given orally.

Work is also continuing at Rothamsted, where Dr Tariq Butt's group is working on the effects of oral dosing and injection of insects with efrapeptins. Bioassays from this work have shown that there is significant toxicity of toxin via both routes in *Plutella* species, *Galleria* species and *Phaedon* species larvae (Butt, pers. comm.).

Colonisation Option: Use of *Tolypocladium* in Biocontrol of Midge Larvae

In many situations it appears that insect colonisation, which takes 36 to 48 hours, is followed by toxin production and that this process accelerates death. This type of insect attack may be suitable for situations where speed of kill is not important, but a lack of toxicity is. Work on a potential practical application of *Tolypocladium cylindrosporum* as a BCA under field conditions is being carried out by Paul van Poppelen at Birkbeck. The targets are non-biting midges that colonise trickling filters in sewage works and which emerge periodically to cause significant nuisance to local business and residents. The three major species are Psychodids *(Psychoda alternata)*, Window Midges *(Sylvicola cinctus)* and Chironomids *(Metriocnemus hygropetricus)*. At least two of these species have previously been reported as being susceptible to *T. cylindrosporum* (Ekbom, 1993) and work carried out at Birkbeck has confirmed that *P. alternata* can be targeted. However the speed of action has been confirmed as being slow (4 to 5 days), appearing to rely on hyphal colonisation of insect larvae rather than toxin kill. Such a mode of action is desirable in the waste water environment because processed water is often released into water courses from which drinking water is abstracted further downstream. If toxin production was involved in the insect control process there would be difficulty in obtaining permission for its use under field conditions without extensive toxicological testing of the effluent water.

Another important consideration in the use of *T. cylindrosporum* in this environment is that the BCA does not persist for an extended period of time. Experimental work has shown that *T. cylindrosporum* spores are capable of survival under the nutritional and physical conditions that exist in trickling filters, notably their survival within a biofilm. However, persistent establishment of *T. cylindrosporum* within a filter is unlikely, and due to avoidance by *P. alternata* larvae of psychodid cadavers, horizontal transmission is only likely to occur under food shortage conditions. Other Nematocera, in particular some of the Chironomid larvae which are carnivorous, may well be affected through such transmission however. This is currently under investigation at Birkbeck.

In designing a BCA for *P. altemata* it is important to understand the details of the life cycle of the midge and the different larval stages it goes through in the filter bed before emergence as an adult fly. Long term observations of lifecycle and behavioural aspects of the larvae have shown that in particular the first three instars are unlikely to be realistic targets, as they have a life span of two days before transformation under standardized temperatures. The fourth instar, with a life span of up to 8 days is a more realistic possibility, as infection and death from *T. cylindrosporum* infection normally take four to five days. Thus part of the research at Birkbeck Coflege has been focused upon the identification of this 'window of opportunity' for the application of *T. cylindrosporum,* based on detailed consideration of life cycle and the ecological niche the target operates in, as well as the mode and speed of operation of the control method.

It is hoped that an understanding of the life cycles and biology of both pests and their diseases will bring rapid results in the development of new and safe controls for many insect pests. Such work requires the study of many different aspects and needs input from many different disciplines. A co-ordinated approach from several different research teams will provide the most effective and speedy solution to such problems.

ACKNOWLEDGEMENTS

Thanks to Dr Tariq Butt and his group for his involvement in these studies.

REFERENCES

Bissett, J., 1983, Notes on *Tolypocladium* and related genera, *Can. J. Bot.* 61: 1311-1329.

Ekbom, B., 1993, *Tolypocladium cylindrosporum* Gams (Deuteromycetes, Moniliaceae) as a biological control agent against the Window Midge *Sylvicola cinctus* Fabr. (Diptera: Anisopodidae), *Biocont. Sci. Technol.* 3: 309-313.

Gillespie, AT. and Claydon, N., 1989, The use of entomogenous fungi for pest control and the role of toxins in pathogenesis, *Pesticide Science* 27: 203-215.

Goettal, M.S., 1987, Preliminary field trials with the entomopathogenic hyphomycete *Tolypocladium cylindrosporum* in Central Alberta, *J. Am. Mosquito Control Ass.* 3: 239-245.

Hajek, A.E. and St Leger, R.J., 1994, Interaction between fungal pathogens and insect hosts, *Ann. Rev. Entomol.* 39: 293-322.

Lam, T.N.C., Goettal, M. and Soares, G.G. Jnr., 1988, Host records for the entomopathogenic hyphomycete *Tolypocladium cylindrosporum, Florida Entomologist* 71: 86-89.

Lisansky, S., Robinson, A. and Coombs, J., 1991, *Green Growers Guide,* CPL Press, U.K.

Roberts, D.W., 1981, Toxins of entomopathogenic fungi, in: *Microbial Control of Pests and Plant Diseases 1970-1980* (H.D. Burges, ed.), pp. 441-464, Academic Press, New York.

Roberts, D.W. and Hajek, A.E., 1992, Entomopathogenic fungi as bioinsecticides, in: *Frontiers in Industrial Mycology* (G.F. Leatham, ed.), pp. 144-159, Chapman & Hall, New York.

Soares, JR., Riba, G.G., Caudal, A. and Vincent, J.J., 1989, Comparative studies of eleven isolates of the fungal entomopathogen *Tolypocladium cylindrosporum* and two isolates of *Tolypocladium extinguens, J. Inv. Pathol.* 46: 115-120.

WWDA, 1997, *The World Wide Directory of Agrobiologicals on CDROM,* CPL Scientific Press, Newbury, Berkshire, UK.

POTENTIAL BIOTECHNOLOGICAL APPLICATIONS OF THERMOPHILIC MOULDS

A. Archana[1] and T. Satyanarayana[2]

[1]Swami Shraddhanand College
Alipur
Delhi-110 036
India

[2]Department of Microbiology
University of Delhi South Campus
Benito Juarez Road
New Delhi-110 021
India

INTRODUCTION

Temperature is one of the extremely important environmental variables that play a key and decisive role in the survival, growth, distribution and diversity of microbes on earth. Brock (1970) stressed that thermophilic microbes are inhabitants of extreme environments where the temperature rises above the prevailing normal temperature in the area. This temperature increment could be caused by geothermal heat, microbial thermogenesis of organic rich materials and solar heating of substrates such as soils, litter and rock.

The biotechnological applications of thermophilic moulds are numerous. Pure culture studies of thermophilic fungi have provided clear evidence that they possess a battery of extracellular enzymes capable of hydrolysing polymers such as starch, protein, pectin, hemicellulose and cellulose. They have also been reported to produce, among others, a number of antibacterial and antifungal substances, extracellular phenolic compounds and organic acids (Satyanarayana et al., 1992). Some thermophilic fungi have already been used in industries involving food processing and bioconversion of organic materials (Mouchacca, 1990). Thermophilic moulds offer several advantages for their use in industry. This is mainly due to their ability to produce thermostable enzymes. Such biocatalysts exhibit a higher degree of stability in the presence of detergents and aqueous organic solvents and reduced loss of activity during storage and purification. Mass cultivation of thermophilic fungi precludes to a large extent contamination problems encountered with mesophiles with concomitant reduction in investment costs (Brock, 1986). Microbial fermentations carried out at elevated temperatures

using thermophilic moulds reduce the viscosity and surface tension of water which allows better mixing of the broth. The metabolic activity of microbes results in heat production, but in the case of thermophilic fermentations the system need not be cooled extensively since elevated temperatures can be tolerated by thermophilic fungi. Consequently, the comparatively lower energy requirement makes the bioprocess more cost-effective.
Thermophilic fungi are also potential candidates for water treatment. The advantages they offer include an enhanced reaction rate and decreased retention time, destruction of pathogenic microbes that might be present in sewage, lower viscosity so that less energy is required for mixing, and easier de-watering of the resulting sludge (Zinder, 1986).

Thermophilic microbes are reported to survive and grow at elevated temperatures on account of rapid resynthesis of essential metabolises, molecular thermostability and ultrastructural thermostability. Survival of thermophilic fungi appears to be a consequence of thermostability and functional permeability of the membranes. Recently, there have been indications that polyamines could be playing a significant role in growth and differentiation of these fungi (Singhania et al., 1991; Magan, 1997). Wright et al. (1983) suggested that growth of the thermophile *Talaromyces thermophilus* could be partially attributed to its inability to synthesize unsaturated fatty acids, which in turn are responsible for growth at lower temperatures.

There has been much work on the occurrence of thermophilic moulds and their role in decomposition of organic matter, production of enzymes, antibiotics, organic acids, phenolic compounds, extracellular polysaccharides, lipids and bioconversions (Tansey and Brock, 1978; Johri, 1980; Johri and Satyanarayana, 1984, 1986; Satyanarayana et al., 1988, 1990, 1992, McHale and Morrison, 1986; Bunni et al., 1989). The enzymes elaborated by thermophilic fungi include thermostable aminopeptidases, carboxypeptidases, lipases, proteases and ribonucleases that show thermostability at temperatures much higher than the growth maxima of the organism (Satyanarayana et al., 1992). Enzymes of the TCA cycle such as glucose-6-P-dehydrogenase, malate dehydrogenase, isocitrate dehydrogenase and succinic dehydrogenase are also thermostable in several thermophilic moulds. Most exopolysaccharases of fungal origin are glycoproteins, and partial removal of the associated carbohydrate renders such enzymes less heat-stable suggesting an important role for the carbohydrate moiety in maintenance of conformation structure at high temperatures (Yoshioka et al., 1981). The natural occurrence of thermophiles in self-heating piles of organic matter such as hay, wood-chips and agricultural composts clearly shows their ability to degrade cellulosic substrates. This has encouraged exploitation of these thermophiles in commercial production of such hydrolases. A growing concern of microbiologists is to be able to convert agricultural and urban wastes to cellobiose and glucose, and in turn allow microbial conversion to single cell protein (Johri, 1980).

PRODUCTION OF THERMOSTABLE ENZYMES

β-Galactosidase

β-galactosidase converts the milk sugar lactose into glucose and galactose, and thereby enhances the effective sweetness of dairy products. Furthermore, it lowers the content of lactose in milk formulated for people suffering from lactose-intolerance. Sorensen and Crison (1974) reported 25 out of 54 thermophilic fungal strains to be producing the enzyme β-galactosidase, commonly called lactase. *Humicola grisea* var. *thermoidea, Humicola lanuginosa, Malbranchea pulchella* var. *sulfurea, Mucor pusillus, M. miehei, Sporotrichum thermophile* and *Torula thermophila* exhibited best growth on semi-solid lactose medium at pH 6.6, and among these *Mucor pusillus* was the best enzyme producer at 61 °C, with good thermostability. Another enzyme from *Thermomyces lanuginosus* was produced in static culture (Prasad and Maheshwari, 1978b).

Satyanarayana et al.(1985) studied β-galactosidase production in 13 thermophilic fungi and found extracellular as well as intracellular activity by *Humicola insolens, Malbranchea pulchella* and *Talaromyces emersonii* while only intracellular activity was seen in *Acremonium alabamensis, Chaetomium thermophile, Humicola lanuginosa* and *Sporotrichum thermophile.* Hence, thermophilic moulds seem to be a potential source for industrial production of ß-galactosidase, whose optimum pH range for activity, 4-8, tallies with that of milk of whey, indicating its suitability for industrial usage. Partial indigestibility and the lower sweetening value of lactose may be overcome by its hydrolysis by ß-galactosidase into its constituent monomeric sugars, glucose and galactose (Fisher et al., 1995). Fisher et al. (1995) purified ß-galactosidase from *Talaromyces lanuginosus*, which was identified as a bulky glucoprotein with a dimeric structure and substantial thermostability.

Phosphatases

The phosphoserine residues in casein can be hydrolysed both by alkaline and acid phosphatases whereby the phosphate content of food can be reduced. Alkaline phosphatases find usage in molecular biology. Production of acid phosphatases was reported in *Thermomyces lanuginosus* (Crison, 1969). Out of 775 thermophilic fungal isolates studied, Bilai et al. (1985) found only fifteen to be producing acid phosphatases. No correlation between the intensity of growth of microbes and the ability to produce acid phosphatases could be established. Satyanarayana et al. (1985) screened thirteen thermophilic fungi and found that only two of them, *Acremonium alabamensis* and *Rhizopus rhizopodiformis,* secreted acid phosphatases whereas the other eleven species produced both types of phosphatases. Spent media from cultures of *Thermoascus crustaceus* contained significant amounts of acid phosphatase activity (Arnold et al, 1988).

Chitinases

The biopolymer chitin is found in the cell walls of fungi and in the exoskeletons of marine invertebrates and arthropods. For its complete hydrolysis, many microbes, plants and animals have an enzyme system composed of two hydrolytic enzymes: chitinase and chitobiase. In cell walls of fungi, chitin may be accompanied by chitosan which is deacetylated chitin (Kauss et al., 1983). Chitinolytic enzymes have been described from thermophilic fungi, viz. *Talaromyces byssochlamydoides* and *T. emersonii* (Adams and Deploey, 1978), *Thermomyces lanuginosus* (Nakaji, 1978), *Mucor miehei* (Kauss et al., 1983), *Talaromyces emersonii* (McCormack et al., 1991). Chitinase of *T. emersonii* was inducible and substantially thermostable (McCormack et al., 1991).

Mannanases

Mannanases are the major components of the hemicellulose fraction of softwoods. Complete hydrolysis of mannans involves the action of endo-acting mannanases and mannosidases. Eight thermophilic fungi were examined for mannanase production by Araujo and Ward (1990) and they found highest enzyme production by *Talaromyces byssochlamydoides* and *T. emersonii.* Cell-free supernatants from *Thielavia terrestris* were resolved into four mannanases and a mannosidase which were quite thermostable (Araujo and Ward, 1990). Such thermostable mannanases show promise as bleaching aids in the paper and pulp industry (Stalbrand et al., 1993).

Xylanases

Xylanases are the most studied enzymes among the hemicellulases since their substrate xylans constitute the largest proportion of hemicelluloses in nature. Among the thermophilic fungi screened by Flannigan and Sellars (1972), *Absidia ramosa, A. fumigatus, H. lanuginosa, M. pusillus* and *T. aurantiacus* actively degraded arabinoxylan. Matsuo et al. (1975) selected *M. pulchella* var. *sulfurea* among several xylanase-producing thermophilic moulds and obtained 93.2% hydrolysis of xylan to xylose. Satyanarayana and Johri (1983) reported a higher xylanolytic activity for *S. thermophile* and *A. alabamensis* among six thermophilic fungal isolates of paddy straw compost. Maheshwari and Kamalam (1985) observed higher xylanase yields in shake flasks than in a fermenter. Durand et al. (1984) demonstrated production of xylanase and ß-xylosidase from *Sporotrichum cellulophilum* and *Thielavia terrestris* using wheat bran as a substrate. Gomes et al. (1994) characterized a highly thermostable xylanase of a wild strain of *Thermoascus aurantiacus*.

Recently *Melanocarpus albomyces* ZIS-68 was shown to produce a xylanase (Jain, 1995). Haq and Deckwer (1995) compared xylanase production by two strains of *Thermomyces lanuginosus*, both of which were found to be free from cellulase activity, making them suitable preparations for biobleaching of paper pulps. Likewise, Dusterhoft et al. (1997) purified and characterized two xylanases from *Humicola insolens*. Submerged fermentation was a better mode of xylanase production by *H. lanuginosa* and *S..thermophile*, while *T. aurantiacus* produced higher titres by solid state fermentation (Grajek, 1987). Purkarthofer et al. (1993) studied production of cellulase-free xylanase by *T. lanuginosus* in both submerged and solid state fermentations. Endoxylanases are known to be induced by their substrate, xylan. Purkarthofer and Steiner (1995) demonstrated enzyme induction in the case of *Thermomyces lanuginosus*.

The complete degradation of branched xylans requires the concerted action of several hydrolytic enzymes, viz. acetyl xylan esterase, α-glucuronidase, α-arabinofuranosidase and ß-xylanosidase, besides endoxylanase (Biely, 1985). Acetyl esterases from *T. reesei* and *S. commune* exerted a synergistic action with endoxylanases in the hydrolysis of acetylated xylan to produce xylooligomers, xylose and acetic acid (Biely et al., 1986). Tuohy et al. (1994) purified two acetyl xylan esterases from *Talaromyces emersonii*. Acetyl esterase of *T. lanuginosus* showed best activity in the acidic range. *Trichoderma reesei* produced α-glucuronidase which acts synergistically with endoxylanases and liberates 4-o-methylglucuronic acid from substituted xylooligomers. α-Arabinofuranosidases hydrolyse non-reducing arabinofuranosyl residues from furanosidase, arabinans, arabinoxylans and arabinogalactans. This enzyme has been characterized from *A. niger* (John et al., 1979), *T. koningii* (Wood and McCrae, 1986), *T. lanuginosus* (Purkarthofer et al., 1993) and *T. emersonii* (Tuohy et al., 1994), and it degrades arabinose residues in heteropolymeric substrates. ß-Xylosidases hydrolyse xylobiose to xylose, and normally exhibit no activity towards xylans and other higher xylooligosaccharides. ß-Xylosidase is produced in appreciable amounts by *M. lanuginosa* (Anand and Vithayathil, 1996) and *A. niger* (Takenishi et al., 1973).

The most important novel biotechnological application of xylanases is to produce rayon grade paper pulps, for which a thermostable cellulase-free xylanase preparation is required (Viikari et al., 1991). Chemical bleaching of paper pulp is a polluting process, whereas xylanase aided bleaching results in a decrease in chlorine consumption and improves paper brightness in an environmentally friendly way (Viikari et al., 1994). Hemicellulose hydrolysis in pulps by these hemicellulases increases the extractability of lignin from kraft pulps in the subsequent bleaching sequences. Other applications of xylanases in the same industry are in debarking, refining of pulp fibres, preparation of dissolving pulps and bioconversion.

Dissolving pulps are used to produce cellulose derivatives such as acetates, cellophanes and rayons, where hemicellulose contamination leads to colour and haze in the product and also to poor cellulose derivation (Paice and Jurasek, 1984). Xylanases find numerous applications in other industries, including clarification of juices and wines, extraction of coffee, plant oils and starch (Wong and Saddler, 1993), improvement of nutritional properties of silage, maceration of cell walls while processing fruits and vegetables, production of food thickeners, fuel and chemical feedstocks (Biely, 1991), preparation of cereal-based diets for poultry and pigs and so on. Besides, xylanase pretreatment of forage crops and other biomass is of great value in improving the nutritional quality and digestibility of rumen feeds or in facilitating composting (Gilbert and Hazlewood, 1993). The enzymatic digestion of industrial wastes as an alternative to landfill deposition may also be useful.

Cellulases

Cellulases include endoglucanases (randomly cleaving internal glycosidic bonds in cellulose, e.g. carboxymethyl cellulose), exoglucanases (removing glucose units from ends of cellulose chains) and ß-glucosidases (integral elements of cellulolytic systems which convert cellulose to glucose). *Chaetomium thermophile* var. *coprophile, C. thermophile* var. *dissitum, Humicola grisea* var. *thermoidea, H. insolens, Myriococcum albomyces, Sporotrichum thermophile* and *Torula thermophila* were observed to be cellulolytic and they could degrade filter paper as well as CMC (Fergus, 1969). *Malbranchea pulchella* var. *sulfurea, Stilbella thermophila* and *Talaromyces thermophilus* could hydrolyse only CMC and not filter paper, whereas *Humicola lanuginosa, H. stellata, Mucor miehei, M. pusillus* and *Thermoascus aurantiacus* were unable to utilize any of these cellulose sources (Satyanarayana et al, 1992).

In contrast, some strains of *T. aurantiacus* (Tong et al, 1980) and *Talaromyces* spp. (Jain et al., 1979, McHale and Coughlan, 1981) were cellulolytic. Roy et al. (1990) reported production of thermostable endocellulase by *Myceliophthora thermophile.* Growth of *Chaetomium cellulolyticum* on many cellulose-rich raw materials indicates the presence of a broad spectrum of enzymes required for bioconversion of such substrates (Ganju et al., 1990). Cellulases and ß-glucosidases were detected in this system on further investigation. Tong et al. (1980) reported production of ß-glucosidase and three cellulases from three-week old *Thermoascus aurantiacus* culture.

Nishio et al. (1981) cultivated *Talaromyces* sp. in solid state on wheat bran and produced cellulases. Grajek (1987) reported greater production of cellulases in solid state fermentation (SSF) than in submerged fermentation. Presence of additives such as Tween-80 was shown to hamper cellulose degradation by *S. thermophile* (Ramanelli et al., 1975), which could be due to its action as a detergent to prevent the adsorption of extracellular cellulase on cellulose. The cell-associated cellulolytic activity reached a maximum by three days, while extracellular activity reached a peak by six days in *Thielavia terrestris* (Breuil et al., 1986). The spores of *S. thermophile* immobilized in agar/polyacrylamide/alginate gels have been shown to grow and produce cellulases (Singh et al., 1989). There have been other instances too where thermophilic fungi such as *T. emersonii* and *S. cellulophilum* have been immobilized for cellulase production (Tamada et al., 1986). Generally immobilized systems are seen to exhibit more cellulase activity than the corresponding free cells. Cellulase production is mostly induced by the presence of cellulose, cellobiose and other related higher sugars, and is repressed by the catabolite, glucose. Efforts to develop hyperproducing cellulose mutants by physical and chemical mutagenesis have resulted in a few cellulose hyperproducers.

Sporotrichum thermophile ATCC 42464 produced two intracellular ß-glucosidases that were mutually very different (Meyer and Canavascini, 1981). Tong et al. (1980) reported the occurrence of a highly glycosylated ß-glucosidase of *Thermoascus aurantiacus* which contributed to its exceptional heat stability. A pair of ß-glucosidases, both intracellular, exhibited relatively less thermostability despite being glycosylated in the cellulolytic system of *Humicola grisea* var. *thermoidea* (Peralta et al., 1990). Filho (1996) obtained a ß-glucosidase from *H. grisea* var. *thermoidea* grown on wheat bran by solid-state culture. This enzyme also was a dimer molecule with two identical subunits and good thermotolerance. A highly thermostable ß-glucosidase was purified and characterized from the thermophilic fungus *H. grisea* var. *thermoidea* (Peralta et al., 1997). ß-Glucosidase plays a crucial role in the process of saccharification of cellulose by removing cellobiose by catalysing hydrolysis of ß-glucosidic linkages.

The potential applications of cellulases are in enhancing the tensile strength of high α-sulfite pulp, recovering agar and the production of sea weed jelly, saccharification of delignified cellulosic wastes, modifying cereals, improving the solubility of raw-materials in brewing, digesting excreta in septic tanks, supplementing enzymes in feed for poultry and pigs, and preparing protoplasts from higher plants and fungi (Satyanarayana et al., 1992).

Amylases

The production of amylases has been reported from several thermophilic fungi viz. *Rhizomucor pusillus* (Somkuti and Steinberg, 1980; Turchi and Bercker, 1987; Kanlayakrit et al., 1987), *Talaromyces emersonii* (Bunni et al., 1989), *Thermomyces lanuginosus* (Jensen et al., 1988), *M. sulfurea* (Gupta and Gautam, 1995) and *T. aurantiacus* (Adams, 1992).

Amylases have numerous applications in biotechnological industries, such as in the starch industry, brewing industry, distilling industry and textile industry. Adams (1992) studied amylase production with respect to mycelial growth of *T. aurantiacus.* The α-amylase of *T. lanuginosus* was maximally active at slightly acidic conditions and 65°C. Near boiling temperatures resulted in its progressive and irreversible denaturation. Sadhukhan et al. (1990) used a synthetic medium for amylase production by *Myceliophthora thermophila* and demonstrated beneficial effects of Ca^{2+} and Mn^{2+} on amylase biosynthesis. Purified amylase preparations of various thermophilic fungi, *Rhizomucor pusillus* (Somkuti and Steinberg, 1980), *Talaromyces emersonii* (Bunni et al., 1989), *Thermomyces lanuginosus* (Jensen et al., 1988) and *M. sulfurea* (Gupta and Gautam, 1995) have been described. Jackson and Seidman (1985) reported production of α-amylase, glucoamylase and α-glucosidase from a thermophilic mould *Humicola grisea (thermoidea),* which formed a strong amylolytic system due to the additive effect of these hydrolysing enzymes, and suggested that this mould has immense industrial potential. Likewise, another mould *Talaromyces emersonii* was reported to be a producer of α-amylase, α-glucosidase and glucoamylase (Bunni et al., 1989; Basawesvara Rao, 1979). Adams (1995) showed extracellular amylase activities of *Rhizomucor pusillus* and *Humicola lanuginosa* at initial stages of mycelial growth. Protoplast fusion in *Malbranchea sulfurea* resulted in enhanced extracellular production of α-amylase (Gupta and Gautam, 1995).

Glucoamylase removes glucose molecules from the polymer chain end (exo-acting) and also forms an important component of the starch hydrolytic system. Campos and Felix (1995) purified and characterized a glucoamylase from *Humicola grisea.* There have been reports of a fungus producing glucoamylase in more than one form. Taylor et al. (1978) reported two forms of this enzyme in *T. lanuginosus,* which had variation in their pH optima for enzyme action. Thermostable glucoamylases from thermophilic fungi are very often glycoproteins, such as in the case of *R.. pusillus* (Kanlayakrit et al., 1987) and *T. lanuginosus* (Basaveswara Rao et al., 1981).

Proteases

The term proteases refers to a complex of protein-degrading enzymes that may include proteinases and peptidases. Species of *Achaetomium, Chaetomium, Humicola, Rhizomucor, Malbranchea, Penicillium, Rhizopus, Sporotrichum, Torula* and *Talaromyces* are good producers of acid, neutral and alkaline protease (Garg and Johri, 1998). Synthesis of many proteases is governed partially by the type **of** (and??) concentration of the available nutrients (Garg and Johri, 1998). Satyanarayana and Johri (1983) reported protease production from *H. lanuginosa, S. thermophile* and *T. aurantiacus.* Acid protease produced from *Rhizomucor miehei* was industrially very significant (Escobar and Barnett, 1995). Thakur et al. (1990) achieved a high milk clotting activity (50,000 Soxhlet U/g mouldy bran) using *Mucor miehei* in SSF. A high ratio of 60:1 between milk clotting and proteolytic activity for this enzyme was recorded. Industrial applications of proteases include those in cheese making, detergents and laundry, tanning industry, baking, brewing, meat tenderization, **digestive and preparation** among others.

Lipases

Thermophilic moulds that have been found to produce lipases include *M. pusillus, M. miehei, H. lanuginosa, H. grisea* var. *thermoidea, T. thermophilus, T. emersonii* and *A. alabamensis* (Satyanarayana and Johri, 1981a; Deploey et al., 1981; Lawson et al., 1994). Arima et al. (1972) reported a high lipase production at 45 °C in a medium containing olive oil, whereas lipases were synthesized by *H. grisea* var. *thermoidea, M. pusillus, T. thermophilus* and *T. crustaceus*, in a medium devoid of lipids, though their addition caused an increase in enzyme production (Ogundero, 1980). Lawson et al. (1994) investigated the nature of substrate binding in *Humicola lanuginosa* lipase through X-ray crystallography and intuitive modelling. In a recent patent filed by Novo Industry, Denmark, the suitability of *Humicola* lipase for the detergent industry was demonstrated.

Invertase

Invertase hydrolyses sucrose into glucose and fructose and hence it is commercially important for production of high fructose corn syrup (HFCS). The ability to produce invertase was studied in four fungi *viz. Humicola hyalothermophila, Chrysosporium thermophilum, Thermomyces lanuginosus* and *Malbranchea pulchella* var. *sulfurea* (Kirillova and Shchepankevich, 1989). Maheshwari et al. (1983) showed that fungal invertase appeared in mycelial suspension only when sucrose or raffinose was added to the medium and rapidly disappeared following its exhaustion. This enzyme was not stable in cell-free extracts and also the enzyme activity was growth dependent. A similar induction effect of sucrose and raffinose on invertase secretion was observed by Palanivelu et al. (1984). Maheshwari and Balasubramanyam (1988) also reported invertase production by *Thermomyces lanuginosus* and *Penicillium dupontii*, and consequently sucrose degradation.

Trehalase

Trehalose (α-D-glucopyranosyl-α-glucopyranoside) is widely distributed in fungi where it is stored as a reserve food supply. In fungi, trehalose is associated with carbon reserves, developmental processes, adaptations to a number of environmental stresses and injuries and in some cases with signal transduction pathways (Neves et al., 1994). The occurrence of trehalose in thermophiles, is thus a clue towards a possible mechanism enabling their survival and growth at extreme temperatures. In order to be able to utilize the trehalose, the corresponding enzyme trehalase is needed. Investigation revealed that thermophilic fungi could indeed produce trehalases. Prasad and Maheshwari (1978a, b) were the first to report the occurrence of trehalase in *Thermomyces lanuginosus*. A high molecular weight extracellular trehalase was purified from conidia of *Humicola grisea* var. *thermoidea* (Zimmermann et al., 1990). Cardello et al. (1994) purified cytosolic trehalase from the conidia of *H. grisea* var. *thermoidea* which was more thermostable than the other known trehalases. The partially-purified, thermally stable trehalase from conidia of *Humicola grisea* was highly specific for trehalose and was free of potentially interfering activities (Neves et al., 1994), and therefore, this was used for the determination of trehalase.

Pectinases

Pectins are heteropolysaccharides that mainly have a structural role in plants and help in maintaining the cell integrity. Craveri et al. (1967) described pectinases from *Penicillium dupontii, Humicola stellata, Humicola lanuginosa, Mucor pusillus* and *Humicola insolens*. Knosel and Resz (1973) investigated thermophilic strains of fungi representing different genera and species for the production of pectinolytic enzymes. The pectinolytic activities of *Thielavia terrestris, Sporotrichum cellulophilum* and *Trichoderma reesei* were mutually compared, which revealed moderate thermostability of these enzymes (Durand et al., 1984). Tuohy et al. (1989) detected pectinolytic activities in solid state cultures of four different strains of thermophilic fungi, namely *C. thermophile, T. emersonii* CBS 814.70, *T. emersonii* UCG 208 and *T. aurantiacus*, whereas pectin lyase was found only in *Chaetomium thermophile*.

Due to the synergistic effect of pectinase and cellulase on the hydrolysis of plant cell biomass Whitehead and Smith (1989) investigated the production of pectinase by *Sporotrichum thermophile* in both static and shake flask cultures. Static culture had a higher yield than shake flask culture. Saccharification of natural substrates with cellulases and hemicellulases may be accentuated using pectinases, since the enzymatic action of the latter would permit better access of the former pair of enzymes to their respective substrates (Tuohy et al., 1989). Commercial pectinases comprising three main activities, e.g. polygalacturonase, pectin lyase and pectin methylesterase, are therefore used to reduce the level of pectic substances in the processing of fruits and juices.

Xylose Isomerase

Thermophilic moulds are known to produce xylanolytic enzymes to utilize xylan (Dubey and Johri, 1987; Gomes et al., 1993; Matsuo and Yasui, 1985; Satyanarayana and Johri, 1983), and to utilize xylose as a carbon source (Reese, 1946; Subramanyam, 1980). Thermostable xylanases from thermophilic moulds could be employed in the hydrolysis of abundantly available hemicellulases in agricultural residues, such as cereal straws, into xylose (Maheshwari and Kamalam, 1985). Banerjee et al. (1994) reported a thermophilic mould, *Malbranchea pulchella* var. *sulfurea* TMD-8 that produced xylanases extracellularly when wheat straw or wheat straw hemicellulose was provided in the medium. Its mycelial extracts contained xylose isomerase, xylose reductase and xylitol dehydrogenase which is an exceptional property. It has been reported that both pathways for the conversion of xylose to xylulose mediated by (i) xylose isomerase, and (ii) xylose reductase and xylitol dehydrogenase occur in thermophilic moulds (Banerjee et al., 1994), mesophilic filamentous fungi such as *Aspergillus* spp., *F. oxysporum* (Suihko et al., 1983), *Gliocladium roseum, Neurospora crassa* (Chiang and Knight, 1960), *Penicillium* spp., *Rhizopus nigrificans*, and in yeasts (Hofer et al., 1971).

Phytase

Thermophilic fungi such as *Sporotrichum thermophile* are known to produce phytases (myo-inositol hexaphosphate phosphohydrolase), a special group of acid phosphatases that hydrolyse phytate to a series of lower phosphate esters of myo-inositol and phosphate. Phytic acid and phytates are common components of plant tissues and hence a major component of plant-derived foods. About 75% of the total phosphorus in cereals, legumes and seeds (Common, 1940; de Boland et al., 1975; Erdman, 1979) exists as phytic acid phosphorus. The phytate phosphorus accounts for between 60 and 90% of total phosphorus present in seeds of both monocot and dicot plants. Phytic acid phosphorus, however, remains largely unavailable to monogastrics such as pigs, poultry and humans as these species are devoid of sufficient, suitable endogenous phosphatase activity that is capable of liberating the phosphate groups from the phytate core structure. This results in excretion of unutilised phosphorus, creating pollution problems in areas of intensive livestock production. The released phosphorus is transported into rivers and lakes where it may cause eutrophication.

Phytate acts as an antinutritional factor since six reactive groups in this molecule chelate minerals such as calcium, magnesium, iron, zinc, cobalt and copper very strongly, thus forming insoluble metal-phytate complexes (Gifford and Clydesdale, 1990). Nwokolo and Bragg (1977) have shown that the presence of phytate in rapeseed causes zinc, calcium and magnesium deficiency in chicken. Phytates reduce the digestibility of proteins (Knuckles et al., 1989) starch and lipids (Nyman and Bjorck, 1989). They are also known to inhibit certain enzymes such as amylases (Sharma et al., 1978; Deshpande and Cheryan, 1984), acid phosphatase (Hayakawa et al., 1990), tyrosinase (Graf, 1986) and trypsin (Deshpande and Damodaran, 1989). Hence, treatment of foods and feeds with phytase or phytase producers could be expected to hydrolyse phytates, and consequently alleviate their antinutritional effects (Ghosh, 1997). Fungal phytases are known to be mostly extracellular (Howson et al., 1983).

Ghosh (1997) reported a fast-growing, well sporulating thermophilic fungus *Sporotrichum thermophile* to be a good phytase producer. A two-day old culture of *S. thermophile* started phytase secretion that reached a peak by eight days, followed by a decline. In this case enzyme biosynthesis was induced by the presence of phytic acid or phytate, amounting to a five-fold enhancement. A very high phytate concentration, however, inhibited phytase production. This pattern is in accordance with many other mesophilic bacteria and fungi (Howson and Davis, 1983; Tambe et al., 1994).

A higher enzyme titre was attained in solid state fermentation than in submerged fermentation, which could be attributed to better nutrient composition of the solid substrate, closer contact between the fungus and the nutrients, and the solid substrate being more similar to the natural growth habit of moulds (Ghosh, 1997). The phytase of *S. thermophile* could be useful in the removal of phytate from soymilk, where a thermostable enzyme would be needed. To meet this purpose as well as in pretreatment of animal feed for removal of antinutritional properties of phytate, a crude enzyme extract may be used. A simpler way could be cultivation of the organism *per se* on the animal feed in SSF to bring about *in situ* phytate reduction. This would also result in protein enrichment of the animal feed.

UPGRADATION OF ANIMAL FEED AND SINGLE CELL PROTEIN PRODUCTION

The cellulolytic thermophilic moulds offer certain advantages over their mesophilic counterparts, such as high rates of cellulose breakdown, good sources of protein, activity over a wider temperature range including higher temperatures, and a higher specific growth rate (Seal and Eggins, 1976). *Chaetomium cellulolyticum* (Chahal and Wang, 1978) and *Sporotrichum pulverulentum* (Thomke et al., 1980) have been used to upgrade animal feeds and to produce single cell protein (SCP). Hecht et al. (1982) reported conversion of cellulose into fungal cell mass, whereas Klingspohn et al. (1993) utilized potato pulp from potato starch processing to meet the same purpose.

Acid pretreatment of the substrate appears to be better than alkali pretreatment if the product is to be used as a ruminant feed in terms of rate and extent of protein production in the insoluble biomass product. Panment et al. (1978) utilised alkali pretreated sawdust to cultivate *Chaetomium cellulolyticum* by solid state fermentation. SCP was produced from various pretreated wood substrates using *C. cellulolyticum* (Moo-Young et al., 1978). Feeding trails suggest that the SCP is nutritious, digestible and non-toxic as an additive in animal feeds. Ghai et al. (1979) used thermophilic moulds *C. cellulolyticum* and *Actinomucor* for producing SCP on canning industry wastes. Similarly, *S. thermophile* converted newsprint waste into protein-rich matter by solid-state fermentation in six days (Zadrazil and Brunnert, 1982). Sen et al. (1979) accomplished conversion of leafy waste material into a brown-coloured product with increased N, P and K contents without a foul smell by using *Sporotrichum* sp. D-14, *Mucor* sp. C-16 and *Humicola* sp. H-37 in 90 days. A two-fold increase in protein content of sugarbeet pulp in 48 hours was obtained using thermophilic moulds, *T. aurantiacus* and *S. thermophile* being the most efficient (Grajek, 1988). In this case, the protein productivity per cubic metre working volume of the fermenter was 0.2 kg/h for *T. aurantiacus* and 0.162 kg/h for *S. thermophile*. Sundman et al. (1981) reported decolourisation of kraft bleach plant effluent by *S. pulverulentum*. Schűgerl and Rosen (1997) used cellulose and nutrient salts as well as potato pulp and potato protein liquor as substrates for cultivating *Chaetomium cellulolyticum* in batch and repeated-batch operations, of which the latter was a more productive process for fungal protein production. Employing thermophilic moulds for SCP production and upgradation of animal feed eliminates the need for heat removal and also reduces contamination problems, and hence makes the technology more cost-effective and simplified.

COMPOSTING

Among the biotechnological applications of thermophilic fungi, composting has been in vogue for the degradation of agroresidues, mushroom production, solid waste management and for understanding the role of fungi in plant litter ecosystems. Composting is a process involving conversion of organic residues into lignoprotein complexes (humus) by thermophiles under optimum moisture and aeration conditions. During this physico-chemical and microbial conversion, CO_2 is evolved and the temperature of the pile may reach 60 to 80°C (Gallardo-Lara and Nogales, 1987). Though the maximum growth temperature of thermophilic moulds is around 60°C (Rosenberg, 1975), their spores may survive at higher temperatures (Falcon et al., 1987). These surviving spores lead to massive fungal growth during the final stage of composting (Straatsma et al., 1989). The effect of these fungi on the growth of mushroom mycelium has been described at three levels. First, the fungi decrease the concentration of ammonia in the compost, which otherwise would counteract the growth of the mushroom mycelium (Ross and Harris, 1983). Second, they immobilise nutrients in a form that apparently is available to the mushroom mycelium (Fermor and Grant, 1985). And third, they may have a growth-promoting effect on mushroom mycelium, as has been demonstrated for *Scytalidium thermophilum* and for other thermophilic moulds (Straatsma et al., 1989).
Scytalidium thermophilum is an important fungus in the production of mushroom compost (Wiegant, 1992). Compost prepared using this fungus had a growth-promoting effect on the mycelia of the edible mushroom *Agaricus bisporus* (Wiegant et al., 1992). However, the hyphal extension rates were not clearly related to the rate of mushroom biomass accumulation. The luxuriant growth of thermophilic moulds in the last phase of the compost preparation

process (Wiegant, 1992) contributes significantly to the quality of compost (Ross and Harris, 1983; Straatsma et al., 1989, 1994) and also, very importantly, such compost is capable of supporting selective growth of the button mushroom, *Agaricus bisporus* (Straatsma et al., 1994).

BIOSORPTION OF HEAVY METALS

The existence of heavy metals and/or radionuclides in the environment, whether derived from natural or anthropological activities, represents a significant environmental hazard. It has been indicated that a potential practical utilization of thermophiles is in bioleaching, which is a faster process than a similar leaching process using mesophiles. Fungal biomass of *Talaromyces emersonii* was reported to be very useful in biosorptive technology (Bengtsson et al., 1995). Faurest et al. (1994) demonstrated the role of pH and cations in the enhancement of heavy metal sorption by fungal mycelial dead biomasses. Similarly Bengtsson et al. (1995) also showed that binding of uranium to the *T. emersonii* biomass decreases significantly when the pH is decreased from 5 to 3. Such thermophilic fungi having the capability to take up heavy metals and radionuclides may be exploited in detoxification of metal-bearing waste-waters, decontamination of radioactive waste-waters, recovery of metals from ore-processing solutions and the recovery of rare metals from sea water (Gadd and White, 1993).

ANTAGONISTIC ATTRIBUTES

The term 'antagonism' refers to any microbial activity that, in some way, adversely affects another organism(s) growing in association with it. It includes lysis, competition and antibiosis. While competing under adverse conditions of temperature regimes, thermophilic moulds are renowned producers of volatile and non-volatile metabolites, lytic enzymes and antimicrobials (Johri and Satyanarayana, 1984, 1986; Johri et al., 1985). These volatile compounds apparently regulate fungal growth *in situ* through fungistatic action and auto-inhibition. Rode et al. (1947) reported penicillin production by *Malbranchea pulchella* var. *sulfurea*. Members of the genera *Thermoascus* and *Malbranchea* were found capable of production of 6-aminopenicillic acid (6-APA) (Kitano et al., 1975), whereas *Mucor pusillus* and *M. miehei* are known to produce extracellular antibiotic peptides namely 'sillurin' and 'miehein' (Somkuti and Greenberg, 1979). Six isolates of *Malbranchea* emanated sporostatic volatile metabolites which inhibited spores of *A. alabamensis, H. lanuginosa, Thermoascus* sp., *T. indicae* and even of itself (Prakash, 1984). Likewise, volatile compounds of *C. thermophile, M. pusillus, S. thermophile* and *T. aurantiacus* inhibited spore germination of *A. fumigatus, H. lanuginosa* and *T. thermophila* (Satyanarayana and Johri, 1981b). Premabai and Narasimhan (1966) reported production of malbranchin A and B' from *M. pulchella* which functioned as an antibacterial.

POLYAMINE BIOSYNTHESIS

A wide range of thermophilic fungi can cause spoilage of bagasse, grain, groundnut, hay, palm, kernels, peat, wood, chips and other agricultural crops (Sharma, 1989). Although microbial colonisation is essential for the disposal of organic waste in nature, and horticulturally the same process is utilized in composting for enriching the soil type and for producing mushroom compost, the biodeterioration of healthy stored grains and wooden articles is indeed undesirable. A strategic prevention of this biodeterioration could be achieved by inhibition of polyamine biosynthesis (Singhania et al., 1991). Thermophilic moulds have been shown to contain polyamines similar to their mesophilic counterparts. These polyamines play a vital role in cellular growth and differentiation (Rajam and Galston, 1985; Walters, 1987) and provide thermal stability for protein synthesis in thermophiles (Oshima and Senshu, 1985).

Singhania et al. (1991) reported a wide distribution of putrescine, spermidine and spermine in fifteen thermophilic moulds belonging to zygomycetes, ascomycetes and deuteromycetes. Besides these, cadaverine and a few unidentified polyamines were also sporadically found. Temperature and age of the culture had a marked effect on the endogenous free pool of polyamines in *Humicola lanuginosa,* and polyamine levels generally declined below or above the optimum temperature of 45°C. Hamana and Matsuzaki (1985) also reported widespread occurrence of putrescine and spermidine in ascomycetes, basidiomycetes and slime moulds, but they did not report any novel polyamines in any of the fungi. Oshima and Senshu (1985) recorded that polyamine levels in *H. lanuginosa* were dependent upon growth temperature and age of the culture.

The inhibitors of polyamines metabolism, particularly ornithine decarboxylase (ODC) inhibitors, may be used as antifungal agents to prevent diseases caused by some thermophilic moulds, and as protectants against biodeterioration caused by these moulds. Difluoromethylornithine (DFMO) effectively inhibited the growth of *H. lanuginosa, T. emersonii* and *M. pusillus* (Singhania et al., 1991). Inhibition of several phytopathogenic fungi has been reported using specific inhibitors of polyamine biosynthesis such as DFMO and cyclohexylamine (Rajam and Galston, 1985; West and Walters, 1989). In contrast, Singhania et al. (1991) showed that difluoromethylarginine (DFMA) had no effect on mycelial growth of the fungi tested except *M. pusillus.* In fact, DFMA caused an enhancement in mycelial growth of *H. lanuginosa.* Therefore, the lack of effectiveness of DFMA suggests the absence of the arginine decarboxylase (ADC) pathway in these thermophilic fungi, and the specificity of action of DFMO on ODC. The main pathway for polyamine biosynthesis is that involving ornithine decarboxylase.

FUTURE PERSPECTIVE

Certain thermophilic moulds have been shown to be the sources of useful enzymes such as glucoamylases, phytases and cellulase-free xylanases. Detailed investigations are needed in exploring the diversity of thermophilic moulds and on the production and characterization of these enzymes. The presence of trehalose and trehalase has been reported in *Humicola* spp. Since trehalose is known to help in the preservation of enzymes, research efforts in understanding enzymes involved in biosynthesis of trehalose are worthwhile. It has not yet been unequivocally established whether only ODC, or ADC in combination with ODC pathways are present in these moulds for biosynthesis of putrescine. This aspect needs further attention. An insight into the production of single cell protein on agro- and food processing industry residues employing these moulds could be useful.

CONCLUSIONS

The biotechnological potential of these moulds is in their ability to degrade organic residues, produce a number of useful enzymes, polysaccharides and antibiotics, transform steroids, bind heavy metals such as uranium and in the production of enriched feeds.

REFERENCES

Adams, P.R., 1992, Growth and amylase production of *Thermoascus aurantiacus* Miehe, *Biotechnol. Appl. Biochem.* 15: 311-313.

Adams, P.R., 1995, Extracellular amylase activities of *Rhizomucor pusillus* and *Humicola lanuginosa* at initial stages of growth, *Mycopathol.* 128: 139-142.

Adams, R. and Deploey, J.J., 1978, Enzymes produced by thermophilic fungi, *Mycologia* 70: 906-910.

Anand, L. and Vithayathil, P.J., 1996, Xylan-degrading enzymes from thermophilic fungus *Humicola lanuginosa*: action pattern of xylanase and ß-glucosidase on xylans, xylooligomers and arabinooligomers, *J. Ferment. Bioeng.* 81: 511-517.

Araujo, A. and Ward, O.P., 1991, Purification and some properties of the mannanases from *Thielavia terrestris, J. Indust. Microbiol.* 6: 269-274.

Arima, K., Liu, W.H. and Beppu, T., 1972, Studies on the lipase production of thermophilic fungus *Humicola lanuginosa, Agr. Biol. Chem.* 36: 893-895.

Arnold, W.N., Garrison, R.G., Mann, L.C. and Wallace, D.P., 1988, The acid phosphatase from *Thermoascus crustaceus,* a thermophilic fungus, *Microbios* 54: 101-112.

Banerjee, S., Archana, A. and Satyanarayana, T., 1994, Xylose metabolism in a thermophilic mould *Malbranchea pulchella* var. *sulfurea* TMD-8, *Curr. Microbiol.* 29: 349-352.

Basawesvara Rao, V., Maheshwari, R., Sastry, N.V.S. and Subba Rao, P.V., 1979, A thermostable glucoamylase from the thermophilic fungus *Thermomyces lanuginosus, Curr. Sci.* 48: 113-115.

Basawesvara Rao, V., Sastry, N.V.S. and Subba Rao, P.V., 1981, Purification and characterization of a thermostable glucoamylase from the thermophilic fungus *Thermomyces lanuginosus, Biochem. J.* 193: 379-387.

Bengtsson, L., Johansson, B., Hackett, T.J., McHale, L. and McHale, A.P., 1995, Studies on the biosorption of uranium by *Talaromyces emersonii* CBS 814.70 biomass, *Appl. Microbiol. Biotechnol.* 42: 807.

Biely, P., 1985, Microbial xylanolytic system, *Trends Biotechnol.* 3: 286.

Biely, P., Mackenzie, C.R., Puls, J. and Schneider, H., 1986, Cooperativity of esterases and xylanases in the enzymatic degradation of acetyl xylan, *Bio/Technol.* 4: 731.

Biely, P., 1991, Biotechnological potential and production of xylanolytic systems free of cellulases, in: *Enzymes in Biomass Conversion* (G.F. Leatham and M.E. Himmel, eds.), ACS Symposium series 460, American Chemical Society, Washington DC.

Bilai, T.I., Chernyagina, T.B., Dorokhov, V.V., Poedinok, N.L., Zakharchenko, V.A., Ellanskaya, I.A. and Lozhkina, G.A., 1985, Phosphatase activity of different species of thermophilic and mesophilic fungi, *Mikrobiologika* 47: 53-57.

Breuil, C., Wojtczak, G. and Saddler, J.N., 1986, Production and localization of cellulases and ß-glucosidase from the thermophilic fungus *Thielavia terrestris, Biotechnol. Lett.* 9: 673-676.

Brock, T.D., 1970, High temperature systems, *Ann. Rev. Ecol. Systemat.* 1: 191-220.

Brock, T.D., 1986, Life at high temperatures, *Science* 230: 132-138.

Bunni, L., McHale, L. and McHale, A.P., 1989, Production, isolation and partial characterization of an amylase system produced by *Talaromyces emersonii* CBS 814.70, *Enz. Microb. Technol.* 11: 370-375.

Campos, L. and Felix, C.R., 1995, Purification and characterization of a glucoamylase from *Humicola grisea, Appl. Env. Microbiol.* 61: 2436.

Cardello, L., Terenzi, H.F. and Jorge, J.A., 1994, A cytosolic trehalase from the thermophilic fungus *Humicola grisea* var. *thermoidea, Microbiol.* 140: 1671-1677.

Chahal, D.S. and Wang, D.I.C., 1978, *Chaetomium cellulolyticum,* growth, behaviour on cellulose and protein production, *Mycol.* 70: 160-170.

Chiang, CL. and Knight, S.G., 1960, Metabolism of D-xylose by moulds, *Nature* 188: 79-81.

Common, R.H., 1940, The phytic acid content of some poultry feeding stuffs, *Analyst* 65: 79-83.

Craveri, R., Craveri, A. and Guicciardi, A., 1967, Research on the properties and activities of enzymes of eumycete thermophilic isolates of soil. *Annal. Microbiol. Enzymol.* 17: 1-30.

Crisan, E.V., 1969, The proteins of thermophilic fungi, in: *Current Topics in Plant Science,* (J.E. Grunckel, ed.), Academic Press, New York, London, pp. 32-33.

deBoland, A.R., Garner, G.B. and O'Dell, B.L., 1975, Identification of properties of phytate in cereal grains and oilseed products, *J. Agric. Food Chem.* 23: 1186-1189.

Deploey, J.J., Fortman, B. and Perva, L., 1981, Daily measurements of *Mucor miehei* lipase activity, *Mycologia* 73: 953-958.

Deshpande, S.S. and Cheryan, M., 1984, Effect of phytic acid, divalent cations, and their interaction on α-amylase activity, *J. Food Sci.* 49: 516-699.

Deshpande, S.S. and Damodaran, S., 1989, Effect of phytate on solubility, activity and confirmation of trypsin and chymotrypsin, *J. Food Sci.* 54: 695-699.

Dubey, A.K. and Johri, B.N., 1987, Xylanolytic activity of thermophilic *Sporotrichum sp.* and *Myceliophthora thermophilum, Proc. Indian Acad. Sci. (Plant Sci.)* 97: 247-255.

Durand, H., Soucaille, P., and Tiraby, G., 1984, Comparative study of cellulases and hemicellulases from four fungi: mesophiles *Trichoderma reesei* and *Penicillium* sp. and thermophiles *Thielavia terrestris* and *Sporotrichum cellulophilum, Enz. Microb. Technol.* 6: 175-180.

Dusterhoft, E.M., Linssen, V.A.J.M., Voragen, A.G.J. and Beldman, G., 1997, Purification, characterization and properties of two xylanases from *Humicola insolens, Enz. Microb. Technol.* 20: 437-445.

Erdman, J.W. Jr., 1979, Oilseed phytates: nutritional implications, *J. Amer. Oil Chem. Soc.* 56: 736.

Escobar, J. and Barnett, S., 1995, Synthesis of acid protease from *Mucor miehei*: integration of production and recovery, *Proc. Biochem.* 30: 695-700.

Falcon, M., Corominas, E., Pere, M.L. and Perestelo, F., 1987, Aerobic bacterial populations and environmental factors involved in the composting of agricultural and forest wastes of the Canary Islands, *Biol. Wastes* 20: 89-99.

Faurest, E., Canal, C. and Roux, J.C., 1994, Improvement of heavy metal sorption by mycelial dead biomasses (*Rhizopus arrhizus, Mucor miehei and Penicillium chrysogenum*): pH control and cationic activation, *FEMS Microbiol. Rev.* 14: 325.

Fergus, C.L., 1969, The cellulolytic activity of thermophilic fungi and actinomycetes, *Mycologia* 61: 120-129.

Fermor, T.R. and Grant, W.D., 1985, Degradation of fungal and actinomycete mycelia by *Agaricus bisporus, J. Gen. Microbiol.* 131: 1729-1734.

Filho, E.X.F., 1996, Purification and characterization of a ß-glucosidase from solid-state cultures of *Humicola grisea* var. *thermoidea, Can. J. Microbiol.* 42: 1-5.

Fischer, L., Scheckermann, C. and Wagner, F., 1995, Purification and characterization of a thermotolerant ß-galactosidase from *Thermomyces lanuginosus, Appl. Env. Microbiol.* 61: 1497.

Flannigan, B. and Sellars, P.N., 1972, Activities of thermophilous fungi from barley kernels against arabinoxylan and carboxymethylcellulose, *Trans. Br. Mycol. Soc.* 58: 338.

Gadd, G.M. and White, C., 1993, Microbial treatment of metal pollution: a working biotechnology, *Trends Biotechnol.* 11: 353-359.

Gallardo-Lara, F. and Nagales, R., 1987, Effect of the application of town refuse compost on the soil-plant system: a review, *Biol. Wastes* 19: 35.

Ganju, R.K., Murthy, S.K. and Vithayathil, P.J., 1990, Purification and functional characteristics of an endocellulase from *Chaetomium thermophile* var. *coprophile, Carbohydr. Res.* 197: 245-255.

Garg, S.K. and Johri, B.N., 1998, Proteolytic enzymes, in: *Thermophilic Moulds in Biotechnology* (T. Satyanarayana, ed.), Narosa Publishing House, New Delhi.

Ghai, S.K., Kahlon, S.S. and Chahal, D.S., 1979, Single cell protein from canning industry waste: Sag waste as substrate for thermotolerant fungi, *Indian J. Exp. Biol.* 17: 789-791.

Ghosh, S., 1997, Phytase of a thermophilic mould *Sporotrichum thermophile* Apinis, *M.Sc. Dissertation*, University of Delhi South Campus, New Delhi.

Gifford, S.R. and Clydesdale, R.M., 1990, Interactions among calcium, zinc and phytate with three protein sources, *J. Food Sci.* 55: 1720-1723.

Gomes, D.J., Gomes, I. and Steiner, W., 1994, Production of a highly thermostable xylanase by a wild strain of thermophilic fungus *Thermoascus aurantiacus* and partial characterization of the enzyme, *J. Biotechnol.* 37: 11-22.

Gomes, J., Gomes, I., Kreiner, W., Esterbauer, H., Sinner, M. and Steiner, W., 1993, Production of high level of cellulase-free and thermostable xylanase by a wild strain of *Thermomyces lanuginosus* using beechwood xylan, *J. Biotechnol.,* 30: 283-297.

Gilbert, H.J. and Hazlewood, G.P., 1993, Bacterial cellulases and xylanases, *J. Gen. Microbiol.* 139: 187-194.

Graf, E., 1986 (ed.), *Phytic Acid Chemistry and Applications*, Pilatus Press, Minneapolis, MN, pp. 1-21.

Grajek, W., 1987, Production of D-xylanases by thermophilic fungi using different methods of culture, *Biotechnol. Lett.* 9: 353-356.

Grajek, W., 1988, Production of protein by thermophilic fungi from sugar beet pulp in solid-state fermentation, *Biotechnol. Bioeng.* 32: 225-260.

Gupta, A.K. and Gautam, S.P., 1995, Improved production of extracellular α-amylase by the thermophilic fungus *Malbranchea sulfurea*, following protoplast fusion, *World J. Microbiol. Biotechnol.* 11: 193-195.

Hamana, K. and Matsuzaki, S., 1985, Distribution of polyamines in prokaryotes, algae, plants and fungi, in: *Polyamines: Basic and Clinical Aspects* (K. Imahori, F. Suzuki, O. Suzuki and U. Bachrach, eds.), VNU Science Press, Utrecht, pp. 105-112.

Haq, M.M. and Deckwer, W.D., 1995, Cellulase-free xylanase by thermophilic fungi: A comparison of xylanase production by two *Thermomyces lanuginosus* strains, *Appl. Microbiol. Biotechnol.* 43: 604-609.

Hayakawa, T., Suzuki, K., Miura, H., Ohno, T. and Igaue, I., 1990, Myoinositol polyphosphate intermediates in the dephosphorylation of phytic acid by acid phosphatase with phytase activity from rice bran, *Agric. Biol. Chem.* 54: 279-286.

Hecht, V., Schügerl, K. and Scheiding, W., 1982, Conversion of cellulose into fungal cell mass, *Eur. J. Appl. Microbiol. and Biotechnol.* 16: 219-222.

70

Höfer, M., Betz, A. and Kotyk, A., 1971, Metabolism of the obligatory yeast *Rhodotorula gracilis*, IV: Induction of an enzyme necessary for D-xylose catabolism, *Biochim. Biophys. Acta* 252: 1-12.

Howson, S.J. and Davis, R.P., 1983, Production of phytate hydrolysing enzyme by some fungi, *Enz. Microb. Technol.* 5: 377-382.

Jackson, L.E. and Seidman, M., 1985, Danish patent application No. 8503417.

Jain, A., 1995, Production of xylanase by thermophilic fungus *Melanocarpus albomyces* ZIS-68, *Proc. Biochem.* 30: 705-710.

Jain, M.K., Kapoor, K.K. and Mishra, M.M., 1979, Cellulase activity, degradation of cellulose and lignin, and humus formation by thermophilic fungi, *Trans. Brit. Mycol. Soc.* 73: 85-89.

Jensen, B., Olsen, J. and Allermann, K., 1988, Purification of extracellular amylolytic enzymes from the thermophilic fungus *Thermomyces lanuginosus*, *Can. J. Microbiol.* 34: 218-223.

John, M., Schmid, B., and Schmid, J., 1979, Purification and some properties of five endoxylanases and ß-D-xylosidases produced by a strain of *Aspergillus niger*, *Can. J. Biochem.*, 57: 125-134.

Johri, B.N., 1980, Biology of thermophilous fungi, in: *Recent Advances in Biology of Microorganisms* (K.S. Bilgrami and K.M. Vyas, eds.), Bishen Singh and Mahendra Pal Singh, Dehradun, India.

Johri, B.N. and Satyanarayana, T., 1984, Ecology of thermophilic fungi, in: *Progress in Microbial Ecology* (K.G. Mukerji, V.P. Agnihotri and R.P. Singh, eds.), Print House, Lucknow.

Johri, B.N. and Satyanarayana, T., 1986, Thermophilic moulds: perspectives in basic and applied research, *Indian Rev. Life Sci.* 6: 75-100.

Johri, B.N., Jain, S. and Chauhan, S., 1985, Enzymes from thermophilic fungi: proteases and lipases, *Proc. Indian Acad. Sci.* (Plant Sci.) 94: 175-196.

Kanlayakrit, W., Ishimatru, K., Nakao, M. and Hayashida, S., 1987, Characteristics of raw starch-digesting glucoamylase from thermophilic *Rhizomucor pusillus*, *J. Ferment. Technol.* 65: 379-385.

Kauss, H., Jeblick, W. and Young, D.H., 1983, Chitin deacetylase from the plant pathogen *Colletotrichum lindemuthianum*, *Plant Sci. Lett.*, 28: 231-236.

Kirillova, L.M. and Shchepankevich, Y.M., 1989, Directed search for invertase-producing strains among the thermophilic micromycetes, *Mikrobiologika* 51: 59-64.

Kitano, K., Kintaka, K., Katamato, K., Nan, K. and Nakao, Y., 1975, Occurrence of 6-aminopenicillinic acid in culture broths of strains belonging to genera *Thermoascus, Gymnoascus, Polypaecium* and *Malbranchea*, *J. Ferment. Technol.* 53: 339-346.

Klingspohn, U., Papsupuleti, P.V. and Schügerl, K., 1993, Production of enzymes from potato pulp using batch operation of a bioreactor, *J. Chem. Technol. Biotechnol.* 58: 19.

Knosel, D. and Resz, A., 1973, *Pilze aus Mullkomport Stadtehygiene* 6: 143.

Knuckles, B.E., Kuzmicky, D.D., Gumbmann, M.R. and Betschart, A.A., 1989, Effect of myoinositol phosphate esters on *in vitro* and *in vivo* digestion of proteins, *J. Food. Sci.* 54: 1348-1350.

Kotwal, S.M., Khan, M.I. and Khire, J.M., 1995, Production of thermostable ß-galactosidase from thermophilic fungus *Humicola* sp., *J. Ind. Microbiol.* 15: 116-120.

Lawson, D.M., Brzozowski, A.M., Rety, S., Verma, C. and Dodson, G.G., 1994, Probing the nature of substrate binding in *Humicola lanuginosa* lipase through X-ray crystallography and intuitive modelling, *Protein Eng.* 7: 543-550.

Magan, N., 1997, Fungi in extreme environments, in: *The Mycota: IV. Environmental and Microbial Relationship* (K. Esser, P.A. Lemke, D.T. Wicklow and B. E. Soderstrom, eds.), Springer-Verlag, Berlin Heidelberg.

Maheshwari, R. and Balasubramanyam, P.V., 1988, Simultaneous utilization of glucose and sucrose by thermophilic fungi, *J. Bacteriol.* 170: 3274-3280.

Maheshwari, R., Balasubramanyam, P.V. and Palanivelu, P., 1983, Distinctive behaviour of invertase in a thermophilic fungus *Thermomyces lanuginosus*, *Archiv. Microbiol.* 134: 255-260.

Maheshwari, R. and Kamalam, P.T., 1985, Isolation and culture of thermophilic fungi, *Melanocarpus albomyces* and factors influencing the production and activity of xylanase, *J. Gen. Microbiol.* 131: 3017-3027.

Matsuo, M. and Yasui, T., 1985, Properties of xylanase of *Malbranchea pulchella* var. *sulfurea* no. 48, *Agric. Biol. Chem.* 49: 839-841.

Matsuo, M., Yasui, T. and Kobayashi, T., 1975, Production and saccharifying action for xylan of xylanase from *Malbranchea pulchella* var. *sulfurea* no. 48, *Nippon Nagaikagaku Kaishi* 49: 263-270.

McCormack, J., Hackett, T.J., Tuohy, M.G. and Coughlan, M.P., 1991, Chitinase production by *Talaromyces emersonii*, *Biotechnol. Lett.* 13: 677-682.

McHale, A. and Coughlan, M.P., 1981, *Biochem. Biophys. Acta.* 662: 145-151.

McHale, A.P. and Morrison, J., 1986, Cellulase production during growth of *Talaromyces emersonii* CBS 814.70 on lactose containing media, *Enz. Microb. Technol.* 8: 749-754.

Meyer, H.P. and Canevascini, G., 1981, Separation and some properties of two intracellular ß-glucosidases of *Sporotrichum (Chrysosporium) thermophile*, *Appl. Environ. Microbiol.* 41: 924-931.

Moo-Young, M., Chahal, D.S. and Vlach, D., 1978, Single cell protein from various pretreated wood substrates using *Chaetomium cellulolyticum*, *Biotechnol. Bioeng.* 20: 107-118.

Mouchacca, J., 1995, Thermophilic fungi in desert soils: a neglected extreme environment, in: *Microbial*

Diversity and Ecosystem Function, CAB International, Oxon, UK.

Nakaji, M.K., 1978, Ultrastructure and physiology of the thermophilic fungus *Humicola lanuginosa*, *Ph.D. Dissertation*, University of Kentucky, USA.

Neves, M.J., Terenzi, H.F., Leone, F.A. and Jorge, J.A., 1994, Quantification of trehalose in biological samples with a conidial trehalose from the thermophilic fungus *Humicola grisea* var. *thermoidea*, *World J. Microbiol. Biotechnol.* 10: 17-19.

Nishio, N., Krisu, H. and Nagai, S., 1981, Thermophilic cellulose production by *Talaromyces* sp. in solid state cultivation, *J. Ferment. Technol.* 59: 407-410.

Nwokolo, E.N. and Bragg, D.B., 1977, Influence of phytic acid and crude fibre on the availability of minerals from four protein supplements in growing chicks, *Can. J. Anim. Sci.* 57: 475-477.

Nyman, M.E. and Bjorck, I.M., 1989, *In vivo* effects of phytic acid and polyphenols on the bioavailability of polysaccharides and other nutrients, *J. Food Sci.* 54: 1332-1335.

Ogundero, V.W., 1980, Lipase activity of thermophilic fungi from mouldy groundnuts in Nigeria, *Mycologia* 72: 118-126.

Oshima, T. and Senshu, M., 1985, Unusual long polyamines in a thermophile, in: *Polyamines: Basic and Clinical Aspects* (K. Imahori, F. Suzuki, O. Suzuki and U. Bachrach, eds.), VNU Science Press, Utrecht, pp. 113-118.

Paice, M.G. and Jurasek, L., 1984, Removing hemicellulose from pulps by specific enzyme hydrolysis, *J. Wood Chem. Technol.* 4: 187-198.

Palanivelu, P., Balasubramanyam, P.V. and Maheshwari, R., 1984, Co-induction of sucrose transport and invertase activities in a thermophilic fungus, *Thermomyces lanuginosus, Arch. Microbiol.* 139: 44-47.

Panment, N., Robinson, C.W., Hilton, J. and Moo-Young, M., 1978, Solid state cultivation of *Chaetomium cellulolyticum* on alkali pretreated sawdust, *Biotechnol. Bioeng.* 20: 1735-1744.

Peralta, R.M., Kadowaki, M.K., Terenzi, H.F. and Jorge, J.A., 1997, A highly thermostable ß-glucosidase activity from the thermophilic fungus *Humicola grisea* var. *thermoidea*: Purification and biochemical characterization, *FEMS Microbiol. Lett.* 146: 291-296.

Peralta, R.M., Terenzi, H.F. and Jorge, J.A., 1990, ß-Glycosidase activities of *Humicola grisea*: biochemical and kinetic characterization of a multifunctional enzyme, *Biochim. Biophys. Acta* 1033: 243-249.

Prakash, A., 1984, Antagonistic attributes of thermophilic fungi and thermophilism, *Ph.D. Thesis*, Bhopal University, Bhopal, 166 pp.

Prasad, A.R.S. and Maheshwari, R., 1978a, Purification and properties of trehalase from the thermophilic fungus *Humicola lanuginosa*, *Biochim. Biophys. Acta.* 525: 162-180.

Prasad, A.R.S. and Maheshwari, R., 1978b., Growth of the trehalase activity in the thermophilic fungus *Thermomyces lanuginosus*, *Proc. Ind. Acad. Sci.* 87B: 231-241.

Premabai, M. and Narasimhan, P.L., 1966, Thermophilic microrganisms Part IV. Elaboration of malbranchins A and B by *Malbranchea pulchella*, *Indian J. Biochem.* 3: 187-190.

Purkarthofer, H. and Steiner, W., 1995, Induction of endo-ß-xylanase in the fungus *Thermomyces lanuginosus*, *Enz. Microb. Technol.* 17: 114-118.

Purkarthofer, H., Sinner, M. and Steiner, W., 1993, Cellulase-free xylanase from *Thermomyces lanuginosus*: Optimisation of production in submerged and solid state culture, *Enz. Microb. Technol.* 15: 677-682.

Rajam, M.V. and Galston, A.W., 1985, The effects of some polyamine biosynthetic inhibitors on growth and morphology and phytopathogenic fungi, *Plant and Cell Physiol.* 26: 683-692.

Reese, E.T., 1946, Aerobic decomposition of cellulose by microorganisms at temperatures above 40°C, *Ph.D. Thesis*, Pennsylvania State College, Pennsylvania.

Rode, L.J., Foster, J.W. and Schuhardt, V.T., 1947, Penicillin production by a thermophilic fungus, *J. Bacteriol.* 53: 565-566.

Romanelli, R.A., Huston, C.W. and Barnett, S.M., 1975, Studies on thermophilic fungi, *Appl. Microbiol.* 30: 276-281.

Rosenberg, S.L., 1975, Temperature and pH optima for 21 species of thermophilic and thermotolerant fungi, *Can. J. Microbiol.* 21: 1535-1540.

Ross, R.C. and Harris, P.J., 1983, The significance of thermophilic fungi in mushroom compost preparation, *Sci. Hortic.* (Amsterdam) 20: 61-70.

Roy, S.K., Dey, S.K., Raha, S.K. and Chakrabarty, S.L., 1990, Purification and properties of an extracellular endoglucanase from *Myceliophthora thermophila* D-14 (ATCC 48 104), *J. Gen. Microbiol.* 136: 1967-1971.

Sadhukhan, R.K., Manna, S., Roy, S.K. and Chakrabarty, S.L., 1990, Thermostable amylolytic enzymes from a cellulolytic fungus *Myceliophthora thermophila* D-14 (ATCC 48 104), *Appl. Microbiol. Biotechnol.* 33: 692-696.

Satyanarayana, T. and Johri, B.N., 1981a, Lipolytic activity of thermophilic fungi of paddy straw compost, *Curr. Sci.* 50: 680-682.

Satyanarayana, T. and Johri, B.N., 1981b, Volatile sporostatic factors of thermophilous fungal strains of paddy straw compost, *Curr. Sci.* 50: 763-766.

Satyanarayana, T. and Johri, B.N., 1983, Variation in xylanolytic activity of thermophilic fungi, *Bionature* 3: 39-41.

Satyanarayana, T., Chavant, L. and Montant, C., 1985, Applicability of apizym for screening enzyme activity of thermophilic moulds, *Trans. Br. Mycol. Soc.* 85: 727-730.

Satyanarayana, T., Jain, S. and Johri, B.N., 1988, Cellulases and xylanases of thermophilic moulds, in: *Perspectives in Mycology and Plant Pathology* (V.P. Agnihotri, A.K. Sarbhoy and D. Kumar, eds.), Malhotra Publishing House, New Delhi, pp. 24-60.

Satyanarayana, T., Johri, B.N. and Klein, J., 1992, Biotechnological potential of thermophilic fungi, in: *Handbook of Applied Mycology, Vol.4* (D.K. Arora, R.P. Elander and K.G. Mukherji, eds.), Marcel Dekker Inc., New York., pp. 729-761.

Schügerl, K. and Rosen, W., 1997, Investigation of the use of agricultural byproducts for fungal protein production, *Proc. Biochem.* 32: 705-7.

Seal, K.J. and Eggins, H.O.W., 1976, The upgrading of agricultural wastes by thermophilic fungi, in: *Food from Wastes* (G.G. Birch, K.J. Parkar and J.T. Worgen, eds.), Appl. Sciences Publish., London, pp. 58-58.

Sen, S., Abraham, T.K. and Chakrabarty, S.L., 1979, Biodeterioration and utilization of city wastes by thermophilic microorganisms, *Indian J. Exp. Biol.* 17: 1284-1285.

Sharma, H.S.S., 1989, Economic importance of thermophilous fungi, *Appl. Microbiol. Biotechnol.* 31: 1-10.

Sharma, H.S.S. and Johri, B.N., 1992, The role of thermophilic fungi in agriculture, in: *Handbook of Applied Mycology, Vol.4* (D.K. Arora, R.P. Elander and K.G. Mukerji, eds.), Marcel Dekker, New York, pp. 707-728.

Sharma, C.B., Goel, M. and Irshad, M., 1978, Myoinositol hexaphosphate as a potential inhibitor of α-amylase of different origins, *Phytochem.* 17: 201-204.

Singhania, S., Satyanarayana, T. and Rajam, M.V., 1991, Polyamines of thermophilic moulds: distribution and effect of polyamine biosynthesis inhibitors on growth, *Mycol. Res.* 95: 915-917.

Somkuti, G.A. and Greenberg, G., 1979, Antimicrobial peptidase of thermophilic fungi, *Dev. Indus. Microbiol.* 20: 661-337.

Somkuti, G.A. and Steinberg, D.H., 1980, Thermoacidophilic extracellular alpha-amylase of *Mucor pusillus*, *Dev. Indus. Microbiol.* 21: 327-337.

Sorensen, S.Y. and Crisan, E.Y., 1974, Thermostable lactase from thermophilic fungi, *J. Food Sci.* 39: 1184-1187.

Stalbrand, H., Sika-aho, M., Tenkanen, M. and Viikari, L., 1993, Purification and characterization of two ß-mannanases from *Trichoderma reesei*, *J. Biotechnol.* 29: 229-234.

Straatsma, G., Gerrits, J.P.G., Augustijn, M.P.A.M., Op den Camp, Vogels, G.D. and var Griendsven, L.J.L.D., 1989, Population dynamics of *Scytalidium thermophilum* in mushroom compost and stimulatory effects on growth rate and yield of *Agaricus bisporus*, *J. Gen. Microbiol.* 135: 751-759.

Straatsma, G., Olijnsma, T.W., Gerrits, J.P.G., Amsing, J.G.M., Camp, H.J.M.O.D. and van Griendsven, L.J.L.D., 1994, Inoculation of *Scytalidium thermophilum* in button mushroom compost and its effect on yield, *Appl. Env. Microbiol.* 60: 3049-3054.

Subramanyam, A., 1980, Studies on *Thermoascus aurantiacus* Miehei, *Acta Mycol.* 26: 121-131.

Suihko, M.L., Suomalainen, I. and Enari, T.M., 1983, D-Xylose catabolism in *Fusarium oxysporum*, *Biotechnol. Lett.* 5: 525-530.

Sundman, G., Kirk, T.K. and Chang, H., 1981, Fungal decolorization of kraft bleach plant effluent, *Tappi* 64: 145.

Takenishi, S., Tsujisaka, Y. and Fukumoto, J., 1973, *J. Biochem.* 73: 335.

Tambe, S.M., Kakli, S.G., Kelkar, S.M. and Parekh, L.J., 1994, Two distinct molecular forms of phytase from *Klebsiella aerogenes:* Evidence for unusally small active enzyme peptide, *J. Ferment. Bioeng.* 77: 23-27.

Tansey, M.R. and Brock, T.D., 1978, Microbial life at high temperatures: Ecological aspects, in: *Microbial Life of Extreme Environments* (D.J. Kushner, ed.), Academic Press, London, pp. 156-215.

Thakur, M.S., Karanth, N.G. and Nand, K., 1990, Production of fungal rennet by *Mucor miehei* using solid state fermentation, *Appl. Microbiol. Biotechnol.* 32: 409-413.

Taylor, P.M., Napier, E.J. and Fleming, I.D., 1978, Some properties of a glucoamylase produced from the thermophilic fungus *Humicola lanuginosa*, *Carbohydr. Res.* 61: 301-308.

Thomke, S., Rundgren, M. and Eriksson, K.E., 1980, Nutritional evaluation of the white-rot fungus *Sporotrichum pulverulentum* as a feed stuff to rats, pigs and sheep, *Biotechnol. Bioeng.* 22: 2285-2303.

Tomada, N., Kasai, N., Kamakura, M. and Kaetsu, I., 1986, Periodical batch culture of the immobilized growing fungus *Sporotrichum cellulophilum* producing cellulase in the non-woven materials, *Biotech. Bioeng.* 28: 1227-1232.

Tong, C.C., Cole, A.L. and Shepherd, M.G., 1980, Purification and properties of the cellulase from the thermophilic fungus *Thermoascus aurantiacus*, *Biochem. J.* 191: 83-94.

Tuohy, M.G., Buckley, R.J., Griffin, T.O., Connelly, I.C., Shanley, N.A., Ximenes, E., Filho, F., Hughes, M.M., Grogan, P. and Coughlan, M.P., 1989, Enzyme production by solid state cultures of aerobic fungi on lignocellulosic substrates, in: *Enzyme Systems for Lignocellulose Degradation* (M.P. Coughlan, ed.), Elsevier Applied Science, London and New York.

Tuohy, M.G., Laffey, C.D. and Coughlan, M.P., 1994, Characterization of the individual components of the xylanolytic enzyme system of *Talaromyces emersonii, Bioresource Technol.* 50: 37-42.

Turchi, S.L. and Becker, T., 1987, Improved purification of alpha-amylase isolated from *Rhizomucor pusillus* by affinity chromatography, *Curr. Microbiol.* 15: 203-206.

Viikari, L., Kantelinen, A., Ratto, M. and Sundquist, J., 1991, Enzymes in pulp and paper processing, in: *Enzyme in Biomass Conversion,* ACS symposium series 469, (G.S. Leatham and M.E. Himmel, eds.), American Chemical Society, Washington, pp. 12-22.

Viikari, L., Komtelinero, A., Sundquist, J. and Linko, M., 1994, Xylanases in bleaching: from an idea to the industry, *FEMS Microbiol. Rev.* 13: 335-350.

Walters, D.R., 1987, Polyamines: the cinderellas of cell biology, *Biologist* 34: 73-76.

West, H.M. and Walters, D.R., 1989, Effects of polyamines biosynthesis inhibitors on growth of *Pyrenophora teres, Gaeumannomyces granuus, Fusarium culmorum* and *Septoria nodorum in vitro, Mycol. Res.* 92: 453-457.

Whitehead, E.A. and Smith, S.N., 1989, Fungal extracellular enzyme activity associated with breakdown of plant cell biomass, *Enz. Microb. Technol.* 11: 736.

Wiegant, W.M., 1992, Growth characteristics of the thermophilic fungus *Scytalidium thermophilum* in relation to production of mushroom compost, *Appl. Env. Microbiol.* 58: 1301-1307.

Wiegant, W.M., Wery, J., Buitenhuis, E.T. and de Bont, J.A.M., 1992, Growth-promoting effect of thermophilic fungi on the mycelium of the edible mushroom *Agaricus bisporus, Appl. Env. Microbiol.* 58: 2654-2659.

Wright, C., Kafkewitz, D. and Somberg, E.W., 1983, Eukaryotic thermophily: Role of lipids in the growth of *Talaromyces thermophilus, J. Bacteriol.* 156: 493.

Wong, K.K.Y. and Saddler, J.N., 1993, Application of hemicellulases in food, feed, pulp and paper industries, in: *Hemicellulose and Hemicellulases* (M.P. Coughlan and G.P. Hazlewood, eds.), Portland Press, London.

Wood, T.N. and McCrae, S.T., 1986, Studies of two low molecular weight endo-1,4-ß-D-xylanase constitutively synthesized by the cellulolytic fungus *Trichoderma koningii, Carbohydr. Res.* 148: 321-330.

Yoshioka, H., Chavanich, S., Nilubol, N. and Hayashida, S., 1981, Production and characterization of thermostable xylanase from *Talaromyces byssochlamydoides* YH-50, *Agric. Biol. Chem.* 45: 579-586.

Zadrazil, F. and Brunnert, H., 1982, Solid state fermentation of lignocellulose containing plant residues with *Sporotrichum pulverulentum* Nov. and *Dichomitus squalens* (Karst.) Reid., *Eur. J. Appl. Microbiol. Biotechnol.* 16: 45-51.

Zimmermann, A.I.S., Terenzi, H.F. and Jorge, J.A., 1990, Purification and properties of an extracellular conidial trehalase from *Humicola grisea* var. *thermoidea, Biochim. Biophys. Acta* 1036: 41-46.

FUNGAL BIOTECHNOLOGY FOR EFFECTIVE CEREAL STRAW MANAGEMENT

Myank U. Charaya

Microbiology Laboratory
Department of Botany
M.M. Postgraduate College
Modinagar-201 204
India

INTRODUCTION

Cereal crops, probably the oldest crops grown by man, still retain a dominant position in world agriculture constituting staple food in most parts of the world. The quantities of nutrients temporarily immobilised in chaff, straw, stubble and roots probably equal the nutrients removed in the grain (Pierce, 1978). Considering the latest world agricultural production records (FAO, 1997), not less than 584 874 x 10^3 MT of wheat straw and 562 259 x 10^3 of rice straw were produced in 1996.

The enormous quantities of straw available allow us to put it to a variety of uses (Staniforth, 1979a,b; Anderson and Anderson, 1980) depending upon our needs and choice: (a) it may be burnt in the soil itself thereby returning minerals immobilized; (b) it may be chopped followed by ploughing or composted for adding to soil to improve fertility; (c) it may be used for feeding animals and for their bedding; (d) it may be used for thatching, packing and making ropes, and for direct combustion to heat public baths, bread ovens, brick kilns and domestic water supplies; (e) it may be used for growing microorganisms for Single Cell Protein, for growing mushrooms and cucumber, for production of xylitol, methane, biogas and alcohol; or (f) it may be used for pulp and paper production. Proper well-thought strategies need to be evolved for effective management of this highly valuable natural resource.

FUNGI AS AGENTS OF STRAW TRANSFORMATION

Straw may be treated in a variety of ways to increase its value. Although mechanical, physical and chemical treatments form a part of some of the strategies for straw utilisation, it is the biological degradation of straw which forms the basis of its most important uses - as livestock feed, for SCP and mushroom growing, for composting and for the production of various chemicals such as methane, biogas and alcohol. Bacon (1979) preferred to use the term 'straw transformation' for the process, laying emphasis on the fact that various constituents of straw are transformed into useful products. The fungi are known to play an important part in this process (Dickinson and Pugh, 1974) and the rise of fungal biotechnology (Charaya and Mehrotra, 1998) has had a positive impact on our efficiency in managing straw.

PROBLEMS ASSOCIATED WITH STRAW MANAGEMENT

Like many agricultural wastes, cereal straw consists largely of lignified cell wall material, cellulose, hemicellulose and lignin being the three primary structural components (Charaya, 1985; Hayn et al., 1993). Though many types of straw exhibit variations in their quantities of cellulose, hemicellulose and lignin, all cereal straws have cellulose as the predominant component followed by hemicellulose - cellulose consisting of more than 40% and hemicellulose more than 20% (Anderson and Anderson, 1980; Charaya, 1985).

A great deal of attention has been paid to the degradation of cellulose and the use of its degradation products. Cellulose, like starch, is a polymer of D-glucose. It differs from starch, however, in that cellulose has $\beta(1\rightarrow4)$ linkages while starch has $\alpha(1\rightarrow4)$ linkages. This difference results in strong hydrogen bonding and gives cellulose its physical strength and rigidity (Anderson and Anderson, 1980). Not only this, an intimate association exists between different components. The lignin, together with the hemicellulose, encrusts the cellulose chains forming a barrier which prevents wetting and access by cellulose-degrading enzymes (Kirk and Haskin, 1973; Kuhad and Singh, 1993). To utilise the cellulose component or the hemicellulose, this association first has to be broken. The loosening of this association, followed by degradation of cellulose, hemicellulose and lignin, and further transformation/ utilisation of the various degradation products constitute the major components of any strategy aimed at the efficient management of straw. Fungal biotechnology holds immense potential to help us overcome any impediments in the effective utilisation of this invaluable resource.

CEREAL STRAW AS RUMINANT FEED

Straw has probably been a component of the ruminant diet since the evolution of the rumen. It is also possible that the rumen evolved to cope with the natural abundance of straw-type feeding material (Morrison, 1979). The rumen contains a large microbial population (about 10^{10} cells/ml). The cellulases produced by these microorganisms hydrolyse cellulose releasing glucose for use by the ruminant. Since the reduction potential of the rumen is -0.4V (the O_2 concentration at this highly reducing potential being 10^{-22} M), anaerobic bacteria dominate the rumen microflora. These bacteria are not known to be efficient in degrading the lignin components, the major barrier for the effective utilisation of the cellulose component. Naturally, the issue of increasing the digestibility of straw by the ruminants has been the target of numerous studies (Harvey and Palmer, 1989; Morrison, 1979; Stewart et al., 1979).

A number of strategies may be employed for achieving this goal, of which at least two may involve the intensive use of fungal biotechnology. These are (i) the modification of rumen microbiota and (ii) delignification of the straw to be used as animal feed.

Modification of the Rumen Microbiota

Of the generally recognised groups of microorganisms, only certain fungi are known to be capable of degrading lignin within natural environments (Kirk et al., 1977). One would naturally be tempted to explore the possibility of introducing lignolytic fungi into the rumen. The environment of the rumen, however, might not prove to be congenial for such a proposition, the aerobic nature of the lignolytic fungi and the requirement of oxygen for the process of lignin degradation itself being the major impediments.

The advent of genetic engineering has raised the possibility of incorporating the gene for lignolytic activity into the rumen microbes. Various laboratories have isolated different cDNA and genomic sequences coding for ligninase (De Boer et al., 1987; Tien and Tu, 1987). An interesting development in this connection is the discovery of an obligately anaerobic fungus, *Neocallimastix frontalis* by Orpin in 1975, from the rumen of a sheep. The fungus is a normal part of the rumen microbiota and appears to play a major role in the breakdown of lignocelluloses in the rumen (Orpin and Joblin, 1988).

A number of other anaerobic fungi have also been isolated from rumen - *Anaeromyces, Caecomyces, Piromyces* and *Orpinomyces* (Hawksworth et al., 1995). It might be more convenient to use these rumen fungi for enhancing the rate of lignin breakdown in the rumen itself by (i) boosting their population and activity through diet modification; (ii) amplification of the gene coding for lignolytic activity; and (iii) incorporating in these the ligninase gene sequence from efficient lignin decomposers like *Phanerochaete* and *Chrysosporium*.

Delignification of the Straw

Various physical, chemical and biological methods to improve the digestibility of the straw have been tried since the mid-nineteenth century. The relative merits and demerits of these have been succinctly summarised by Staniforth (1979a). The physical treatments have mainly focussed on increasing the surface area available for enzyme action. However, these do not break down the lignin bonds in the straw; instead, the fine particles formed as a result of grinding have the tendency to form a slurry which passes through the rumen more quickly, shortening the contact time between the feed particles and the rumen microflora and thereby decreasing the nutritive value (Anderson and Anderson, 1980).

The various chemical methods to remove lignin are based on (i) solubilising the lignin using alkali (Han et al., 1978; Katrib et al., 1988), ammonia (Morrison, 1988) or hot acid (Katrib et al., 1988; Morrison, 1988); (ii) oxidising the lignin component to break it down into simple compounds using alkaline-H_2O (Gould et al., 1989), peracetic acid (Gharpuray et al., 1983), or acidified sodium chlorite (Katrib et al., 1988). The chemical modification of straw, however, has not found favour on several counts: (i) it dictates the use of high concentrations of chemicals and treatment under pressure for better efficiency (Gould et al., 1989) leading to lower palatability and a mere 40% increase in digestibility; (ii) it is necessary to wash the treated straw with huge quantities of water to remove residual acid/alkali leading to a loss of up to 25% of the organics solubilised during hydrolysis (Israilides et al., 1978; Gould et al., 1989); and (iii) it also leads to higher expenses from drying wet straw and the generation of large amounts of effluents leading to either heavy environmental pollution or heavy expenses on waste treatment (Lonsane and Ghildiyal, 1989).

In the midst of debates regarding the feasibility of different physical and chemical methods for improving the feeding value of straw, a considerable interest has developed during the last three decades for using biological means for the purpose. It is hoped that a biologically structured technology with attributes of greater specificity, lower energy requirements and lower pollution generation might be developed in the near future (Harvey and Palmer, 1989). Initial attempts with this strategy were not very encouraging, for example, Harley et al. (1974) obtained an increase in the *in vitro* digestibility of barley straw from 46 to 70 using *Polystichus sanguineus* plus NaOH while from NaOH alone it was increased to 64.

During the last two decades, however, considerable progress has been made in the understanding of the mechanism of lignin degradation by the fungi. A number of lignolytic fungi have been identified, the enzyme ligninase has been characterised and genomic sequences for the ligninase have been isolated, thereby improving the prospects of delignification of straw through fungal biotechnology.

Of the lignolytic groups of fungi, the white rot basidiomycetes are probably the most efficient lignin-degraders (Haider and Trojanowski, 1980). Under proper environmental conditions, these completely degrade all structural components of lignin, with ultimate formation of CO_2 and H_2O (Cowling, 1961). There are several hundred species of white rot fungi belonging to a variety of fungal families including the Agaricaceae, Hydnaceae, Corticiaceae, Polyporaceae and Thelephoraceae (Kirk, 1971). Solid state fermentation of cereal straw using a number of these fungi has been attempted, and a significant increase in digestibility has been reported by several investigators (Agosin and Odeir, 1985; Neelakantan and Sondhi, 1989; Zadrazil, 1979). Among the white rot fungi, the species belonging to the genus *Pleurotus* are reported to be most efficient (Detroy et al., 1980; Zadrazil, 1979).

However, the white rot fungi remove the cellulose and hemicellulose along with lignin. As a result, a good proportion of cellulose, which would have been available to the ruminant, is lost. If this is allowed to happen, the purpose of biological treatment itself is defeated. Hence, a number of workers are trying to employ different strategies to tackle this problem. Zadrazil and Brunnert (1980, 1981) and Zadrazil (1985) have tried to work out the cultural conditions suitable for maximum delignification by white rot fungi with minimum loss of polysaccharides. It is hoped that this will be achieved by controlling temperature, moisture, dissolved oxygen, CO_2 content in the air phase, incubation period and supplementing with appropriate nutrients. Another approach is to obtain cellulase-less mutants of white rot fungi which would degrade lignin using hemicellulose as a sole carbon source leaving cellulose for the ruminants (Eriksson et al., 1980).

Meanwhile, efforts have been made in screening white rot fungi for selective delignification, i.e. the ability to remove lignin with lesser removal of cellulose. A number of preferential lignin degraders have been identified (Blanchette, 1984; Otjen et al., 1987). These include *Phanerochaete chrysosporium*, *P. rimosa* and *Ceriporiopsis subvermispora*. There is an urgent need to screen more white rot fungi for preferential delignification. A fast method for the purpose has already been developed by Nishida et al. (1988). The modern techniques of genetic engineering may also be used for developing better strains of fungi with the capacity to degrade lignin preferentially.

CEREAL STRAW AS A SUBSTRATE FOR THE PRODUCTION OF FUNGAL PROTEINS

Due to their rapid growth rate and high (sometimes more than 50%) protein content, fungi might prove to be an ideal source of protein. Large amounts of fungal protein can be produced in a small area using cheap waste products. Yaaris (1977) described a procedure adopted for growing yeasts on straw in Western Australia. Straw is chopped into pieces of 6 to 25 mm length, acid-sprayed and pressure-cooked at 100,000 kNm^{-2} for 30 minutes. The hydrolysed straw (with some of the polysaccharides converted to glucose) is then ammoniated, inoculated with a suitable yeast and fermented. Yeasts grown on straw can be filtered off and dried; these may contain up to 50% protein (Worgan, 1974). In addition to the use of *Saccharomyces* species, a number of other yeasts have been found useful for the purpose. *Candida utilis* is one such yeast, which can grow well on straw hydrolysate (Anderson and Anderson, 1980).

Filamentous fungi have been used for protein production since the 1920s (Thatcher, 1954). The earliest attempts to produce fungal proteins were made in 1920 by Pringsheim and Lichenstein when they fermented straw with *Aspergillus fumigatus* for animal feed (Litchfield, 1968). The last three decades have witnessed greater interest in this aspect of fungal biotechnology. The ability of filamentous fungi to grow at a fast rate, the fibrous nature of the finished product for easy conversion into textured food, greater retention time in the digestive system, higher protein content, greater digestibility, lesser nucleic acid content, acceptable mushroom-like odour, ease with which their biomass can be processed and greater penetrating power (hence suitability for solid state fermentation of lignocellulosic material), place these fungi in a much more advantageous position than the single-celled yeasts (Chahal, 1982).

Pietersen and Andersen (1976) demonstrated the feasibility of using straw for the production of mycoprotein by *Trichoderma viride* while Han et al. (1976) investigated the possibilities of cultivating *Aureobasidium pullulans (Pullularia pullulans)* on straw hydrolysate. In fact, a considerable amount of research has been devoted to growing *Trichoderma* spp. on cheap cellulosic waste products (Solomons, 1985). The species of *Trichoderma* - *T. viride, T. lignorum* and *T. koningi* - are among the very few fungi which can degrade native cellulose (Berghem and Pettersson, 1973; Eveleigh, 1987). Another approach is to use *T. viride* for degrading straw to simple sugars which are then used by yeasts such as *Saccharomyces cerevisiae:* the protein content of the mixture so obtained is quite high (Wainwright, 1992). In the last two decades, another fungus, *Chaetomium cellulolyticum,* has drawn the attention of biotechnologists for the production of proteins from lignocelluloses including cereal straw. The fungus is highly cellulolytic, has high protein productivity on alkali-treated straw, and can even utilise xylose as well as other hemicellulose sugars. The suitability of the fungus for SCP production has been aptly projected by Chahal (1982). He also proposed an integrated plan for the production of food, feed and fuel from lignocelluloses (including crop residues) using *Chaetomium cellulolyticum* as the main fermenting agent.

The rate at which a given lignocellulosic substrate is utilised by a fungus is limited by the type of enzyme it produces. This limitation may be overcome by using mixtures of two or more fungi which produce different but complementary enzyme systems for more efficient utilisation of the resource. Experiments to examine the feasibility of such an approach have met with considerable success for the production of *Sporotrichum pulverulentum* and *Aspergillus niger* on citrus wastes (Wainwright, 1992). Co-cultivation of *Aspergillus ellipticus* and *A. fumigatus* for solid state fermentation of lignocellulosic wastes has also been attempted by Gupta and Mudamwar (1997).

The new tools of genetic engineering have further brightened the prospects of enhanced production of microbial proteins using 'engineered' microbes. For example, enhancement of cellobiose activity in *Trichoderma reesei* has been achieved through amplification of its cellobiase gene (Barnett et al., 1991).

ETHANOL FROM STRAW

The use of biomass for manufacturing alcohol dates back to the 1930s and the Second World War, when many countries converted agricultural products to alcohol for use in automobiles. Ethanol is an anti-knock fuel and has the power to withstand high compression ratios. Its use as such or in the form of ethanol-gasoline blends would stretch the available fuel supplies for meeting energy crises (Mathur, 1985). Due to the rapid rise of crude oil prices in the 1970s, the research and development of fuel ethanol from biomass received considerable attention from countries all over the world. There is an urgent need to develop an alternative fuel. The elimination of lead from gasoline and the use of ethanol as an octane enhancer in heavy vehicular engines may help reduce air pollution by curtailing CO_2 emissions (Nguyen, 1993). In Brazil, neat alcohol (>98%) is used directly, while in North America, ethanol-gasoline blends are preferred (Saddler, 1993). Incidentally, almost all the transportation ethanol utilised today is derived from starch or sugar substitutes like corn or sugar-cane. However, excessive diversion of these crops for ethanol production is not desirable. In many countries like India, it may not be wise to displace sugar-cane and/or tapioca as food crops at all (Khoshoo, 1984). Hence the conversion of lignocellulosic crop residues like straw holds immense potential as raw materials for the production of ethanol.

Pre-treatment of the Straw

Various physical and chemical barriers greatly inhibit the accessibility of the cellulose and hemicellulose to the hydrolytic enzymes. Thompson et al. (1991) have shown that the surface area available to the cellulase enzyme complex was perhaps the most important factor governing the initial stages of cellulose hydrolysis. A number of reviews have appeared on this aspect (Chahal and Overend, 1982; Schell et al., 1991). Different types of pre-treatment methods are available which may be broadly grouped into three categories - physical, chemical and biological. The biological methods are based upon delignification through white rot fungi. We have already discussed this approach in connection with the upgrading of straw as feed for the ruminants. Of a number of physical and chemical methods (Dunlop et al., 1976), Saddler (1993) has found steam pre-treatment to be the preferred method so far since (i) it provides recovery of most of the lignocellulosic components, (ii) it enhances hydrolysis of the cellulosic residues, and (iii) steam is the only operating cost.

Fungal biotechnology needs to be developed to achieve the objects of pre-treatment. Research during the last two decades has revealed that it is possible to disrupt the lignin shield without removing the lignin from the substrate simply by removing the hemicelluloses (Converse, 1993). Hemicellulolytic activity is widespread among the fungi (Charaya, 1985) and a number of hemicellulases produced by fungi have been purified (Viikari et al., 1993).

Saccharification of Cellulose

The pre-treated straw is subjected to either acid hydrolysis or enzymatic hydrolysis using cellulases for the conversion of lignocellulose into monomeric sugars. The cellulase-producing potential of the fungi is the basis of the utilisation of fungal biotechnology for the enzymatic hydrolysis of cellulose. Also, as compared to the acid hydrolysis process, the bioconversion processes (using cellulases) promise higher glucose yields and lower levels of degradation by-products (Nguyen, 1993; Wayman et al., 1979).

Due to their low specific activity or turnover number, large quantities of cellulases are required for the hydrolysis; these can be produced from the submerged cultures of *Trichoderma reesei* (San Martin et al., 1986; Webb et al., 1986). In fact, the cellulolytic fungi serve as a highly useful resource for the cellulases for various reasons: (i) they can be easily grown on simple media, (ii) most of them secrete cellulase extracellularly when grown on cellulose as a source of carbon, and (iii) it is easy to separate the enzyme from the mycelial mass (Chahal and Overend, 1982). However, although cellulolytic ability has been demonstrated to exist in quite a large number of fungi, only very few fungi are able to produce an enzyme system which can degrade native cellulose. Some of these are: *Trichoderma reesei, T. lignorum, T. koningi* (Berghem and Pettersson, 1973; Wood, 1972; Wood and McCrae, 1972), *Sporotrichum pulverulentum* (Streamer et al., 1975), *Penicillium funiculosum* (Selby, 1968), *Penicillium iriensis* (Boretti et al., 1973), *Polyporus adustus* (Eriksson, 1975), *Myrothecium verrucaria* (Updegraff, 1971), *Fusarium solani* (Wood, 1972) and *Chaetomium thermophile* var. *dissitum* (Goksoyr et al., 1975).

Furthermore, the ability of a fungus to degrade cellulosic substrates and to convert the substrate into its biomass does not mean that its cell-free cellulase system also will have high hydrolytic potential. Rather, a fungus which is able to convert most of the cellulose substrate into enzyme protein with high hydrolytic potential must be the most suitable organism for cellulase production (Chahal and Overend, 1982). On these counts, *Trichoderma reesei* is believed to be the most efficient producer of cellulase at present; it can be grown easily in submerged aerobic cultures where, under proper conditions, it excretes the complete complex of enzymes required for the hydrolysis of crystalline cellulose (Pourquie and Warzywoda, 1993). Considerable attention has been devoted, therefore, to biochemical characterisation, structure and production of cellulase from *Trichoderma reesei* (Allen and Roche, 1989; Coughlan, 1992; Pourquie and Warzywoda, 1993; Warzywoda et al., 1992).

Since the cost of enzymes is quite high, it is advisable to produce the required cellulase on site instead of purchasing it from suppliers. For this process, hemicellulose-rich sugar solutions or glucose-rich hydrolysates of lignocelluloses produced on site can be used as these provide a cheap source of substrate. Incidentally, it has been found that crude water extracts of steam-exploded straw support good cellulose production by *T. reesei* (Pourquie and Warzywoda, 1993). It is also worth mentioning that the wild type strains of fungi produce the cellulase in amounts too low to support an economic industrial process, and hence require genetic improvement. *T. reesei* is no exception. Using the well-established strain improvement programmes involving a succession of mutagenesis and screening for hyperproduction of cellulases, considerable improvements in the production of celluloses by *T. reesei* have been obtained. Durand et al. (1984) isolated a hyperproducing strain *T. reesei* CL 847 which was better than all strains isolated since then, although attempts at further improvement of the strain did not succeed (Pourquie and Warzywoda, 1993).

However, using the tools of genetic engineering, a number of breakthroughs have been achieved. Barnett et al. (1991) have been able to enhance cellobiase activity by amplification of the cellobiase gene in *T. reesei*. Transformation systems and plasmid vectors useful for *T. reesei* have been developed (Berges and Barreau, 1991) which have been used to clone and express an invertase gene from *Aspergillus niger* in *T. reesei*. The resultant transformant strains can use sucrose as a sole carbon source while the wild type strains would not. It is thus possible now to clone and express other genes coding for polysaccharide hydrolases into *T. reesei*. If it is accomplished, the saccharification efficiency of *T. reesei* might be improved significantly.

The enzyme solution is mixed with pre-treated straw. The mixture is allowed to react in hydrolysis reactors which are basically tall cylindrical vessels with mixers. Using 1 tonne of steam pre-treated straw, 29 kg of hexoses per 100 kg of original wheat straw have been obtained (Hayes et al., 1993).

The hydrolysis of cellulose results in a gradual accumulation of sugars which, in turn, cause end product inhibition of the cellulases, leading to a gradual decrease in the rate of cellulose hydrolysis with time, as observed by Fan et al. (1987). The *Trichoderma* enzyme system is strongly inhibited by glucose and cellobiose at concentrations much lower than 1% (Grohmann, 1993), while starting concentrations of fermentable sugars higher than 8% are required for efficient recovery of ethanol (Phillips and Humphrey, 1983). A number of methods have been proposed to overcome this problem. One of these is to use high concentrations of enzymes by increasing the ß-glucosidase and cellulase loadings several-fold (Ishihara et al., 1991) but this would increase enzyme cost. Another approach which is finding more favour among biotechnologists is to add fermenting organisms to the saccharification vessel so that the sugars are taken up for fermentation to ethanol as soon as these are formed. This would keep the sugar concentrations low so that saccharification might proceed uninhibited. This approach is called Simultaneous Saccharification and Fermentation (SSF) as compared to the usual SHF (Separate Hydrolysis and Fermentation). The strengths and weaknesses of both SHF and SSF have been discussed by Nguyen (1993).

Fermentation of Hydrolysate into Ethanol

The hexoses obtained by the hydrolysis of cellulose are subjected to fermentation. Yeasts are the most important agents which bring about the fermentation of sugar to alcohol. The genera *Saccharomyces* and *Schizosaccharomyces* are the classic sources of industrial yeasts, while *Candida, Kluyveromyces* and some other genera of yeasts may also provide strains for the purpose. The ability to ferment cellobiose to ethanol is found mainly in yeasts belonging to *Brettanomyces, Candida, Debaryomyces, Hansenula* and *Pichia*. Efforts are required to develop ethanol-tolerant, high-yielding yeasts which are also thermotolerant. The use of thermotolerant strains would decrease the need for water for cooling fermenters. It is possible to develop ethanol-tolerant strains of yeasts which can carry out fermentation at 39-40°C through hybridisation (Seki et al., 1983).

Fermentation of Pentoses to Alcohol

Since hemicelluloses form a sizeable fraction of the straw, an economically viable process for the production of ethanol from straw must also include the fermentation of pentoses. Though *Fusarium* sp. has been in use for ethanol production from pentoses since the 1940s (Nord and Mull, 1945), it was the identification of a xylose-fermenting yeast, *Pachysolen tannophilus* (Schneider et al., 1981) which stimulated research and development in that direction. A number of other yeasts have been found to have the ability to ferment xylose. These include *Candida blankii, C. famata, C. fructus, C. guilliermondii, C. shetae, Clavispora* sp., *Kluyveromyces cellobiovorus, K. marxianus, Pachysolen tannophilus, Pichia segobiensis, P. stipitis* and *Schizosaccharomyces combe*. The non-yeast fungi with xylose fermenting ability are *Aureobasidium pullulans, Fusarium avenaceum, F. chlamydosporum, F. graminearum, F. lycopersici, F. oxysporum, F. solani, F. tricinetum, Monilia* sp., *Mucor* sp., *Neurospora crassa* and *Paecilomyces* sp.

The pentose-fermenting yeasts have a lower tolerance to ethanol as compared to that of *Saccharomyces cerevisiae* (Skoog and Hahn-Hägerdal, 1988). Efforts for developing ethanol-tolerant strains have been reviewed by Esser and Karsch (1984) and Lynd (1989).

The biotechnologies for producing ethanol from hexoses and pentoses are based on different organisms for the two different groups of monomers. One would naturally wish to have an ideal organism which can ferment both hexoses and pentoses into ethanol, but none of the main hexose fermenters (like *Saccharomyces cerevisae*) can ferment pentoses to ethanol. Hallborn et al. (1991) have reported the cloning of pentose-fermenting genes into *S. cerevisae*, and the prospects of engineering an organism having the ability to produce ethanol from pentoses as well as hexoses have improved.

MUSHROOM COMPOST FROM STRAW

Agaricus bisporus and *Volvariella volvacea* are the most generally grown mushrooms using straw as the main ingredient for cultivation, though *Pleurotus ostreatus* and *P. cystidiosus*, also grown on straw, are gaining popularity (Wainwright, 1992). For cultivating *A. bisporus,* traditionally horse manure and/or chicken manure, and wheat straw are pre-treated by a process called 'composting' before being inoculated with spawn. Composting is a form of pasteurisation or fermentation specific to mushrooms. The process (i) renders the substrate mixture unsuitable for the growth of unwanted microorganisms and (ii) modifies the nutritional status of the substrate such that it becomes particularly favourable for the growth and development of the mushrooms. Mixed cultures of fungi, actinomycetes and bacteria play an essential role in natural as well as traditional composting systems (Poincelot, 1974). The microbiology and enzymology of mushroom composts, along with other aspects, have been discussed by Fermor and Wood (1979) and Hayes and Lim (1979).

Composting is the most critical part of mushroom growing and consists of two phases - Phase I in the open air; and Phase II which is indoors and temperature-controlled (Wiegant,

1992). Nowadays, biotechnologists are exploring the feasibility of using individual fungi or mixed cultures of fungi for what is called industrial composting (Wainwright, 1992). In this connection, the widespread use of thermophilic fungi in the near future seems likely since these contribute significantly to the quality of the compost (Straatsma et al., 1989). Straatsma et al. (1994) used *Scytalidium thermophilum* to prepare a compost of invariably high quality that does not emit ammonia or odour into the environment; moreover, they observed a two-fold increase in the yield of mushrooms on the compost inoculated with *S. thermophilum* as compared to the uninoculated one. Dr. Aneja and his coworkers at Kurukshetra University (India) have obtained similar results with *Torula thermophila* and *Malbranchea sulfurea* (personal communication). Biotechnological applications of thermophilic moulds in mushroom compost preparation have been discussed elsewhere in this volume by Dr. R. Kumar and Dr. K.R. Aneja.

Some bonus benefits also accrue when the straw is utilised for growing mushrooms. The spent compost after harvesting the mushrooms can be used as a fertiliser for soil conditioning as well as a good substrate for growing the *Pleurotus* mushrooms (Change and Li, 1982).

USING STRAW FROM IMPROVING SOIL FERTILITY

The immense utility of cereal straw provides enough grounds for its substantial diversion from the fields. However, complete diversion of crop residues for industrial or off-farm use could result in soil deterioration and destruction of soil resources as were observed in North Africa, the Near East and Korea (Pierce, 1978). Cereal straw, when mature, contains approximately 0.5% N, 0.6% P_2O_5 and 15% K_2O (Gaur, 1992). A substantial fraction of nutrients immobilised in the straw need to be returned to the soil to maintain soil fertility. The burning of straw may restore some of the nutrients in the form of ash, but the entire organic matter is lost; the soil is thus deprived of the benefits of organic matter conditioning. This loss may be avoided by the addition of well-decomposed compost to the soil - an age-old practice to restore soil structure. Another method of adding organic matter to the soil is direct ploughing in of the straw.

Straw Compost for Soil Application

The use of straw (as well as other organic residues) to prepare compost manure has been practised for a very long time in many parts of the world and has been the subject of innumerable writings (Acharya, 1940a,b; Fowler, 1930; Gaur, 1992; Howard, 1935; Poincelot, 1974). Composting is basically a microbiological process - the materials undergoing composting are decomposed by the microorganisms to be converted, in about three to six months, into an amorphous, brown to dark brown humified material called compost. Here again, fungal biotechnology might be immensely useful in (i) reducing the time for composting, and (ii) enrichment of the compost. A number of strains of cellulolytic and lignolytic microorganisms have already been isolated (Bhardwaj and Gaur, 1985; Gaur, 1987). An attempt to use fungal inoculants as compost accelerators by Gaur and his coworkers at IARI, New Delhi has yielded encouraging results. Several types of wastes, chopped to 5 to 6 cm size, were placed in pits. Homogenised cultures of the fungi *Trichurus spiralis, Paecilomyces fusisporus, Trichoderma viride* and *Aspergillus* sp. were added at 300 g per tonne of material. Rock phosphate was added at 1% and moisture was initially maintained at 100%. Every fortnight, the composting mass was turned upside down. Within 8 to 10 weeks, a good quality compost was prepared from paddy straw. The compost contained about 1.7% N and a C:N ratio of 12:3. Earlier in 1980, Yadav and Rao also found that inoculation with the fungi *Trichoderma viride, Chaetomium abuanse, Myrothecium roridum, Aspergillus niger* and *A. terreus* (i) accelerated composting, (ii) brought down the C:N ratio, and (iii) brought about an increase in the humic acid content of the composted material (Subba Rao, 1992). The nutrient content of the compost can also be improved by introducing efficient microbial inoculants including *Aspergillus awamori* (Gaur, 1992).

Application of Uncomposted Straw to the Soil

A number of workers in the past have recommended direct ploughing in of crop residues/straw in soil. According to Sloneker (1976), if properly incorporated into the soil, crop residues improve soil tillage and water absorption during rains, consequently reducing soil erosion. Sugars in the residues supply energy for the abundant microbial system that releases carbon dioxide which accelerates chemical weathering of minerals. The refractory organic material remaining, i.e. the humus, changes the soil pH, chelates heavy metal ions and affects the physical conditions and water-holding capacity of the soil. The carbohydrates in such residues can be used by certain microorganisms for fixing nitrogen (Mortensen, 1963).

Application of uncomposted straw to the soil has been found to be more beneficial to crops than composts prepared from straw (Dhar, 1968; Kavimandan, 1980) because direct ploughing produces more nitrogen fixation in the soil (Dhar, 1968). The ploughing in of wheat straw as such for increasing soil fertility is practised already in Italy (Verona and Lepidi, (1972). Gaur and Mathur (1979) and Gaur et al. (1980) have also reported significant favourable effects of ploughing in wheat straw on maize yield and addition of paddy straw and stubble on wheat and green gram respectively. One of the most important properties of a manure is its ability to decompose quickly and mix with the soil. This can be achieved artificially by inoculating the straw, in the field itself, with microorganisms which can quickly decompose it (Subba Rao, 1982).

A number of studies have been carried out by different workers in the past to study the fungi which decompose cereal straw in nature as well as the ability of individual fungi to decompose the straw *in vitro* (Bowen and Harper, 1989; Broder and Wagner, 1988; Charaya, 1985; Dkhar and Mishra, 1991; Johri and Satyanarayana, 1977; Robinson et al., 1994; Singh et al., 1979). Charaya (1985) found *Emericella nidulans, Myrothecium verrucaria, Alternaria alternata, Penicillium oxalicum, Fusarium semitectum, Cladosporium cladosporioides* and *Aspergillus niger* to be good decomposers of wheat straw. *Aspergillus flavus, Emericella nidulans, Alternaria alternata, Chaetomium globosum, Aspergillus fumigatus* and *Cladosporium cladosporioides* were efficient decomposers of paddy straw. It is suggested that these fungi may be used as inoculants for accelerating the rate of decomposition of the straw in the field itself. Suitable carriers for the cellulose-decomposing organisms to be used as inoculants have already been developed by Rasal et al. (1987).

Charaya (1985) suggested that the selection of a fungal strain to be used as an inoculant must be done carefully. Tolerance to the pH and temperature to which it would be exposed, and the pre-existing competitors or antagonists it would have to deal with, are among the major factors which would determine the success of the inoculant decomposer.

One problem which the fungal inoculants have to face is that of C:N ratio of the straw. Straw, on average, contains only 0.5% nitrogen and 40% carbon. When subject to colonisation by fungi, it has only 0.5 units of nitrogen to satisfy them while 1.2 to 1.6 units are required. So a deficit of 1.2 to 1.6 units of nitrogen appears in the environment (Alexander, 1977). In order to utilise straw for direct ploughing, therefore, the problem of nitrogen deficiency must be tackled. At present there are two alternatives available to us - (i) to treat the straw beforehand with nitrogenous fertiliser, and (ii) to inoculate with those fungi which have the potential to decompose the straw even at low nitrogen levels. Efforts to search for such fungi are being made in this laboratory at Modinagar. Already, Jain (1989) has found that

Aspergillus fumigatus could be used as an inoculant to decompose paddy straw without any amendment. However, if low nitrogen pre-treatment is provided, *Cladosporium cladosporioides* may prove to be equally effective.

Another practical problem which farmers might face is that the incorporation of straw into the soil might have deleterious effects on the crops (Elliott et al., 1978). Charaya and coworkers also observed a steep depression in the soil microflora during the first week of incorporation of fresh wheat straw in the soil (Charaya, 1985). However, such a depression was not discernible if stored straw was used or if the straw was first allowed to rot for one week prior to its incorporation into the soil. Thus, if stored straw is incorporated into the soil, there is little likelihood of any damage to the crop or to the soil microflora. Moreover, in many parts of the world, including Northern India, the current agricultural practice is to keep the land unused for many days after wheat has been harvested. Thus, it is quite feasible to plough in straw some fifteen days in advance of sowing the next crop.

OTHER USES OF STRAW

There are many other fields where fungal biotechnology might be helpful in the better management of cereal straw. For example, biogas yields from the mycostraw are reported to be twice that of untreated straw (Baisaria et al., 1983; Muellar and Troesch, 1986).

Another area is biomechanical pulping (Jurasek and Paice, 1988) and fungal bleaching of straw. Attempts at developing cellulase-less mutants of lignolytic fungi or searching for preferential lignin-decomposers have been discussed elsewhere in this article. The use of immobilised wood-decomposing fungi like *Phanerochaete chrysosporium* for bleaching has been advocated (Wainwright, 1992). The development of fungal technology for pulping cereal straw may help reduce the consumption of wood.

CONCLUSIONS

Cereal straw represents a great resource for the production of food, feed, manure, fuel and other products. The technology does exist for its conversion into many of these useful products. Fungal biotechnology needs to be strengthened further not only for yielding a greater variety of products from cereal straw but also for making the existing biotechnological processes economically more profitable and competitive.

REFERENCES

Acharya, C.N., 1940a, Composts and soil fertility, *Indian Farming* 1: 66-68.

Acharya, C.N., 1940b, Composts and soil fertility, *Indian Farming* 1: 121-125.

Agosin, E. and Odier, E., 1985, Solid state fermentation, lignin degradation and resulting digestibility of wheat straw fermented by selected white rot fungi, *Appl. Microbiol. Biotech.* 21: 397.

Alexander, M., 1977, *Introduction to Soil Microbiology* (2nd Edn.), John Wiley & Sons Inc., Reprinted by Wiley Eastern Ltd., New Delhi.

Allen, A.L. and Roche, C.D., 1989, Effects of strain and fermentation conditions on production of cellulase by *Trichoderma reesei, Biotech. Bioeng.* 33: 650-656.

Anderson, A.W. and Anderson, J.F., 1980, On finding a use for straw, in: *Utilisation and Recycle of Agricultural Wastes and Residues* (M.L. Shuler, ed.), pp. 237-272, CRC Press Inc., Florida.

Bacon, J.S.D., 1979, What is straw decay? Some retrospective comments, in: *Straw Decay and its Effects on Disposal and Utilisation* (E. Grossbard, ed.) pp. 227-236, John Wiley & Sons, Chichester, New York, Brisbane, Toronto.

Baisaria, R., Madan, M. and Mukhopadhyay, 1983, Production of biogas from the residues from mushroom cultivation, *Biotech. Lett.* 5: 811-812.

Barnett, C.C., Berka, R.M. and Fowler, T., 1991, Cloning and amplification of the gene encoding an extra-cellular glucosidase from *Trichoderma reesei*: evidence for improved rates of saccharification of cellulosic substrates, *Biotech.* 9: 562-567.

Berges, T. and Barreau, C., 1991, Isolation of uridine auxotrophs from *Trichoderma reesei* and efficient transformation with the cloned ura 3 and ura 5 genes, *Current Genetics* 19: 359-365.

Berghem, L.E.R. and Pettersson, L.G., 1973, The mechanism of enzymatic cellulase degradation. Purification of cellulolytic enzyme from *Trichoderma viride* active on highly ordered cellulose, *Eur. J. Biochem.* 37: 21-32.

Bhardwaj, K.R. and Gaur, A.C., 1985, *Recycling of Organic Wastes,* ICAR, New Delhi, India.

Blanchette, R.A., 1984, Screening wood decayed by white rot fungi for preferential lignin degradation, *Appl. Environ. Microbiol.* 48: 647-653.

Boretti, G., Gurafano, L., Montecucci, P. and Spalla, C., 1973, Cellulase production with *Penicillium iriensis* (n.sp.), *Archives Mikrobiologie* (N.S. Subba Rao, ed.), pp. 551-584, Oxford & IBH Publishing Co., New Delhi.

Bowen, R.M. and Harper, S.H.T., 1989, Fungal populations on wheat straw decomposing in arable soils, *Mycol. Res.* 22: 401-406.

Broder, M.W. and Wagner, G.H., 1988, Microbial colonisation and decomposition of corn, wheat and soyabean residues, *Soil Sci. Soc. Am. J.* 52: 110-117.

Chahal, D.S., 1982, Bioconversion of lignocelluloses into food and feed rich in protein, in: *Advances in Agricultural Microbiology* (N.S. Subba Rao, ed.), pp. 551-584, Oxford & IBH Publishing Co., New Delhi.

Chahal, D.S. and Overend, R.P., 1982, Ethanol fuel from biomass, in: *Advances in Agricultural Microbiology* (N.S. Subba Rao, ed.), pp. 585-641, Oxford & IBH Publishing Co., New Delhi.

Chahal, D.S., Vlach, D. and Moo-Young, M., 1983, Upgrading the protein feed value of lignocellulosic materials using *Chaetomium cellulolyticum* in solid state fermentation, *Advances Biotech.* 2: 327-332.

Chang, S.T. and Li, S.F., 1982, Mushroom culture, in: *Advances in Agricultural Microbiology* (N.S. Subba Rao, ed.), pp. 677-691, Oxford & IBH Publishing Co., New Delhi.

Charaya, M.U., 1985, Taxonomical, ecological and physiological studies on the mycoflora decomposing wheat and paddy crop residues, *Ph.D. Thesis*, Meerut University, Meerut, India.

Cheshire, M.V., Sparling, G.P. and Inkson, R.H.E., 1979, The decomposition of straw in soil, in: *Straw Decay and its Effects on Disposal and Utilisation* (E. Grossbard, ed.), pp. 65-71, John Wiley & Sons, Chichester, New York, Brisbane, Toronto.

Converse, A.O., 1993, Substrate factors limiting enzymatic hydrolysis, in: *Bioconversion of Forest and Agricultural Plant Residues* (J.N. Saddler, ed.), pp. 93-106, CAB International, Wallingford, Oxon, U.K.

Coughlan, M.P., 1992, Enzymatic hydrolysis of cellulose: an overview, *Bioresource Tech.* 39: 107-115.

Cowling, E.B., 1961, Comparative biochemistry of decay of sweetgum sapwood by white rot and brown rot fungi, *USDA Technical Bulletin* 1258.

De Boer, H.A., Zhany, Y.Z., Collins, C. and Reddy, C.A., 1987, Analysis of nucleotide sequences of two ligninase cDNAs from a white rot filamentous fungus, *Phanerochaete chrysosporium*, Gene 60: 93-102.

Detroy, R.W., Lindenfelser, L.A., St. Julian, Jr. G. and Orton, W.L., 1980, Saccharification of wheat straw cellulose by enzymatic hydrolysis following fermentative and chemical pretreatment, *Biotechnology and Bioengineering Symposium* 10: 135-148.

Dhar, N.R., 1968, The value of organic matter, phosphates and sunlight in nitrogen fixation and fertility improvements in world soils, in: *Remaine D'Étude sur le theme Matiére Organique et Fertilité du Sol,* April 22-27 1968, North Holland Publishing Co., Amsterdam & Wiley Interscience Div., John Wiley & Sons Inc., New York, pp. 243-260.

Dickinson, C.H. and Pugh, G.J.F. (eds.), 1974, *Biology of Plant Litter Decomposition, Vols. I & II,* Academic Press, London and New York.

Dkhar, M.S. and Mishra, R.R., 1991, Decomposition of maize (*Zea mays* L.) crop residues, *J. Indian Bot. Soc.* 70: 135-138.

Dunlop, C.E., Thomson, I. and Chiang, L.C., 1976, Treatment processes to increase cellulose microbial digestibility, *Proceedings of AIChE Symposium on Energy, Renewable Resources and New Foods* 158: 58-63.

Durand, H., Tiraby, G. and Pourquie, J., 1984, Amélioration génétique de *Trichoderma reesei* en vue d'une production industrielle de cellulase, in: *Génétique des Microorganismes Industriels,* pp. 39-50, 9éme Colloque de la Société Français de Microbiologie, 15-16 March 1994, Société Français de Microbiologie, Paris.

Elliott, L.F., McCalla, T.M. and Waiss, A., 1978, Phytotoxicity associated with residue management, in: *Crop Residue Management Systems* (Oschwald, ed.), pp. 131-146, American Society of Agronomy, Crop Science Society of America, Soil Science Society of America, Madison.

Eriksson, K.E., Grunewald, A. and Vallander, L., 1980, Studies of growth conditions in wood for three white rot fungi and their cellulase-less mutants, *Biotech. Bioeng.* 22: 363-376.

Esser, K. and Karsch, T., 1984, Bacterial ethanol production: advantages and disadvantages, *Process Biochem.* 19: 116-121.

Eveleigh, D.E., 1987, Cellulase: a perspective, *Phil. Trans. Royal Soc. London A* 321: 435-447.

Fan, L.T., Gharpuray, M.M. and Lee, Y.H., 1987, *Cellulose Hydrolysis*, Springer-Verlag, Berlin, Germany.

Fermor, T.R. and Wood, D.A., 1979, The microbiology and enzymology of wheat straw mushroom compost production, in: *Straw Decay and its Effect on Disposal and Utilisation* (E. Grossbard, ed.), pp. 105-112, Chichester, New York, Brisbane, Toronto.

Fowler, G.J., 1930, Recent experiments on the preparation of organic manure: a review, *Indian Agriculture and Livestock* 7: 711-712.

Gaur, A.C., 1987, Recycling of organic wastes by improved techniques of composting and other methods, *Resource and Conservation* 13: 154-174.

Gaur, A.C., 1992, Bulky organic manures and crop residues, in: *Fertilisers, Organic Manures, Recyclable Wastes and Biofertilisers* (H.L.S. Tandon, ed.), pp. 36-51, Fertiliser Development and Consultation Organisation, New Delhi, India.

Gharpuray, M.M., Lee, Y.H. and Fan, L.T., 1983, Structural modification of lignocellulosics by pretreatments to enhance enzymatic hydrolysis, *Biotech. Bioeng.* 25: 157-172.

Goksoyr, J., Edisa, G., Eriksen J. and Osmundsvag, K., A comparison of cellulases from different microorganisms, in: *Proceedings of Symposium on Enzymatic Hydrolysis of Cellulose* (M. Bailey, T.M. Enari and M. Linco, eds.), pp. 217-230, Aulanko, Finland, SITRA, Helsinki.

Gould, J.M., Jasberg, B.K., Fahey, G.C. and Berger, L.L., 1989, Treatment of wheat straw with alkaline hydrogen peroxide in a modified extruder, *Biotech. Bioeng.* 33: 233-236.

Grohmann, K., 1993, Simultaneous saccharification and fermentation of cellulosic substrates to ethanol, in: *Bioconversion of Forest and Agricultural Plant Residues* (J.N. Saddler, ed.), pp. 183-209, CAB International, Wallingford, Oxon.

Gupta, A. and Mudamwar, D., 1997, Solid state fermentation of lignocellulosic waste for cellulase and ß-glucosidase production by cocultivation of *Aspergillus ellipticus* and *Aspergillus fumigatus, Biotech. Progress* 13: 166-169.

Haider, K. and Trojanowski, J., 1980, A comparison of the degradation of ^{14}C labelled DHO and corn stalk lignins by macrofungi and by bacteria, in: *Lignin Biodegradation - Microbiology, Chemistry and Potential Applications* Vol. 1 (T.K. Kirk, T. Higuchi and H.M. Chang, eds.), pp. 111-134, CRC Press, Florida.

Hallborn, J., Walfridsson, J., Airaksinen, U., Ojamo, H., Hahn-Hägerdal, B., Penttila, M. and Keränen, S., 1991, Xylitol production by recombinant *S. cerevisae, Bio/Technology* 9: 1090-1095.

Han, Y.W., Cheek, P.R., Anderson, A.W. and Likprayoon, C., 1976, Growth of *Aureobasidium pullulans* on straw hydrolysate, *Appl. Environ. Microbiol* 32: 799.

Han, Y.W., Yu, P.L. and Smith, S.K., 1978, Alkali treatment and fermentation of straw for animal feed, *Biotech. Bioeng.* 20: 1915-1926.

Hartley, R.D., Jones, E.C., King, N.J. and Smith, G.A., 1974, Modified wood waste and straw as potential components of animal feed, *J. Sci. Food Agric.* 25: 433-437.

Harvey, P.J. and Palmer, J.M., 1989, Lignin degradation - a biotechnological perspective, in: *Proceedings National Seminar on Biotechnology of Lignin Degradation* (R. Singh, ed.), pp. 4-13, Indian Veterinary Research Institute, Izatnagar, India.

Hawksworth, D.L., Kirk, P.M., Sutton, B.C. and Pegler, D.N. (eds.), 1995, *Ainsworth and Bisby's Dictionary of Fungi,* 8th ed., International Mycological Institute, Egham, U.K.

Hayes, W.A. and Lim, W.G., 1979, Wheat and rice straw composts and mushroom production, in: *Straw Decay and its Effect on Disposal and Utilisation* (E. Grossbard, ed.), pp. 85-93, John Wiley & Sons, Chichester, New York, Brisbane, Toronto.

Hayn, M., Steiner, W., Klinger, R., Steinmuller, H., Sinner, M. and Esterbauer, H., 1993, Basic research and pilot studies on the enzymatic conversion of lignocellulosics, in: *Bioconversion of Forest and Agricultural Plant Residues* (J.N. Saddler, ed.), pp. 33-72, CAB International, Wallingford, Oxon.

Howard, A., 1935, The waste products in agriculture, *J. Royal Soc. Arts* 1935: 84-120.

Ishihara, M., Uemura, S., Hayashi, N. and Shimizu, K., 1991, Semicontinuous enzymatic hydrolysis of lignocelluloses, *Biotech. Bioeng.* 37: 948-954.

Israilides, C.J., Grant, G.A. and Han, Y.W., 1978, Sugar level, fermentability and acceptability of straw treated with different acids, *Appl. Environ. Microbiol.* 36: 43-46.

Jain, S.C., 1989, A study of the potential of fungi to decompose paddy straw in relation to varying nitrogen levels, *Ph.D. Thesis,* Dept. of Botany. M.M. Postgraduate College, Modinagar, India.

Johri, B.N. and Satyanarayana, T., 1977, Role of thermophilic and mesophilic fungi in the decomposition of paddy straw. Proceedings of the International Symposium on Microbial Ecology, Dunedin, New Zealand, 1977(B): 35.

Jurasek, L. and Paice, M.G., 1988, Biological treatment of pulps, *Biomass* 15: 103-108.

Katrib, F.A., Chambat, G. and Joseleau, P., 1988, Organic solvent pretreatment to enhance enzymic saccharification of straw, *J. Sci. Food Agric.* 43: 309-317.

Kavimandan, S.K., 1980, *In situ* composting of wheat straw/weeds for improving crop yields and soil fertility, in: *Recycling Residues of Agriculture and Industry* (M.S. Kalra, ed.), pp. 281-289, Proceedings of

Symposium held at the Punjab Agricultural University, Ludhiana, India.

Khoshoo, T.N., 1984, Bioenergy: scope and limitations, in: *Bioenergy from Waste and Wasteland* (R.N. Sharma, O.P. Vimal and P.D. Tyagi, eds.), pp. 1-11, *Bioenergy Society of India* (Dept. of Non-conventional Energy Sources), New Delhi, India.

Kirk, T.K., 1971, Effect of microorganisms on lignin, *Ann. Rev. Phytopathol.* 9: 185-210.

Kirk, T.K., Connors, W.J. and Zeikus, J.G., 1977, Advances in understanding the microbial degradation of lignin, *Rev. Adv. Phytopathol.* 11: 369-394.

Kirk, T.K. and Haskin, J.M., 1973, Lignin biodeterioration and the bioconversion of wood, *American Society of Chemical Engineering Symposium Series* 69: 124-126.

Kuhad, R.C. and Singh, A., 1993, Lignocellulose biotechnology: current and future prospects, *CRC Crit. Rev. Biotechnol.* 13: 151-172.

Litchfield, J., 1968, The production of fungi, in: *Single Cell Protein* (R.I. Mateles and S.R. Tannenbaum, eds.), pp. 304-329, MIT Press, Cambridge.

Lonsane, B.K. and Ghildial, N.P., 1989, Upgradation of straw for improving digestibility and nutritive quality: design and economics of farm and large scale solid state fermentation plants, in: *Proceedings National Seminar on Biotechnology of Lignin Degradation* (R. Singh, ed.), pp. 78-87, Indian Veterinary Research Institute, Izatnagar, India.

Lynd, L.R., 1989, Production of ethanol from lignocellulosic materials using thermophilic bacteria: critical evaluation of potential and review: production and utilisation, in: *Bioenergy from Waste and Wasteland* (R.N. Sharma, O.P. Vimal and P.D. Tyagi, eds.), pp. 178-190, Bioenergy Society of India (Dept of Non-conventional Energy Sources), New Delhi.

Morrison, I.M., 1979, The degradation and utilisation of straw in the rumen, in: *Straw Decay and its Effect on Disposal and Utilisation* (E. Grossbard, ed.), pp. 237-246, John Wiley & Sons, Chichester, New York, Brisbane, Toronto.

Morrison, I.M., 1988, Influence of chemical and biological pretreatments on the degradation of lignocellulosic material by biological systems, *J. Sci. Food Agric.* 42: 295-304.

Mortensen, J.L., 1963, Decomposition of organic matter and mineralisation of nitrogen in Brookston silt loam and alfalfa green manure, *Plant and Soil* 19: 374-384.

Mueller, H.W. and Troesch, W., 1986, Screening of white rot fungi for biological pretreatment of wheat straw for biogas production, *Appl. Microbiol. Biotechnol.* 24: 180-185.

Neelakantan, S. and Sondhi, M.S., 1989, Bioconversion of wheat straw by selected white rot, soft rot and other fungi, in: *Proceedings National Seminar on Biotechnology of Lignin Degradation* (R. Singh, ed.), pp. 40-46, Indian Veterinary Research Institute, Izatnagar, India.

Nguyen, Q.A., 1993, Economic analyses of integrating a Biomass-to-Ethanol plant into a pulp/saw mill, in: *Bioconversion of Forest and Agricultural Plant Residues* (J.N. Saddler, ed.), pp. 321-340, CAB International, Wallingford, Oxon, U.K.

Nishida, T., Kashino, Y., Mimura, A. and Takahara, Y., 1988, Lignin biodegradation by wood-rotting fungi, *Mokuzai Gakkaishi* 34: 530-536.

Nord, F.F. and Mull, R.P., 1945, Recent progress in the biochemistry of Fusaria, *Advances in Enzymology* 5: 165-205.

Orpin, C.G. and Joblin, K.N., 1988, The rumen anaerobic fungi, in: *The Rumen Microbial Ecosystem* (P.N. Hobson, ed.), pp. 129-150, Elsevier Applied Science, London.

Otjen, L., Blanchette, R., Effland, M. and Leatham, G., 1987, Assessment of 30 white rot basidiomycetes for selective lignin degradation, *Holzforschung* 41: 343-349.

Phillips, J.A. and Humphrey, A.E., 1983, An overview of process technology for the production of liquid fuels and chemicals feedstocks via fermentation, in: *Organic Chemicals from Biomass* (D.L. Wise, ed.), pp. 249-304, Benjamin/Cummings, Menlo Park, CA.

Pierce, R.G., 1978, Crop residues not a waste, *Agric. Res.* 27: 6-11.

Pietersen, N. and Andersen, B., 1976, Fermentation of barley straw by *Trichoderma viride*: recovery and nutritive value of the fermentation product, *Food, Pharmaceuticals and Bioengineering* (American Institute of Chemical Engineers), Ser. 74: 100.

Poincelot, R.P., 1974, A scientific examination of the principles and practice of composting, *Compost Sci.* 15: 24-31.

Pourquie, J. and Warzywoda, M., 1993, Cellulase production by *Trichoderma reesei*, in: *Bioconversion of Forest and Agricultural Plant Residues* (J.N. Saddler, ed.), pp. 107-116, CAB International, Wallingford, Oxon, U.K.

Rasal, P.H., Kalbhor, H.B., Shingle, V.V. and Patil, P.L., 1987, A study on the carriers for cellulolytic fungi, *Madras Agric. J.* 74: 556-558.

Robinson, C.H., Dighton, J., Frankland, J.C. and Roberts, J.D., 1994, Fungal communities on decaying wheat straw of different resource qualities, *Soil Biol. Biochem.* 26: 1053-1058.

Saddler, J.N., 1993, Introduction, in: *Bioconversion of Forest and Agricultural Plant Residues* (J.N. Saddler, ed.), pp. 1-9, CAB International, Wallingford, Oxon, U.K.

San Martin, R., Blanch, H.W., Wilke, C.R. and Sciamanna, A.F., 1986, Production of cellulase enzymes and hydrolysis of steam exploded wood, *Biotech. Bioeng.* 28: 564-569.

Schell, D.J., Torget, R., Power, A., Walter, P.J., Grohmann, K. and Hinman, H.D., 1991, A technical and economic analysis of acid-catalysed steam explosion and dilute sulfuric acid pretreatments using wheat straw or aspen wood chips, *Appl. Biochem. and Biotech.* 28 & 29: 87-97.

Schneider, H., Wang, P.Y., Chan, Y.K. and Maleszka, R., 1981, Conversion of D-xylose into ethanol by the yeast Pachysolen tannophilus, *Biotech. Lett.* 3: 89-92.

Seki, T., Myoga, S., Limtong, S., Uedono, S., Kumnuanta, J. and Taguchi, U., 1983, Genetic construction of yeast strains for high ethanol production, *Biotech. Lett.* 5: 351-356.

Selby, K., 1968, Mechanism of biodegradation of cellulose, in: *Biodeterioration of Materials* Vol. 1 (A.H. Walter and J.J. Elphick, eds.), pp. 62-78, Applied Science Publications Ltd, London.

Singh, K.P., Gupta, S.R. and Edward, J.C., 1979, Fungi associated with different types of decomposing organic matter with special reference to cellulose decomposition, *The Allahabad Farmer* 50: 259-269.

Skoog, K. and Hahn-Hägerdal, B., 1988, Xylose fermentation, *Enzymes and Microbial Technology* 10: 66-80.

Sloneker, J.H., 1976, Agricultural residues, including feedlot wastes, *Biotechnology and Bioengineering Symposium* 6: 235-250.

Solomons, G.L., 1985, Production of biomass by filamentous fungi, in: *Comprehensive Biotechnology*, Vol. 3 (M. Moo-Young, ed.), pp. 483-505, Pergamon Press, Oxford.

Staniforth, A.R., 1979a, *Cereal Straw*, Clarendon Press, Oxford.

Staniforth, A.R., 1979b, Industrial uses for straw and the relevance of decay to such uses, in: *Straw Decay and its Effect on Disposal and Utilisation* (E. Grossbard, ed.), pp. 209-216, John Wiley & Sons, Chichester, New York, Brisbane, Toronto.

Stewart, C.S., Dinsdale, D., Chang, K.J. and Paniagua, C., 1979, The digestibility of straw in the rumen, in: *Straw Decay and its Effect on Disposal and Utilisation* (E. Grossbard, ed.), pp. 123-130, John Wiley & Sons, Chichester, New York, Brisbane, Toronto.

Straatsma, G., Gerrits, J.P.G., Augustijn, M.P.A.M., Opden Camp, H.J.M., Vogels, G.D. and van Griensven, L.J.D., 1989, Population dynamics of *Scytalidium thermophilum* in mushroom compost and stimulatory effects on growth rate and yield of *Agaricus bisporus, J. Gen. Microbiol.* 135: 751-759.

Straatsma, G., Olijnsma, T.W., Gerrits, J.P.G., Amsing, J.G.M, Opden Camp, H.J.M. and van Griensven, L.J.D., 1994, Inoculation of *Scytalidium thermophilum* in Button Mushroom compost and its effect on yield, *Appl. Environ. Microbiol.* 60: 3049-3054.

Streamer, M., Eriksson, K.E. and Pettersson, B., 1975, Extracellular enzyme system utilised by the fungus *Sporotrichum pulverulentum (Chrysosporium lignorum)* for the breakdown of cellulose, *Eur. J. Biochem.* 59: 607-613.

Subba Rao, N.S., 1982, Utilisation of farm wastes and residues, in: *Advances in Agricultural Microbiology* (N.S. Subba Rao, ed.), pp. 509-521.

Thatcher, F.S., 1954, Food and feeds from fungi, *Ann. Rev. Microbiol.* 3: 449-472.

Thompson, D.N., Chen, H.C. and Grethlien, H., 1991, Comparison of pretreatment methods on the basis of available surface area, *Bioresource Tech.* 39: 155-164.

Tien, M. and Tu, D., 1987, Cloning and sequencing of a cDNA for ligninase from *Phanerochaete chrysosporium, Nature* 326: 522-523.

Updegraff, D.M., 1971, Utilisation of cellulose from waste paper by *Myrothecium verrucaria, Biotech. Bioeng.* 13: 77-79.

Verona, O. and Lepidi, A.A., 1972, Decomposition of outward waxy film of wheat straw by soil microorganisms, *Symp. Biol. Hung.* 11: 55-58.

Viikari, L., Tenkanen, M., Buchert, J., Rättö, M., Bailey, M., Siika-aho, M. and Linko, M., 1993, Hemicellulases for industrial applications, in: *Bioconversion of Forest and Agricultural Plant Residues* (J.N. Saddler, ed.), pp. 131-182, CAB International, Wallingford, Oxon, U.K.

Wainwright, M., 1992, *An Introduction to Fungal Biotechnology,* John Wiley & Sons, Chichester, U.K.

Warzywoda, M., Larbre, E. and Pourquie, J., 1992, Production and characterisation of cellulolytic enzymes from *Trichoderma reesei* grown on various carbon sources, *Bioresource Tech.* 39: 125-130.

Wayman, M., Lora, J.H. and Gulbinas, E., 1979, Material and energy balances in the production of ethanol from wood, *Chemistry for Energy* (ACS Symposium Series) 90: 183-201.

Webb, C., Fukuda, H. and Atkinson, B., 1986, The production of cellulase in a spouted bed fermenter using cells immobilised in biomass support particles, *Biotech. Bioeng.* 28: 41-50.

Wiegant, W.M., 1992, Growth characteristics of the thermophilic fungus, *Scytalidium thermophilum* in relation to production of mushroom compost, *Appl. Environ. Microbiol.* 58: 1301-1307.

Wood, T.M., 1972, The C_1 component of the cellulase complex, in: *Proceedings of the Fourth International Fermentation Symposium, Fermentation Technology Today* (G. Terui, ed.), pp. 711-718, Society of Fermentation Technology, Osaka.

Wood, T.M. and McCrae, S.I., 1972, The purification and properties of the C_1 components of *Trichoderma koningii* cellulase, *Biochem. J.* 128: 1183-1192.

Worgan, J.T., 1974, Single cell protein, *Plant Food Man* 3: 99-112.

Yaaris, L.C., 1977, Feed from straw, *Agric. Res.* 25: 3-5.

Zadrazil, F., 1979, Screening of basidiomycetes for optimal utilisation of straw (production of fruiting bodies and feed), in: *Straw Decay and its Effect on Disposal and Utilisation* (E. Grossbard, ed.), pp. 139-146, John Wiley & Sons, Chichester, New York, Brisbane, Toronto.

Zadrazil, F., 1985, Screening of fungi for lignin decomposition and conversion of straw into feed, *Agnew. Botanik.* 59: 433.

Zadrazil, F. and Brunnert, H., 1980, The influence of ammonium nitrate supplementation on degradation and *in vitro* digestibility of straw colonised by higher fungi, *Eur. J. Appl. Microbiol. Biotechnol.* 9: 37.

Zadrazil, F. and Brunnert, H., 1981, Investigation of physical parameters important for the solid state fermentation of straw by white rot fungi, *Eur. J. Appl. Microbiol. Biotechnol.* 11: 183.

BIOTECHNOLOGY FOR THE PRODUCTION AND ENHANCEMENT OF MYCOHERBICIDE POTENTIAL

K.R. Aneja

Department of Botany
Kurukshetra University
Kurukshetra-136 119
India

INTRODUCTION

Pest species cause billions of dollars of damage annually, displace native species of plants and animals, and cause other types of environmental and social damage (Delfosse and Moorhouse, 1992). There is a worldwide effort to move toward the use of ecologically safe 'environmentally friendly' methods of protecting crops from pests and pathogens (Mehrotra, Aneja and Aggarwal, 1997; Thompson, 1993). Biological pest-control offers us a tremendous opportunity to provide agriculture with effective tools for abundant crop production while minimizing impacts on health and the environment (Panetta, 1992). Biological control with plant pathogens is an effective, safe, selective and practical means of weed management that has gained considerable importance (Charudattan, 1990a).

Interest in exploitation of fungal plant pathogens as weed control agents is increasing partly because this approach offers an exploitable biotechnology and is an effective supplement to conventional weed control based on chemical and mechanical methods (Charudattan, 1988, 1990a; Hasan, 1988, McWhorter and Chandler, 1982; TeBeest, Yang and Cisar, 1992, Yang and TeBeest, 1992). The technique of controlling weeds with concerted applications of large doses of inoculum has been regarded as 'inundative control' or 'bioherbicidal strategy' (Charudattan, 1988). When the living microorganism applied in the bioherbicidal strategy is a fungus, the bioherbicide is referred to as a mycoherbicide (Auld and Morin, 1995). Over 100 fungal pathogens have been assessed for potential as mycoherbicides between 1980 and 1990 (Templeton, 1992) and this has resulted in the development of over ten mycoherbicides - DeVine, Collego, Casst, BioMal, Biochon, Velgo, Luboa, *Cercospora rodmanii* (ABG 5003), *Alternaria eichhorniae* and *Colletotrichum orbiculare.*

Progress in the development of mycoherbicides is slow due to a wide range of biological, economic and regulatory constraints (Auld and Morin, 1995). Some of these could be overcome by the application of recent advances in biotechnology and genetic engineering. The interaction between a biological control fungus and its target organism is complex and likely to be under the control of a large number of genes. The recombination of whole genomes, therefore, rather than the manipulation of single genes, is required for the generation of new strains (Hocart and Peberdy, 1989).

Protoplast fusion provides a procedure for promoting recombination of whole genomes, even between incompatible strains. Protoplast fusions with fungi, both intraspecific and interspecific, have been demonstrated to be feasible with fusants possessing characteristics of both parents (TeBeest and Templeton, 1985). It is a method used to combine the pathogenicity of two separate plant pathogenic fungi, a technology recently being used for the development and improvement of mycoherbicides.

One of the greatest stimulators of research in the rapid commercialization of mycoherbicides has been the advances in technology associated with the selection, culture and formulation of fungi in relation to specific targets. To overcome the biological factors - for example virulence, stability, host-range (specificity), environmental dew requirements, geographic biotypes of the weeds - that constrain the development of mycoherbicides into products, genetic manipulation of the fungal pathogens is being practised to enhance mycoherbicide potential (Auld and Morin, 1995; Charudattan, 1990b; Templeton and Heiny, 1989). It has been predicted that 30 weeds might be controlled by mycoherbicides by the year 2000 (Templeton and Heiny, 1989) and advances in biotechnology may one day be able to create on request fungal pathogens with the desired characteristics as commercial mycopesticides (Trujillo, 1992).

This review deals with the history of the use of fungi as biocontrol agents of weeds, biological weed control strategies, mycoherbicide concept, advantages of mycoherbicides, characteristics of potential mycoherbicide pathogens, status, development and commercialization of mycoherbicides, and some anticipated benefits of biotechnology for improvements of fungi to enhance mycoherbicide potential.

HISTORY OF THE USE OF MYCOHERBICIDES

Fungal diseases are one of the worst threats to cultivated plants. Both the farmer and the amateur gardener know to their cost that fungal infections can damage and destroy their plants. It is perhaps surprising, considering awareness of the potential of plant diseases to cause crop failures (e.g. late blight of potato by *Phytophthora infestans,* blast of rice by *Pyricularia oryzae,* coffee rust by *Hemileia vastatrix*) for a century or more, that scientists had overlooked the extent to which wild plants are also vulnerable to fungal infections, thus presenting an opportunity to turn fungal diseases to the advantage of the farmers as weed killers.

Although plant pathogens as biological weed control agents were suggested as early as 1893 (Wilson, 1969), their utilization for the control of weeds actually only started in the 1970s when Arkansas rice growers posed a question to scientists - why not use weed diseases as weed killers (Templeton et al., 1988)? Testing of this approach began in 1969 with the discovery of a disease affecting curly indigo (northern jointvetch), *Aeschynomene virginica* (L.) B.S.P., an important weed of rice and soybean fields of Arkansas, USA. This disease was caused by a fungal pathogen, later identified as *Colletotrichum gloeosporioides* (Penz.) Sacc. f. sp. *aeschynomene* (Smith et al., 1973a,b; Ternpleton, 1982; Templeton et al., 1988). The causal agent was isolated and evaluated for specificity to the weed host, virulence to kill the weed in a rice field and reproductive ability in culture media. The Arkansas researchers, in cooperation with the farmers and Cooperative Extensive Service Personnel, succeeded in getting the fungus commercialized as Collego for use by Arkansas rice and soybean growers (Templeton, 1982). This approach of using indigenous fungal pathogens to control weed populations has been termed the mycoherbicide approach (Templeton et al., 1979).

U.S. scientists in Florida succeeded in controlling strangler (milk weed) vine *(Morrenia odorata* Lindl.), a problematic weed of citrus orchards, with the fungus *Phytophthora palmivora* (Butler) Butler. This formulation, registered as DeVine in 1981 was marketed by Abbott Laboratories, North Chicago, Illinois (Kenney, 1986). Thus DeVine [R] in 1981 and Collego[R] in 1982 were the first two mycoherbicides which appeared in the early 1980s on the U.S. market.

The early work at Florida and Arkansas stimulated a host of other studies on a range of pathogen-weed associations and led to the isolation of many new strains of fungi from a wide variety of hosts. Between 1980 and 1990 over 100 fungal pathogens have been assessed for potential as mycoherbicides (Templeton, 1992) and eight fungal pathogens are registered as mycoherbicides worldwide (Table 1).

Table 1. Mycoherbicides presently being used all around the world.

Fungal pathogen	Weed	Mycoherbicide	Year of Regn.	Reference(s)
Phytophthora palmivora	*Morrenia odorata* (milkweed vine)	DEVINE	1981	Ridings (1986), Ridings et al. (1976)
Colletotrichum gloeosporioides f. sp. *aeschynomene*	*Aeschnomene virginica* (northern jointvetch)	COLLEGO	1982	Te Beest and Templeton (1985), Templeton (1986), Templeton et al. (1984)
Alternaria cassiae	*Cassia obtusifolia* (sickle pod)	CASST	1987	Bannon (1988), Charudattan et al. (1986), Walker (1983), Walker and Boyette (1985)
Colletotrichum gloeosporioides f. sp. *malvae*	*Malva pusilla* (round leaf marrow)	BIOMAL	1992	Auld and Morin (1995), Makowski (1987), Mortensen (1988)
Chondrostereum purpureum	*Prunus serotina* (black cherry)	BIOCHON	1997	de Jong (1997)
Colletotrichum coccodes	*Abutilon theophrasti* (velvet leaf)	VELGO		Wymore et al. (1988)
C. gloeosporioides f. sp. *cuscutae*	*Cuscutta* spp.	LUBOA	1963	Wang (1990)
Cercospora rodmanii	*Eichhornia crassipes* (water hyacinth)	ABG 5003		Charudattan (1984, 1986), Charudattan et al. (1985), Conway (1976), Conway et al. (1978)

BIOLOGICAL WEED CONTROL STRATEGIES

Biological weed control with fungal plant pathogens is approached from one of two strategies depending upon the pathogen discovered: the classical strategy (Bruckart and Dowler, 1986; Templeton, 1982; Watson, 1991) and the mycoherbicide strategy (Charudattan, 1991; TeBeest and Templeton, 1985; Templeton et al., 1988, 1979).

According to the classical strategy, a fungus is simply introduced or released into a weed population to establish, in time, an epiphytotic requiring no further manipulation. In a severe epidemic, the weed is killed or stressed such that its population is reduced to economically acceptable levels. The pathogens used in this strategy are generally rust and other fungi capable of self-dissemination through airborne spores. The most significant examples of the classical strategy are given in Table 2. Sands and Miller (1993) stated there will always be room for a classical approach to the problem, ending in either a public or private release of a host-specific, exotic pathogen. The strength of the classical approach is that it can result in control of a single species of pest, the development costs are low to moderate, and it may have a long term residue and, consequently, long-term control.

According to the inundative (or mycoherbicidal) strategy usually large doses of inundative indigenous fungal pathogens are applied to specific weed-infested fields to infect or kill susceptible weeds (Charudattan, 1988; Daniel et al., 1973; Templeton, 1982, 1986; Templeton et al., 1979). Computer simulation has shown that the dynamics of plant-pathogen interactions of annual weeds can be determined by the properties of the pathogen. Pathogens with high levels of virulence may exist in nature in low frequencies due to high extinction rates and are suitable for the mycoherbicide strategy (Yang and TeBeest, 1992). On the other hand, pathogens with a low level of virulence are frequent, may coexist stably with their host pathogens with intermediate pathogenicity, and are good candidates for the classical strategy, maintaining a stable interaction and a high control efficiency. The probability of extinction of a pathogen increases when pathogenicity is greater than a critical value at the intermediate range.

A third approach - the manipulated mycoherbicide strategy - has been used by Sands and Miller (1993). In this strategy, lethal broad host-range pathogens are genetically modified to permit their safe release. Either they are rendered host-specific or they are given a chemical dependency that prevents their spread or long-term survival. This genetic-manipulative approach offers numerous and diverse scenarios for biocontrol of weeds and may open the door to larger-scale corporate development and perhaps also to larger-scale public development.

THE MYCOHERBICIDE CONCEPT

In 1973, Daniel and coworkers first introduced the mycoherbicide concept. They demonstrated that an endemic fungal pathogen might be rendered completely destructive to its weed host by applying a massive dose of inoculum at a particularly susceptible stage of weed growth. Since this initial definition of the concept, the term mycoherbicide has been redefined by Templeton et al. (1986, 1988, 1979), TeBeest and Templeton (1985) and Auld and Morin (1995). Mycoherbicides are simply plant-pathogenic fungi developed and used in the inundative strategy to control weeds the way chemical herbicides are used (TeBeest and Templeton, 1985). They are highly specific disease-inducing fungi which are isolated from weeds, cultivated in fermentation tanks and sprayed on fields to control biologically a specific weed without harm to the crop or any non-target species in the environment (Templeton et al., 1988). Figure 1 depicts a summary of the steps in the development of a fungus as a mycoherbicide.

Table 2. Successful examples of biocontrol fungal pathogens used to control weeds using the 'classical strategy'

Fungal pathogen	Introduced to control weed	Pathogen native to country/continent	Introduced into country/continent	Reference(s)
Puccinia chondrillina	*Chondrilla juncea* (skeleton weed)	Europe	Australia, 1971 Western USA, 1976	Adams and Line (1984), Cullen (1984), Cullen et al. (1973), Supkoff et al. (1998)
Phragmidium violaceum	*Rubus constrictus* and *R. ulmifolius* (blackberries)	Europe	Chile, 1973 Australia, 1984	Oehrens (1977), Oehrens and Gonzales (1974)
Entyloma compositarum	*Ageratina riparia (=Eupatorium riparium)* (hamakua pamakani)	Jamaica	Hawaii, USA, 1974	Trujillo (1976, 1984), Trujillo et al. (1988)
Puccinia carduorum	*Carduus nutans* (musk thistle)	Turkey	North-eastern USA, 1987	Bruckart et al. (1988), Politis et al. (1984)
Puccinia abrupta var. *partheniicola*	*Parthenium hysterophorus* (congress grass)	Mexico	Australia, 1991	Evans (1987), Tomley and Evans (1995)
Phyllactinia sp.	*Passiflora triparta* var. *triparta* (banana poka vine)	Southern Colombia & Ecuador	Hawaii, 1992	Trujillo (1992)
Maravalia cryptostegiae (Rust)	*Cryptostegia grandiflora* (rubber-vine weed)	Madagascar	Australia, 1993	Evans (1993, 1995a,b), Evans and Fleureau (1993), Evans and Tomley (1994)
Diabole cubensis, Sphaerulina (Phloeospora) mimosae-pigrae	*Mimosa pigra* (giant sensitive plant)	South and Central America, Mexico	Australia, 1993	Evans (1995a), Evans et al. (1995)
Uredo eichhorniae	*Eichhorniae crassipes* (water hyacinth)	Brazil	S. Africa	Morris (1997)

ADVANTAGES OF MYCOHERBICIDES

Mycoherbicides have several advantages over conventional chemical pesticides (Auld and Morin, 1995; Ayers and Paul, 1990). These are: (i) They are cost effective. For example in developing a new herbicide, an agrochemical company screens up to 13,000 compounds, more than half of these picked at random. In contrast, the search for mycoherbicides is more directed towards reducing the costs of development. (ii) They could be used as alternatives to herbicides when the weed has developed resistance. (iii) They are more selective than most

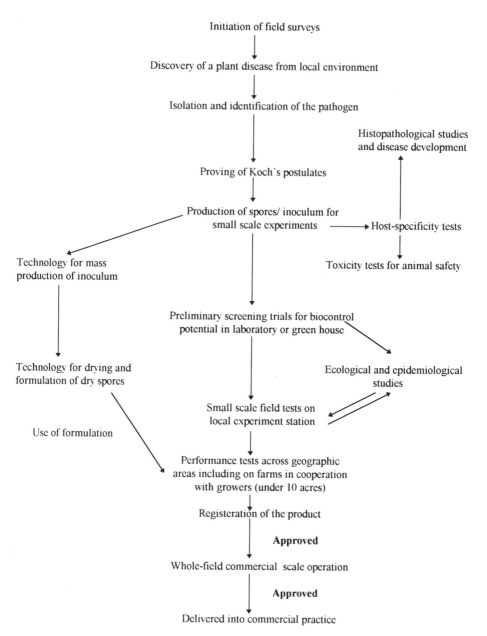

Initiation of field surveys

Discovery of a plant disease from local environment

Isolation and identification of the pathogen

Histopathological studies
and disease development

Proving of Koch's postulates

Production of spores/ inoculum for
small scale experiments ⟶ Host-specificity tests

Technology for mass
production of inoculum

Toxicity tests for animal safety

Preliminary screening trials for biocontrol
potential in laboratory or green house

Technology for drying and
formulation of dry spores

Ecological and epidemiological
studies

Small scale field tests on
local experiment station

Use of formulation

Performance tests across geographic
areas including on farms in cooperation
with growers (under 10 acres)

Registeration of the product

Approved

Whole-field commercial scale operation

Approved

Delivered into commercial practice

Figure 1. Protocol for the development of a fungal biocontrol agent into a mycoherbicide.

96

chemicals, making them a better choice to control weeds that are close relatives to the crop that they infest. (iv) They do not vaporise and are biodegradable, unlike many chemical herbicides which are persistent in the environment. (v) They do not cause pollution problems and are safe for humans, animals, and other plants. (vi) Mycoherbicides are the only solutions for some specific weed species or situations for which no suitable herbicides have been found. For example, parasitic plants and vine weeds which scramble over crops provide a particularly difficult target for even the most selective herbicides. (vii) Mycoherbicides may be preferred to chemical herbicides for aquatic weeds to avoid chemical contamination of water.

CHARACTERISTICS OF POTENTIAL MYCOHERBICIDE PATHOGENS

Weeds, like all plants, may be attacked by a number of diseases but the pathogens that cause them are not all of equal potential as biological control agents (Templeton, 1982). In a few cases, these diseases have yielded pathogens which have been developed into commercial bioherbicides because few plant pathogens are both lethal and specific enough to be effective weed control agents (Sands et al., 1990).

The characteristics of fungal plant pathogens that make them desirable candidates as biological control agents have been extensively reviewed by Daniel et al. (1973), Freeman et al. (1973), Freeman (1977), Templeton et al. (1979, 1986), Charudattan (1990a), Trujillo (1992), Sands and Miller (1993), and Aneja (1997). Fungal pathogens that could be used as potential biological control agents of weeds have the following characteristics: (i) the disease-causing organisms must be host-specific or restricted in host range; (ii) they must be readily cultured in artificial media and able to provide infective units (e.g. spores) readily in culture; (iii) they must be capable of aggressive pathogenicity resulting in effective control of the weed; (iv) they must have high levels of virulence; (v) they must have inefficient natural distribution mechanisms; (vi) the agent must be efficacious under sufficient different environmental conditions; (vii) the inoculum must be capable of abundant production using conventional methods such as liquid fermentation; (viii) the final product (inoculum) formulation must be stable; (ix) storage (shelf-life), handling and methods of application of formulation must be compatible with current agricultural application and practices; (x) the biocontrol agent must be genetically stable; (xi) the agent must cause no pollution and pose no risk to human health; and (xii) the agent must provide quick, complete and easy weed control.

CURRENT STATUS OF MYCOHERBICIDES

Mycoherbicide research and development over the past three decades has resulted in the commercialization of several indigenous fungal pathogens for weed control, for example DeVine, Collego, Casst, BioMal, Biochon, Luboa, Velgo and *Cercospora rodmanii* (see Table 1). Over thirty are in an advanced stage of development and should be available on the market for commercial use by the end of this century.

DEVELOPMENT AND COMMERCIALIZATION OF NOTABLE MYCOHERBICIDES

DeVine™

DeVine™ is a liquid suspension ('wet-pack') consisting of chlamydospores of *Phytophthora palmivora* (Butler) Butler (*=Phytophthora citrophthora* (R.E. Sm. and EH.

Sm.) Leonian). It was the first bioherbicide registered in the United States in 1981 for use as a post-emergent directed spray to control strangler (milkweed) vine *(Morrenia odorata* Lindl.) in Florida citrus orchards. It is manufactured and marketed by Abbott Laboratories, North Chicago, Illinois, USA (Burnett et al., 1974; Charudattan, 1997; Ridings, 1986; Ridings et al., 1976, 1977). DeVine™ gives over 90% control of the weed and control lasts for at least two years after application (Kenney, 1986).

P. citrophthora was isolated from a dying strangler vine by the Department of Plant Industries, Florida Department of Agriculture in 1973-1974. It was later identified as *P. palmivora* (Feichtenberger et al., 1981, 1984). It controls the weed by causing a root and stem rot and girdling the stems of infested plants near the soil surface, resulting in the death of seedlings and older plants. Young seedlings can be killed within one week of inoculation, and larger vines, which often over-grow mature trees, can be killed within four to six weeks. Inoculum (chlamydospores or zoospores) incites the disease whether applied to the seedlings or the soil (TeBeest and Templeton, 1985). The protocol for the development of DeVine™ is given in Table 3.

Table 3. Steps in the development of DeVine *(Phytophthora palmivora)* as a mycoherbicide to control strangler vine *(Morrenia odorata).*

Steps	Year	References
Discovery of the disease	1973	Burnett et al. (1974)
Isolation and identification of the pathogen *as Phytophthora citrophthora*	1973-1974	Burnett et al. (1974)
Host specificity tests	1973-1976	Ridings et al. (1976)
Efficacy tests by the formulation prepared by Abbott Laboratories, North Chicago, IL	1973-1981	Burnett et al. (1974), Ridings et al. (1976, 1977), Woodhead (1981)
Reidentification of the pathogen as *Phytophthora palmivora* (Butler) Butler	1981-1984	Feichtenberger et al. (1981, 1984)
Registration of the pathogen as mycoherbicide DeVine as a liquid suspension (wet pack) consisting of chlamydospores of the fungus by the USEPA	1981	Kenney (1986), Ridings (1986), Te Beest and Templeton (1985)

The production of DeVine™ was discontinued because of the limited market size and the high cost of registration (Evans, 1995) but it is now back on the market produced and marketed by Abbott Laboratories. Abbott is selling the product on a 'made-to-order' basis (Charudattan, 1997).

Collego™

Collego™, a wettable powder formulation of *Colletotrichum gloeosporioides* (Penz.) Sacc. f. sp. *aeschynomene*, was registered in 1982 for the selective control of northern jointvetch *(Aeschynomene virginica* (L) B.S.P.), a leguminous weed whose black seeds contaminate harvests of rice and soybeans in Arkansas, Louisiana and Mississippi, USA

(Daniel et al., 1973; Templeton, 1982, 1986). The pathogen *cga* was isolated from the weed by United States Department of Agriculture (USDA) and University of Arkansas scientists (Smith et al., 1973a,b; Templeton, 1982). The Upjohn Company of Kalamazoo, Michigan, USA developed the production protocols and formulations of Collego™. Collego™ was the first commercially available herbicide for use on an annual weed in annual crops in the United States (Bowers, 1986; Templeton, 1982). The steps for the development of Collego™ as a mycoherbicide to control northern jointvetch are given in Table 4.

Table 4. Steps in the development of Collego *(Colletotrichum gloeosporioides* f. sp. *aeschynomene)* as a mycoherbicide to control northern jointvetch *(Aeschynomene virginica).*

Steps	Year	References
Discovery of the disease	1969	Daniel (1972), Daniel et al. (1973)
Isolation and identification of the pathogen	1970	Daniel et al. (1 973)
Host specificity tests	1970-1973	Daniel (1972), Daniel et al. (1973)
Production of spores for field tests	1973	Daniel et al. (1973)
Efficacy tests in grower field	1973-1981	Smith, Jr. (1978, 1986), Templeton et al. (1981)
Toxicity tests for animal safety	1975	Beasley et al. (1975)
Technology for drying and formulation of dry spores	1976	Bowers (1982, 1986), Churchill (1982), Templeton et al. (1981)
Histopathological studies and disease development mechanism	1978	TeBeest et al. (1978)
Successful use of dried spore formulations on a commercial scale by TUCO division of the Upjohn Company	1982	Quimby and Walker (1982)
Registration of the product COLLEGO by the US Environmental Protection Agency (EPA) as a wettable powder formulation of dried spores	1982	Bowers (1986), Smith Jr. (1986), Te Beest and Templeton (1985), Templeton et al. (1984)
Approval of the Benlate resistant -mutated strain of the pathogen from EPA	1987	Templeton et al. (1988)

Collego™ consists of 15% living spores and 85% inert ingredients (w/w) which are rehydrated and resuspended in a sugar solution before being mixed with water in an applicator's spray tank. It is applied post-emergence, aerially or with land-based sprayers. Within a week or two after application the vetch plants begin to show lesions that gradually encircle the stem. Most of the plants die within five weeks. Collego has provided consistently high levels of weed control (92% on average over 10 years of experimental and commercial use [TeBeest and Templeton, 1985; Templeton, 1986]).

Collego fungus (i.e. *cga*) is sensitive to Benlate fungicide which is used for the control of rice blast. To overcome the reduction in the efficacy of Collego due to Benlate in short season varieties where both pesticides may need to be applied within a short span, an improved strain of the fungus, tolerant to Benlate, has been developed by mutation which has all the other desirable traits of the parent strain. This new strain produced excellent weed control in plots sprayed immediately before and immediately after Benlate application. Approval for the Benlate-resistant strain was obtained from EPA in 1987 (Templeton et al., 1988). Growers have taken enthusiastically to Collego and the fungus has reduced the input of pesticides by nearly 500 000 litres since its introduction (Ayres and Paul, 1990).

The production of Collego was discontinued but Encore Technologies, Minnesota (USA) were successful at getting this bioherbicide back on the market for the 1997 growing season (TeBeest, 1997).

Casst[(R)]

Sicklepod (*Cassia obtusifolia* L.) (Caesalpinaceae, Leguminosae), an annual, non-nodulating legume, is an exotic weed that is difficult to control using conventional weed control practices. It is a problem in agronomic crops, especially soybean (*Glycine max*), peanut (*Arachis hypogea*) and cotton (*Gossypium hirsutum*) in the south-eastern United States. It is the fifth most troublesome weed and sixth most common weed in the southern region of the United States (Charudattan et al., 1986).

In 1982, *Alternaria cassiae* Jurair and Khan, a foliar blight pathogen discovered from Pakistan (Jurair and Khan, 1960) was isolated and identified from seeds and diseased seedlings of sicklepod collected in Mississippi, USA (Walker, 1982). Subsequently, the pathogen was discovered to be naturally occurring in Florida and North Carolina. Walker and Riley (1982) reported *A. cassiae* to be a safe and effective mycoherbicide candidate for sicklepod. It has a narrow host range and is capable of controlling three economically important leguminous weeds - sicklepod (*Cassia obtusifolia*), coffee-senna (*Cassia occidentalis*) and showy crotalaria (*Crotalaria spectabilis*) in soybean and peanuts (Boyette, 1988; Walker, 1983; Walker and Boyette, 1985).

Standardized preparations of conidia for field trials were provided by Abbott Laboratories, Chicago, Illinois and the USA Southern Weed Science Laboratory, Stoneville, Mississippi. An inoculum level of 1 x 10^6 conidia per ml of spray was used. The spray mixture consisted of conidia of *A. cassiae* in 0.02 to 0.04% (v/v) of aqueous surfactant sterox and the spray volume used was 935 litres of water per hectare. *A. cassiae* provided 95% (one application) and 100% (two applications) control in the first year and 95 and 97% control in the second years, in comparison. The toxaphene (chlorinated camphene, 67 to 69% chlorine), an insecticide used to control sicklepod (but presently banned as a pesticide) treatments yielded 78% (one application) and 100% control (two applications) in the first year, indicating the better performance of the fungus than the chemical control (Charudattan, 1990b, Charudattan et al., 1986).

Casst[(R)], a wettable powder formulation of dried spores of *A. cassiae*, has been developed by Mycogen Corporation, Santiago, California for commercial use (Bannon, 1988; Bannon and Hudson, 1988). Protocol for the development of Casst[(R)] as a mycoherbicide is given in Table 5.

BioMal[TM]

BioMal[TM], a formulation of *Colletotrichum gloeosporioides* (Penz) Sacc. f. sp. *malvae*, is a post-emergent mycoherbicide to control round-leaved mallow (*Malva pusilla* Sm.= *M. rotundifolia*) in wheat, lentil and flax in Canada (Makowski and Morrison, 1989; Mortensen, 1988). It gives over 90% control of the weed. BioMal is the first bioherbicide registered in Canada. It has been developed by Philom Bios, a Biotechnology Company based in

Saskatchewan, Canada in collaboration with Agriculture Canada (Makowski and Mortensen, 1992). The fungal pathogen was originally isolated from diseased round-leaved mallow tissues in 1982 by scientists at the Agriculture Canada Research Station, Regina, Saskatchewan (Makowski, 1987; Mortensen, 1988). Under controlled environmental conditions, *M. pusilla* was effectively controlled with *c.g.m.* at a concentration of 2 x 10^6 spores/ml, with a minimum of 20 hours of dew or repetitive dew periods of 16 hours, at temperatures below 30°C, and at all growth stages (Makowski, 1987). In field trials, excellent control of round-leaved mallow was obtained when the fungus was applied as a spore suspension in wheat and lentil at a concentration of 60 x 10^6 spores/m^2 in 150 l/ha water volume under favourable conditions (Mortensen and Makowski, 1990). Moisture, a 12 to 15 hour dew period following application, a temperature around 20°C and an overcast sky following application of the spray are the most important factors responsible for successful control under natural conditions (Makowski and Mortensen, 1990). BioMal^{TM} is marketed in Western Canada by Dow Elanco Canada Inc. (Leggett and Gleddie, 1995).

Table 5. Steps in the development of Casst *(Alternaria cassiae)* as a mycoherbicide for the biological control of three weeds - sicklepod, showy crotolaria and coffee senna - in agricultural fields.

Steps	Year	References
Discovery of the pathogen from Pakistan	1960	Jurair and Khan (1960)
Reisolation & identification of the pathogen from Mississippi, USA	1977-1981	Walker (1982)
Evaluation of the pathogen for biocontrol potential in greenhouse and small field plots	1982	Walker (1982) Walker and Riley (1982)
Host specificity tests	1982	Walker (1982, 1983)
Preparation of the inoculum for field tests by Abbott Lab., North Chicago, Illinois (Code No. 5005)	1982	Charudattan et al. (1986)
Preparation of the inoculum by USDA-ARS, Southern Weed Science Lab, Stoneville, Mississippi	1983	Charudattan et al. (1986)
Evaluation of the pathogen as a myco-herbicide in soybeans and peanuts	1982	Walker (1982) Walker and Boyette (1985) Bannon et al. (1988)
Development of *A. cassiae* product for field testing under EUP	1986	Anonymous (1986)
Evaluation of *A. cassiae* as a mycoherbicide for sicklepod in field tests by 5005 formulation provided by Abbott Lab	1986	Charudattan et al. (1986)
Commercial formulation of mycoherbicide by Mycogen Corporation San Diego, California, as wettable powder formulation of dried spores	1988	Bannon (1988), Bannon and Hudson (1988)

Biochon

Biochon is a recently developed mycoherbicide based on the basidiomycetous fungus *Chondrostereum purpureum* (Pers. ex Fr.) Pouzar to control black cherry (*Prunus serotina* Erhr., Rosaceae) in the Netherlands. It is marketed by Koppert Biological Systems, Netherlands (de Jong, 1997).

Prunus serotina was introduced from North America into the Netherlands to improve the understorey of forests and to ameliorate forest soil by its litter (Bakkar, 1963). It attained the status of a noxious weed in coniferous forests because of competition with newly planted forest trees and the natural vegetation in the understorey of forests (Scheepens and Hoogerbrugge, 1990).

C. purpureum is a native facultative parasite of *P. serotina* and several other deciduous tree species causing silver leaf disease. The fungus was isolated by inverting parts of fresh basidiocarps over malt extract plates and allowing the basidiospores to fall on the agar for several hours and then transferring mycelial mats to new malt agar plates. For inoculation experiments, 2-3 week old mycelial cultures on malt agar, or 2-3 week old liquid cultures grown in malt extract aerated with magnetic stirrers, were used (de Jong et al., 1982). In field experiments conducted at the Centre for Agro Biological Research (CABO), Wageningen, the Netherlands between 1980 and 1986, 90% control was achieved when stumps were treated with agar cultures or suspensions of fragmented mycelium of *C. purpureum* in both spring and autumn seasons. The lowest dose found to be effective was 20 μg mycelial dry weight/stump (de Jong et al., 1982; Scheepens and Hoogerbrugge, 1990; Scheepens and Van Zon, 1982). Various adjuvants (e.g. glycerine, transfilm, soybean and corn oil) had no effect on *C. purpureum* growth, thus these adjuvants may be tested for their role in modifying the efficacy of the bioherbicide (Prasad, 1993).

Lubao

Colletotrichum gloeosporioides Penz., a fungal pathogen isolated from dodder (*Cuscuta chinensis* and *Cuscuta australis*) on soybean in Jinan, Shandong Province (China) in 1963 was developed as a mycoherbicide called 'Lubao No. 1'. This formulation showed good potential and has been used since its discovery in 1963 to control this parasitic weed on soybeans in the People's Republic of China (Liu and Zhu, 1980). However, there was a decline in the use of Lubao No. 1 because of serious limitations such as loss of virulence, a requirement for high humidity in treated fields, limited shelf life and technical problems associated with fermentation and commercialization (Li, 1985). A granular mixture form of Lubao No. 1 is being produced in two County Factories in the Ningxia Autonomous Region, China (Wang, 1990).

A selected strain of the pathogen, *C. gloeosporioides* f. sp. *cuscutae,* that gave better control than Lubao No. 1 was developed as Lubao No. 2. Virulence of the pathogen has been positively correlated with spore size. Spore concentrations of 2 x 10^7 spores/ml are applied with a hand sprayer until run-off occurs. For best results spraying is done at 16:00 to 17:00 hours on days when humidity is high, usually in late July to early August (Templeton and Heiny, 1989). Although Lubao was the first mycoherbicide developed in the world, it is still the only known fungal formulation used to control weeds in China.

Velgo[R]

Velgo is a formulation of *Colletotrichum coccodes* (Wallr.) Hughes for the control of velvet leaf *(Abutilon theophrasti* Medik., Malvaceae), a vigorous terrestrial weed in corn and soybeans in the USA and southern Ontario, Canada. The fungus killed 46% of the plants when the formulation was applied at a rate of 1 x 10^9 spores m^{-2} at the two- to three-leaf stage. Lack of complete control of the weed is due to premature shedding of infected leaves and regrowth of new leaves (Wymore et al., 1987). The adjuvant effect of thidiazuron (N-phenyl N-1,2,3-thiadiazol-5-ylurea), which promotes defoliation and retards plant regrowth, in improving the effectiveness of *C. coccodes* was observed and might be used to control A. *theophrasti* (Hodgson et al., 1988).

Cercospora rodmanii as a Mycoherbicide for Waterhyacinth

Cercospora rodmanii Conway has been developed as a mycoherbicide to control waterhyacinth in Florida, USA (Freeman and Charudattan, 1984). Waterhyacinth, *Eichhornia crassipes* (Mart.) Solms, a native of South America, now occurs worldwide mainly in tropical and sub-tropical regions. It is a major threat to waterways wherever it occurs, blocking irrigation canals and impeding hydroelectric schemes (Center et al., 1990; Evans, 1987; Pieterse and Murphy, 1990). It has been called the world's worst aquatic weed and ranked the second most serious weed (Aneja, 1996).

 C. rodmanii, was first isolated in 1973 from diseased waterhyacinth plants found in the Rodman reservoir, Florida, USA (Conway, 1976a) and subsequently from Kurukshetra, India (Aneja and Srinivas, 1990). The pathogen is host-specific and safe for use. It is grown for three weeks on potato-dextrose broth containing 5% yeast extract, blended and applied at the rate of 1.1 g net weight/m^2. Following inoculation of the plants grown in buckets, disease stress caused a significant reduction in the net rate of leaf production (Conway, 1976a, b; Conway and Freeman, 1977; Conway et al., 1978a,b; Freeman and Charudattan, 1984). Effective control of the weed is obtained when *C. rodmanii* is used with multiple applications of inoculum early in the growing season or under conditions favouring low to moderate host rates, including combining with sublethal rates of chemical herbicides or insect biocontrol agents (Charudattan, 1986; Charudattan et al., 1985).

 The use of *C. rodmanii* as a biological control agent of waterhyacinth was patented by the University of Florida, Gainesville, Florida (Conway et al., 1978a). The University signed an agreement with Abbott Laboratories, USA to produce a commercial product of the fungus. Subsequently, Abbott Laboratories was licenced to develop the fungus as a microbial herbicide for commercial use. Abbott developed wettable powder formulations of the fungus (ABG-5003) and obtained a U. S. Environmental Protection Agency (EPA) Experimental Use Permit (EUP) to evaluate it as a microbial herbicide (Freeman and Charudattan, 1984). The protocol for the development of *C. rodmanii* as a biocontrol agent of waterhyacinth is given in Table 6.

Colletotrichum orbiculare as a Mycoherbicide to Control *Xanthium spinosum*

Xanthium spinosum L. (Asteraceae), Bathurst burr, spiny cockleburr or spiny clotburr, an annual, native to south America, is widespread as a weed especially in temperate regions. It has been recorded in 34 countries as a weed in 13 crops (Holm et al., 1977). In Australia, Bathurst burr is an exotic weed and has been regarded as a major weed since the early 1850s (Hocking and Liddle, 1986). It is principally a weed of grazing land and a wool contaminant and occasionally a problematic weed of horticulture and irrigated summer crops in Australia (Auld et al., 1988, 1990a, 1992).

 Colletotrichum orbiculare (Berk. and Mont.) v. Arx (=*Gloeosporium orbiculare* (Berk. and Mont.) is known as a pathogen of the Cucurbitaceae (Sutton, 1980). The pathogen was isolated from *X. spinosum* in 1984 from southeastern Australia (Nikandrow et al., 1990). Controlled environment studies (Auld et al., 1988; McRae and Auld, 1988) and field experiments (Auld et al., 1990a,b) have demonstrated that *C. orbiculare* is a potential mycoherbicide for the control of Bathurst burr (Klein and Auld, 1995a). Auld et al. (1988, 1990a) and McRae and Auld (1988) demonstrated successful control of the weed with this fungus under controlled environmental conditions using 10^6 spores/ml and in the field with 10^7 spores/ml (Auld *et al.*, 1990b). At concentrations of 10^7 spores/ml, the fungus killed six-week old plants within 14 days. The fungus showed considerable potential as a control agent provided inoculations were made in the late afternoon or evening and provided a period of dew occurs following inoculation (Auld et al., 1988; McRae and Auld, 1988). In the absence of dew, vegetable oil suspension emulsions containing *C. orbiculare* conidia gave significantly better anthracnose development on Bathurst burr than aqueous suspension; in other words, the use of vegetable oil emulsions reduced the dependence on dew for infection (Auld, 1993).

Table 6. Steps in the development of *Cercospora rodmanii* as a biocontrol agent of waterhyacinth.

Steps	Year	References
Discovery of the disease from the Rodman Reservoir, Florida, USA	1973	Conway et al. (1974)
Identification and description of the pathogen	1975	Conway (1976a) Conway and Cullen (1978)
Preliminary greenhouse and field test of efficacy (conducted in small lake in Fish Prairie)	1973-1976	Conway et al. (1979)
Host range testing in the green house and the field	1973-1977	Conway and Freeman (1977)
Advanced field studies in Florida and Louisiana	1973-1976	Addor (1977), Conway (1976b), Conway and Freeman (1976) Freeman et al. (1981,1982)
A Cercospora rodmanii use patent obtained (Patented by University of Florida)	1978	Conway et al. (1978a)
Industrial production of *C. rodmanii* as wettable powder formulation, ABG-5003 (Abbott Lab.)	1978	Charudattan (1991), Charudattan et al. (1985), Conway et al. (1978b), Freeman and Charudattan (1984)
An EPA Experimental use permit obtained	1979	Freeman and Charudattan (1984)
Successful large-scale aerial application of Abbott formulation	1980	Theriot (1980, 1982)
EUP studies continued	1980	Freeman and Charudattan (1984)
Registration cancelled by Abbott Lab. due to the following reasons: -Lack of acceptable level of efficacy -Technical difficulties in production and marketing of commercially acceptable formulation -Competition from chemical herbicides -Unprofitable market	1984	Charudattan (1991)

Vegetable (Canola and soybean) and mineral (Caltex summer oil and Ampool A) oils improved mycoherbicide activity in comparison with spores applied in water only (Klein et al., 1995). Better control of the weed was obtained when the mycoherbicide was applied to young, actively growing plants during long dew periods in spring or early summer (Klein and Auld, 1995a). Higher spore dose rates were associated with increased plant kill (Klein and Auld, 1995b). Moreover, wounding could increase the efficacy of *C. orbiculare* as a mycoherbicide for Bathurst burr, if the formulation was applied immediately after injury (Klein and Auld, 1996). Protocol for the development of *C. orbiculare* as a mycoherbicide for the control of *X. spinosum* is given in Table 7.

Table 7. Steps in the development of *Colletotrichum orbiculare* as a mycoherbicide for Bathurst burr or spiny cocklebur *(Xanthium spinosum* L.) in Australia.

Steps	Year	References
Discovery of the disease in eastern Australia	1984	Nikandrow et al. (1990)
Isolation and identification of the pathogen	1984	Nikandrow et al. (1990)
Production of spores on solid media and submerged culture	1987	Auld et al. (1988)
Testing of potential of *Colletotrichum orbiculare* as a mycoherbicide under controlled environmental conditions	1987-1989	Auld et al. (1988), McRae and Auld (1988)
Efficacy test in fields	1988-1989	Auld et al. (1990a,b)
Study of effect of environmental factors on disease development	1988-1990	Auld et al. (1990a,b), McRae and Auld (1988)
Dried preparation of spores of the pathogen in Kaolin (hydrated aluminium silicate) by Sandoz Agro, Switzerland for field testing	1991	Klein et al. (1995)
Evaluation of oil suspension emulsions with mycoherbicide under controlled environmental conditions and in the fields	1991-1993	Auld (1993), Klein et al. (1995)
Assessment of influence of spore dose and water volume on the efficacy of mycoherbicide in field trials	1991-1993	Klein and Auld (1995a)
Effect of wounding on mycoherbicide's efficacy	1992-1993	Klein and Auld (1996)

Alternaria eichhorniae as a Biocontrol Agent of Waterhyacinth

Alternaria eichhorniae Nag Raj and Ponnappa was first described in 1970 from Bangalore and Assam, India, as the causal agent of a leaf-spot disease of waterhyacinth and was suggested as a suitable biocontrol agent of this weed (Nag Raj and Ponnappa, 1970). It causes discrete or blotchy necrotic leaf spots with dark centres and brownish black margins, often with a thin yellow halo surrounding the spots resulting in severe leaf blights. This disease of waterhyacinth is not widespread in the world, restricted to India, Pakistan, Indonesia, Thailand and Egypt (Aneja, 1996; Charudattan, 1984, 1990c; Gopal, 1987; Shabana et al., 1995a,b).

Shabana et al. (1995a) have described *A. eichhorniae* as an effective and safe bioherbicide candidate in Egypt. Pellet and powder formulations of the fungus were tested in greenhouse trials. Two months after four sequential applications of the formulation supplemented with a hydrophilic polyacrylamide, disease severity on fungus-treated plants increased by 93% and the biomass (fresh weight) decreased by 89% compared to the control, thus showing its ability to curtail waterhyacinth growth in the greenhouse (Shabana et al., 1995b).

A. eichhorniae is a potential mycoherbicide for the control of waterhyacinth in India. It has a narrow host range. Under experimental conditions, complete death of the laminae resulted two months after inoculation of the laminae in cemented tanks (Aneja, 1992). Inoculum of the pathogen for experimental studies was prepared on waterhyacinth dextrose agar medium at 25 °C for 10 days as 25 °C was the best temperature for sporulation.

ANTICIPATED BENEFITS OF BIOTECHNOLOGY TO MYCOHERBICIDE DEVELOPMENT

Mycoherbicides have still not realized their potential, mainly owing to an array of constraints, biological (e.g. host variability and host range, resistance mechanisms and interaction with other microorganisms), environmental (e.g. temperature and dew requirement), technological (e.g. mass production and formulation), commercial (e.g. market potential, customer demand, cost of production) and regulatory (Auld and Morin, 1995; Charudattan, 1990b; Evans 1995a; Leggett and Gleddie, 1995; Templeton and Heiny, 1989). But with the application of recent advances in biotechnology, mycoherbicide development will gain momentum and these will play an increasingly significant role in agricultural ecosystems. Virulent fungal pathogens with broad host ranges offer the most commercial potential, if these can be tailored to suit particular crop systems (Evans, 1995b). Charudattan (1990b) suggested that in order to advance in the area of mycoherbicide development, genetic improvement of microbial herbicide should be achieved for: (i) increased virulence for improved efficacy; (ii) phytotoxin production for improved efficacy and host specificity; (iii) modification for the host range or increased or decreased host specificity; and (iv) resistance to chemical pesticides to enable integration of the microbial agent with chemical pesticides.

The feasibility of improving bioherbicide candidates through genetic manipulation has received considerable attention in recent years (Auld and Morin, 1995; Charudattan, 1985; Greaves et al., 1989; Kenney, 1986, Templeton, 1990; Templeton and Heiny, 1989; Turgeon and Yoder, 1985; Yoder et al., 1989). For example, genetic alteration of the Collego fungus by treatment with methane sulphonate resulted in Benomyl resistance in spores of the fungus (TeBeest, 1984); in other words, development of a new and improved Collego with pesticide resistance. Thus Collego could be used with Benomyl for the control of blast of rice caused by *Pyricularia oryzae* or for control of soybean anthracnose disease (Smith, 1986).

The genetic manipulation of broad host range fungal pathogens to restrict their host range and limit survival has been attempted (Miller et al., 1989a,b; Sands and Miller, 1993a,b; Sands et al., 1990). Genetic improvement to restrain broad host-range is genetically possible in *Sclerotinia sclerotiorum* (Lib.) deBary, *Phymatotrichopsis omnivorous* Hennebert, *Rhizoctonia solani* Kuhn, and *Sclerotium rolfsii* Sacc. For example, three classes of induced mutants of *Sclerotinia sclerotiorum*, a highly virulent and aggressive pathogen of several weeds, were obtained which meet criteria for delimitation, i.e. auxotrophic mutants (primidine auxotrophs) that only attack plants when applied concomitantly with an exogenous source of required nutrient; mutants unable to form sclerotia; and mutants with reduced virulence and/or host ranges (Sands et al., 1990).

The use of recombinant DNA technology, in contrast to mutagenesis technique, may offer a much more precise and directed approach of altering the DNA of a pathogen to enhance effectiveness and modify host specificity of pathogens (Auld and Morin, 1995; Sands and Miller, 1993b; Templeton and Heiny, 1989). A successful mating in culture and plant tissues has been reported by TeBeest et al. (1992) between *Colletotrichum gloeosporioides* f. sp. *aeschynomene* and strains of the same fungus, i.e. *C.g.* infecting winged waterprimrose (*Ludwigia decurrens* Walt.) and peacon *(Carya illinoensis* (Wangenh.) C. Koch.).

Genetic characterisation using RAPD technology has been used to identify various isolates and species in the genus *Colletotrichum* for the purpose of assessing species relatedness and for correlation with isolate pathogenicity and phytotoxin production. Genetic transformation of biocontrol agents with a cutinase gene derived from *Colletotrichum gloeosporioides* and *Fusarium solani* to improve pathogen penetration is under investigation. Transformants have been obtained by protoplast fusion with BioMal, a mycoherbicide used to control round-leaved mallow in Canada (Boyetchko and Bailey, 1997).

CONCLUSIONS

Farmers and consumers, both concerned about the potential health and environmental consequences of chemical applications, are in search of alternatives. To meet the demand for alternatives to chemical control of weeds, microbial control has emerged as a possible option. The goal of biological weed control is the use of mainly host specific arthropods and pathogens to reduce the population density of a weed to below its economic or ecological injury level (Schroeder, 1995). It is perhaps surprising, considering the age-old awareness of the potential of plant diseases to cause crop failures, that only recently fungi have been seriously evaluated as a means of controlling weeds. Of the various types of disease-causing microorganisms such as fungi, bacteria, viruses and mycoplasmas, fungi may be preferred as potential biocontrol agents of weeds because they possess most of the desirable characteristics which an agent should have. To control a weed, before it becomes obnoxious, work should be initiated using all three biocontrol strategies, i.e. classical, mycoherbicidal and manipulated mycoherbicidal, simultaneously. Much research is currently underway worldwide to determine the extent to which mycoherbicides may be employed to control weeds.

Commercial mycoherbicides first appeared on the market in the USA in the early 1980s with the release of the product DeVine (a formulation of *Phytophthora palmivora)* in 1981 to control milkweed vine in Florida citrus orchards, and in the next year the release of the product Collego (a formulation of *Colletotrichum gloeosporioides* f. sp. *aeschynomene),* to control northern jointvetch, a leguminous weed in rice. This early work at Arkansas and Florida stimulated a host of other studies on a range of pathogen-weed associations. Progress made in recent years to explore fungi as biocontrol agents of weeds in the developed countries of the world like the USA, Canada, Australia, UK and the Netherlands has resulted in the production and commercial use of eight mycoherbicides, i.e. DeVine, Collego, Casst, Velgo, *Cercospora rodmanii* (in USA), BioMal (in Canada), Biochon (in Netherlands) and Lubao (in China). There are several others which are in an advanced stage of development such as *Colletotrichum orbiculare* for controlling *Xanthium spinosum* (Bathurst burr) in Australia and *Alternaria eichhorniae* for controlling waterhyacinth in Egypt and India.

The pace of development in the area of commercialisation of mycoherbicides is still slow because of an array of biological, economic and regulatory constraints. To overcome biological constraints - such as virulence, stability, producing sufficient concentration of spores to be economically viable, host range, environmental conditions, dew requirements of the pathogens and geographic biotypes of the weeds - the use of fungal biotechnology by genetic manipulation of fungal pathogens through protoplast fusion, improvement of the fermenters, spraying techniques and modification of the carriers to the inoculum is being currently pursued in order to enhance the development of mycoherbicides into products. Advances in biotechnology may one day be able to create on request fungal pathogens with desirable characteristics as commercial mycoherbicides.

In spite of the better potential of fungi to control obnoxious weeds in the tropical climate than in temperate countries, researchers in the Indian subcontinent are still in the pioneering stages of using fungi as biocontrol agents of weeds. The single greatest constraint to the development of mycoherbicides is adequate field explorations for suitable new mycoherbicide candidates. Considering the rich biodiversity, and good prospects for discovering and developing biocontrol agents for many types of weeds, progress made in this area in the developing countries has been almost nil. It seems the major constraint is the lack of available of funds. For example, for the development of a potential biocontrol agent into a mycoherbicide, the first step is its identification. Scientists in developing countries do not have enough funds to pay for authentic identification and unless the identification is confirmed, it is not possible to go ahead with publication and further evaluation for efficacy and development and commercialisation of the agent. Moreover, private enterprises in developing countries are unwilling to give assistance for any research project unless they are guaranteed one hundred per cent economic returns on their investment.

FUTURE STRATEGIES

The strategies discussed earlier demonstrate that the future development of mycoherbicides is dependent on fundamental knowledge of biological interactions at the organism and ecosystem level. Intensive and long term research is needed for finding out and understanding the unique biology of specific biocontrol agents, target combinations, environmental impact and their potential in the management of a weed. The future of mycoherbicides appears to be promising, if proper and immediate attention is paid, especially in the developing countries, to the following:

1. Conducting systematic field surveys for the identification of the major endemic diseases of the major weeds;
2. Isolation and establishment of stable cultures of the causal organisms;
3. Selection of pathogens, particularly those which are highly pathogenic;
4. Developing methods for mass production of stable inoculum;
5. Understanding the disease cycle and the weed patho-system;
6. Understanding the genetics of the pathogen or molecular basis of the disease that may eventually help in selecting and establishing pathogens with greater virulence ('super pathogens') or may even assist to increase virulence through hybridisation, parasexuality or genetic manipulation, i.e. through the application of biotechnology on a wider scale.

To appreciate fully the value of mycoherbicides the farmers must be taught the limitations and benefits of the new technologies prior to or during market launch of any new biological product. Awareness campaigns, aimed at both potential sponsors as well as the farmers and public in general should be initiated to stress the menace posed by obnoxious weeds and the environmental pollution caused by the poisonous chemicals used for controlling the pests.

ACKNOWLEDGEMENTS

I am thankful to my students - Dr. Raj Kumar, Mr. Shahalam and Mrs. Shailja Kaushal for assistance in the preparation of the manuscript. My special thanks to the UK Department for International Development (DFID) for providing financial assistance for the RNRRS Research

Project (1996-99).

REFERENCES

Adams, E.B. and Line, R.F., 1984, Epidemiology and host morphology in the parasitism of rush skeletonweed by *Puccinia chondrillina, Phytopathology* 74: 745-748.

Addor, E.E., 1977, A field test of selected insects and pathogens for control of waterhyacinths. Report 1, Preliminary results for the 1975-76 season, *Misc. Pap.* A-77-2, pp. 1-44, U.S. Army Waterways experiment station, Vickasburg MS.

Aneja, K.R., 1996, Exploitation of fungal pathogens for biocontrol of waterhyacinth, in: *Some Facets of Biodiversity* (R.K. Kohli, N. Jerath and D. Batish, eds.), pp. 141-156, Research Periodicals and Book Services, New Delhi.

Aneja, K.R., 1997, Discovery and development of mycoherbicides for biological control of weeds, in: *New Approaches in Microbial Ecology* (J.P. Tewari, G. Saxena, N. Mittal, I. Tewari and B.P. Chamola, eds.), pp. 517-555, Aditya Books Pvt. Ltd., New Delhi.

Aneja, K.R. and Srinivas, B., 1990, Leaf spot disease of waterhyacinth *Eichhornia crassipes* - a new disease record from India, *Trop. Pest Manag.* 36: 76.

Anonymous, 1986, *Alternaria cassiae,* mycoherbicide for the control of sicklepod, Mycogen Corporation, San Diego, CA.

Auld, B.A., 1993, Vegetable oil suspension emulsions reduce dew dependence of a mycoherbicide, *Crop Prot.* 12: 477.

Auld, B.A. and Morin, L., 1995, Constraints in the development of bioherbicides, *Weed Tech.* 9: 638-652.

Auld, B.A., McRae, C.F. and Say, M.M., 1988, Possible control of *Xanthium spinosum* by a fungus, *Agric. Ecosyst. Environ.* 21: 219.

Auld, B.A., McRae, C.F. and Say, M.M., 1990a, Potential for *Xanthium spinosum* control by *Colletotrichum orbiculare* as a mycoherbicide, in: *Proc. VII Int. Symp. Biol. Contr. Weeds* (E.S. Delfosse, ed.), pp. 435-443, CSIRO Publications, Australia.

Auld, B.A., Say, M.M., Ridings, H.I. and Andrews, J., 1990b, Field applications of *Colletotrichum orbiculare* to control *Xanthium spinosum, Agric. Ecosyst. Environ.* 32: 315-323.

Auld, B.A., Talbot, H.E. and Radburn, K.B., 1992, Host range of three isolates of *Alternaria zinniae, a* potential biocontrol agent for *Xanthium* spp., *Plant Prot. Quart.* 7(3): 114.

Ayers, P. and Paul, N., 1990, Weeding with fungi, *New Scientist* 1: 36-39.

Bakkar, G., 1963, De Ontwikkelingsgeschiedenis van *Prunus serotina* Erhr. in Nederland, *RIVON-mededcling* 147: 201-206.

Bannon, J.S., 1988, CASST herbicide (*Alternaria cassiae*): a case history of a mycoherbicide, *Amer. J Alt. Agric. USA* 3 (213): 73-76.

Bannon, J.S. and Hudson, R.A., 1988, The effect of application timing and lighting intensity on efficacy of CASST™ herbicide (*Alternaria cassiae*) on sicklepod (*Cassia obtusifolia*), *Weed Sci. Soc. Am. Abstr.* 28: 143.

Bannon, J.S., Hudson, R.A., Stowell, L. and Glatzhofer, J., 1988, Combinations of herbicide/plant growth regulator with CASST™, *Proc. South Weed Sci. Soc.* 41: 268.

Beasley, J.N., Patterson, L.T., Templeton, G.E. and Smith, R.J., Jr., 1975, Response of animal to a fungus used as a biological herbicide, *Arkansas Farm Research* 24(6): 16.

Boyette, C.D., 1988, Biocontrol of three leguminous weed species with *Alternaria cassiae, Weed Tech.* 2: 414-417.

Bowers, R.C., 1982, Commercialization of microbial biological control agent, in: *Biological Control of Weeds with Pathogens* (R. Charudattan and H.L. Walker, eds.), pp. 157-173, John Wiley and Sons, New York.

Bowers, R.C., 1986, Commercialization of Collego™ - an industrialist's view, *Weed Sci.* 34 (Suppl.1): 24-25.

Boyetchko, S. and Bailey, K., 1997, Bioherbicide research - status reports, *IBG News* 6(1):9.

Bruckart, W.L., Baudoin, A.B., Abad, R. and Kok, L.T., 1988, Limited field evaluation of *Puccinia carduorum* for biological control of musk thistle, *Phytopathology* 78: 1593.

Bruckart, W.L. and Dowler, W.M., 1986, Evaluation of exotic rust fungi in the United States for classical biological control of weeds, *Weed Sci.* 34 (Suppl. 1): 11-14.

Burnett, H.C., Tucker, D.P.H. and Ridings, W.H., 1974, *Phytophthora* root and stem rot of milkweed vine, *Plant Dis. Rep.* 58: 355-357.

Center, T.D., Cofrancesco, A.F. and Balciunas, J.K., 1990, Biological control of aquatic and wetland weeds in the southeastern United States, in: *Proc. VII Int. Symp. Biol. Contr. Weeds* (E. S. Delfosse, ed.), pp. 239-262, CSIRO Publications, Australia.

Charudattan, R., 1984, Role of *Cercospora rodmanii* and other pathogens in the biological and integrated controls of waterhyacinth, in: *Proc. of the Int. Conf. on Waterhyacinth* (G. Thyagarajan, ed.), pp. 834-859, U.N. Environ. Prog. Nairobi, Kenya.

Charudattan, R., 1985, The use of natural and genetically altered strains of pathogens for weed control, in: *Biological Control in Agricultural IPM Systems* (M.A. Hoy and D.C. Herozog, eds.), pp. 347-372, Academic Press Inc., Orlando, FL.

Charudattan, R., 1986, Integrated control of waterhyacinth with a pathogen, insects and herbicides, *Weed Sci.* 34 (Suppl. 1): 26-30.

Charudattan, R., 1988, Inundative control of weeds with indigenous fungal pathogens, in: *Fungi in Biological Control Systems* (M.N. Burge, ed.), pp. 86-110, Manchester University Press, Manchester.

Charudattan, R., 1990a, Pathogens with potential for weed control, in: *Microbes and Microbial Products as Herbicides* (R.E. Hoagland, ed.), pp. 132-154, ACS Symposium Series 439, American Chemical Society, Washington, DC.

Charudattan, R., 1990b, Prospects for biological control of weeds by plant pathogens, *Fitopathol. Bras.* 15(1): 13-19.

Charudattan, R., 1990c, Biological control of aquatic weeds by means of fungi, in: *Aquatic Weeds: The Ecology and Management of Nuisance Aquatic Vegetation* (A.H. Pieterse and K.J. Murphy, eds.), pp. 186-200, Oxford University Press, New York.

Charudattan, R., 1991, The mycoherbicide approach with plant pathogen, in: *Microbial Control of Weeds* (D.O. TeBeest, ed.), pp. 24-57, Chapman and Hall, New York.

Charudattan, R., 1997, Status of DeVine and Collego, *IBG News* 6(1):8.

Charudattan, R., Lenda, S.B., Kluepfel, M. and Osman, Y.A., 1985, Biocontrol efficacy of *Cercospora rodmanii* on waterhyacinth, *Phytopathology* 75: 1263-1269.

Charudattan, R., Walker, H.L., Boyette, C.D., Ridings, W.H., TeBeest, D.O., Van Dyke, C.G. and Worsham, AD., 1986, Evaluation of *Alternaria cassiae* as a mycoherbicide for sicklepod (*Cassia obtusifolia*) in regional field tests, *Southern Cooperative Series Bulletin* 317, pp. 1-19, Alabama, Agricultural Experiment Station, Auburn University.

Churchill, B.W., 1982, Mass production of microorganisms for biological control, in: *Biological Control of Weeds with Plant Pathogens* (R. Charudattan and H.L. Walker, eds.), pp. 139-156, John Wiley and Sons, New York.

Conway, K.E., 1976a, *Cercospora rodmanii*, a new pathogen of waterhyacinth with biological control potential, *Can. J. Bot.* 54: 1079-1083.

Conway, K.E., 1976b, Evaluation of *Cercospora rodmanii* as a biological control of waterhyacinth, *Phytopathology* 66: 914-917.

Conway, K.E. and Cullen, R.E., 1978, The effect of *Cercospora rodmanii*, a biological control for waterhyacinth, on the fish *Gambusia affinis*, *Mycopathologia* 66: 113-116.

Conway, K.E., Cullen, R.E., Freeman, T.E. and Cornell, J.A., 1979, Field evaluation of *Cercospora rodmanii* as a biological control for waterhyacinth. Inoculum Rate Studies, *Misc. Pap.* A-79-6, pp. 1-46, U.S. Army Waterways Experiment Station, Vicksburg, MS.

Conway, K.E. and Freeman, T.E., 1976, The potential of *Cercospora rodmanii* as a biocontrol for waterhyacinth, in: *Proc. IV Int. Symp. Biol. Contr. Weeds* (T.E. Freeman, ed.), pp, 207-209, University of Florida, Gainesville.

Conway, K.E. and Freeman, T.E., 1977, Host specificity of *Cercospora rodmanii*, a potential biological control of waterhyacinth, *Plant Dis. Rep.* 61: 262-266.

Conway, K.E., Freeman, T.E. and Charudattan, R., 1974, The fungal flora of the waterhyacinth in Florida, Publ. No. 30, pp. 1-11, Water Resources Research Center, University of Florida, Gainesville.

Conway, K.E., Freeman, T.E. and Charudattan, R., 1978a, Method and composition for controlling waterhyacinth, U.S. Patent 4, 097, 261, p.7.

Conway, K.E., Freeman, T.E. and Charudattan, R., 1978b, Development of *Cercospora rodmanii* as a biological control for *Eichhornia crassipes*, in: *Proc. EWRS Vth Symp. on Aquatic Weeds*, pp. 225-230, Wageningen, The Netherlands.

Cullen, J.M., 1984, Bringing the cost benefit analysis of biological control of *Chondrilla juncea* up to date, in: *Proc. VI Symp. Biol. Contr. Weeds* (E.S. Delfosse, ed.), pp. 145-152, Agriculture Canada, Ottawa.

Cullen, J.M., Kable, P.F. and Katt, M., 1973, Epidemic spread of a rust imported for biological control, *Nature* 244: 462-464.

Daniel, J.T., 1972, Biological control of northern jointvetch in rice with a newly discovered *Gloeosporium*, *MS Thesis*, University of Arkansas, Fayetteville.

Daniel, J.T., Templeton, G.E., Smith, R.J. Jr. and Fox, W.T., 1973, Biological control of northern jointvetch in rice with an endemic fungal disease, *Weed Sci.* 21: 303-307.

de Jong, M.D., 1997, New commercial bioherbicides, *IBG News* 6(1):8.

de Jong, M.D., Bulder, C.J.E.A., Weijers, C.A.G.M. and Scheepens, P.C., 1982, Myceliumproduktie van *Chondrostereum purpureum* in vloeistofcultures, *Intern. Rep.* CABO, Wageningen, p. 12.

Delfosse, E.S., 1992, The biological control regulatory process in Australia, in: *Regulation and Guidelines:*

Critical Issues in Biological Control (R. Charudattan and H.W. Browning, eds.), pp. 135-141, Proc. USDA/CSRS National Workshop, IFAS, University of Florida, Gainesville.

Delfosse, E.S. and Moorhouse, J., 1992, Biological control of Paterson's curse in Australia: After the injunction, in: *Proc. VI Int. Symp. Biol. Contr. Weeds* (E.S. Delfosse, ed.), pp. 249-292, Agriculture Canada, Ottawa.

Evans, H.C., 1987, Fungal pathogens of some subtropical and tropical weeds and the possibilities for biological control, *Biocontrol News Inf.* 8: 7-30.

Evans, H.C., 1993, Studies on the rust *Maravalia cryptostegiae,* a potential biological control agent of rubber-vine weed *Cryptostegia grandiflora* (Asclepadaceae: Periplocoideae) in Australia, I: Life Cycle, *Mycopathologia* 124: 163-174.

Evans, H.C., 1995a, Fungi as biocontrol agents of weeds: a tropical perspective, *Can. J. Bot.* 73 (Suppl. I): S58-S64.

Evans, H.C., 1995b, Pathogen-weed relationships: The practice and problems of host range screening, in: *Proc. of the VIII Int. Symp. Biol. Contr. Weeds* (E.S. Delfosse and R.R. Scott, eds.), pp. 539-551, DSIR/CSIR, Melbourne.

Evans, H.C., Carrion, G. and Ruiz-Belin, F., 1995, Mycobiota of the giant sensitive plant, *Mimosa pigra sensu lato* in the Neotropics, *Mycol. Res.* 99: 420-428.

Evans, H.C. and Fleureau, L., 1993, Studies on the rust *Maravalia cryptostegiae,* a potential biological control agent of rubber-vine weed, *Cryptostegia grandiflora* (Asclepiadaceae: Periplocoideae) in Australia, II: Infection, *Mycopathologia* 124(3): 175-184.

Evans, H.C. and Tomley, A.J., 1994, Studies on the rust *Maravalia cryptostegiae,* a potential biological control agent of rubber-vine weed *Cryptostegia grandiflora* (Asclepiadaceae: Periplocoideae) in Australia, III: Host Range, *Mycopathologia* 126: 93-108.

Feichtenberger, E., Zentmyer, G.A. and Menge, J.A., 1981, Identity of *Phytophthora* isolates from milkweed vine *(Morrenia odorata), Phytopathology* 71: 215.

Feichtenberger, E., Zentmyer, G.A. and Menge, J.A., 1984, Identity of *Phytophthora* isolated from milkweed vine, *Phytopathology* 74: 50-55.

Freeman, T.E., 1977, Biological control of aquatic plants with plant pathogens, *Aquatic Botany* 3: 145-184.

Freeman, T.E. and Charudattan, R., 1984, *Cercospora rodmanii* Conway - a biocontrol agent for waterhyacinth, *Bulletin 842 (technical),* Agr. Exp. Sta. Institute of Food and Agricultural Sciences, University of Florida, Gainesville.

Freeman, T.E., Charudattan, R., Conway, K.E., Cullen, R.E., Martyn, R.D., McKinney, D.E., Olexa, M.T. and Reese, D.F., 1981, Biological control of aquatic plants with pathogenic fungi, *Technical Report* A-81-1, pp. 1-47, U.S. Army Waterways Experiment Station, Vicksburg MS.

Freeman, T.E., Charudattan, R., Cullen, R.E. and Kenney, D.S., 1982, Biological control of waterhyacinth with the fungal pathogen *Cercospora rodmanii,* in: *Biological Control of Weeds with Plant Pathogens* (R. Charudattan and H.L. Walker, eds.), pp. 239-240, John Wiley and Sons, New York.

Freeman, T.E., Charudattan, R. and Zetter, F.W., 1973, Biological control of water weeds with plant pathogens, *Water Resources Research Publication* No. 23, University of Florida, Gainesville.

Gopal, B., 1987, *Waterhyacinth - Biology, Ecology and Management,* Elsevier Science, Amsterdam.

Greaves, M.P., Bailey, J.A. and Hargreaves, J.A., 1989, Mycoherbicides: opportunities for genetic manipulation, *Pestic. Sci.* 26: 93-101.

Hasan, S., 1988, Biocontrol of weeds with microbes, in: *Biocontrol of Plant Diseases* (K.G. Mukerji and K.L. Garg, eds.), pp. 109-151, CRC Press, Florida.

Hocart, M.J. and Peberdy, J.F., 1989, Protoplast technology and strain selection, in : *Biotechnology of Fungi for Improving Plant Growth* (J.M. Whipps and R.D. Lumsden, eds.), pp. 235-258, Cambridge University Press, Cambridge.

Hocking, P.J. and Liddle, M.J., 1986, Biology of Australian weeds: 15. *Xanthium occidentale* Bertol complex and *Xanthium spinosum* L., *J. Aust. Inst. Agric. Sci.* 52: 191.

Hodgson, R.H., Wymore, L.A., Watson, A.K., Snyder, R.H. and Collette, A., 1988, *Weed Tech.* 2: 437-480.

Holm, L.G., Plucknett, D.L., Pancho, J.V. and Herberger, J.P., 1977, *The World's Worst Weeds: Distribution and Biology,* University of Hawaii Press, Honolulu.

Jurair, A.M.M. and Khan, A., 1960, A new species of *Alternaria* on *Cassia holosericea* Fresen., *Pakistan J. Sci. Ind. Res.* 3: 71-72.

Kenney, D.S., 1986, DeVine[R] - the way it was developed: an industrialist's view, *Weed Sci.* 34(Suppl. I): 15-16.

Klein, T.A. and Auld, B.A., 1995a, Influence of spore dose and water volume on a mycoherbicide's efficacy in field trials, *Biol. Control* 5: 173-178.

Klein, T.A. and Auld, B.A., 1995b, Evaluation of Tween 20 and glycerol as additives to mycoherbicide suspension applied to Bathurst burr, *Plant Prot. Quart,* 10(1): 14-16.

Klein, T.A. and Auld, B.A., 1996, Wounding can improve efficacy of *Colletotrichum orbiculare* as a mycoherbicide for Bathurst burr, *Aust. J. Exp. Agric.* 36: 185-187.

Klein, T.A., Auld, B.A. and Fang, W., 1995, Evaluation of oil suspension emulsions of *Colletrotrichum orbiculare* as a mycoherbicide in field trials, *Crop Prot.* 14 (3):193-197.

Leggett, M.E. and Gleddie, S.C., 1995, Developing biofertilizer and biocontrol agents that meet farmers' expectations, *Advances in Plant Pathol.* 2: 59-74.

Li, Z.Y., 1985, Lubao No. 1 and its application, *Leafl. 1st Symp. Biol. Contr. Weeds,* Yangzhou, China (In Chinese).

Liu, Z.H. and Zhu, Q.R., 1980, A plant pathogen Lubao No. 1, *Shandong Sci. and Tech. Press.*

Makowski, R.M.D., 1987, The evaluation of *Malva pusilla* Sm. as a weed and its pathogen *Colletotrichum gloeosporioides* (Penz.) Sacc. f. sp. *malvae* as a bioherbicide, *Ph.D. Dissertation,* University of Saskatchewan, Saskatoon, Canada.

Makowski, R.M.D. and Morrison, I.N., 1989, The biology of Canadian weeds, 91. *Malva pusilla* Sm. *(=M. rotundifolia L.), Can. J. Plant Sci.* 69: 861-879.

Makowski, R.M.D. and Mortensen, K., 1990, *Colletotrichum gloeosporioides* f. sp. *malvae* as a bioherbicide for round-leaved mallow *(Malva pusilla):* Conditions for successful control in the field, in: *Proc. VII. Int. Symp. Biol. Contr. Weeds* (E.S. Delfosse, ed.), pp. 513-522, CSIRO Publications, Australia.

Makowski, R.M.D. and Mortensen, K., 1992, The first mycoherbicide in Canada: *Colletotrichum gloeosporioides* f. sp. *malvae* for round-leaved mallow control, in: *Proc. First Int. Weed Contr. Congr.* Vol. 2 (R.G. Richardson, ed.), pp. 298-300, Weed Science Society of Victoria Inc., Melbourne, Victoria.

Mehrotra, R.S., Aneja, K.R. and Aggarwal, A., 1977, Fungal control agents, in: *Environmentally Safe Approaches to Crop Disease Control* (N.A. Rechcigl and J.E. Rechcigl, eds.), pp. 111-137, CRC Lewis Publishers, New York.

McWhorter, G.C. and Chandler, J.M., 1982, Conventional weed control technology, in: *Biological Control of Weeds with Plant Pathogens* (R. Charudattan and H.L. Walker, eds.), pp. 5-27, John Wiley and Sons, New York.

McRae, C.F. and Auld, B.A., 1988, The influence of environmental factors on anthracnose of *Xanthium spinosum, Phytopathology* 78: 1182.

Millar, R.V., Ford, E.J. and Sands, D.C., 1989a, A nonsclerotial pathogenic mutant of *Sclerotinia sclerotiorum, Can. J. Microbiol. 35:* 517-520.

Millar, R.V., Ford, E.J., Zidack, N.J. and Sands, D.C., 1989b, A pyrimidine auxotroph of *Sclerotinia sclerotiorum* for use in biological weed control, *J. Gen. Microbiol.* 135: 2085-2091.

Morris, M., 1997, Classical biological control of weeds with plant pathogens, *IBG News,* 6(1)*:* 11.

Mortensen, K., 1988, The potential of an endemic fungus, *Colletotrichum gloeosporioides,* for biological control of round-leaved mallow *(Malva pusilla)* and velvet leaf *(Abutilon theophrasti), Weed Sci.* 36: 473-478.

Mortensen, K., and Makowski, R.M.D., 1990, Field efficacy at different concentrations of *Colletotrichum gloeosporioides* f. sp. *malvae* as bioherbicide for round-leaved mallow *(Malva pusilla), in: Proc.VII Int. Symp. Biol. Contr. Weeds* (E.S. Delfosse, ed.), pp. 523-530, CSIRO Publications, Australia.

Nag Raj, T.R. and Ponnappa, K.M., 1970, Blight of waterhyacinth caused by *Alternaria eichhorniae* sp. Nov., *Trans. Brit. Mycol. Soc.* 55(1): 123-130.

Nikandrow, A., Weidemann, G.J. and Auld, B.A., 1990, Incidence and pathogenicity of *Colletotrichum orbiculare* and a *Phomopsis* sp. on *Xanthium* spp., *Plant Dis.* 74: 796-799.

Oehrens, E., 1977, Biological control of blackberry through the introduction of rust, *Phragmidium violaceum* in Chile, *FAO Plant Protect. Bull,* 25: 26-28.

Oehrens, E.B. and Gonzales, S.M., 1974, Introduction of *Phragmidium violaceum* (Schulz.) Winter as a biological control agent for zarzamora *(Rubus constrictus* Lef. and M. and *R. ulmifolius* Schott,) (in Spanish), *Agro. Sur.* 2: 30-33.

Panetta, J.D., 1992, An industry perspective of a streamlined regulatory process for biological pest control, in: *Regulation and Guidelines: Critical Issues in Biological Control* (R. Charudattan and H.W. Browning, eds.), pp. 93-97, Proc. USDD/CSRS National Workshop, IBAS, University of Florida, Gainesville.

Pieterse, A.H. and Murphy, K.J., 1990, *Aquatic Weeds, The Ecology and Management of Nuisance Aquatic Vegetation,* Oxford University Press, New York.

Politis, D.J., Watson, A.K. and Bruckart, W.L., 1984, Susceptibility of musk thistle and other composites to *Puccinia carduorum, Phytopathology* 74: 687-691.

Prasad, R., 1993, Role of adjuvants in modifying the efficacy of a bioherbicide on forest species: Compatibility studies under laboratory conditions, *Pestic. Sci,* 38(2-3): 273-275.

Quimby, P.C. and Walker, H.C., 1982, Pathogens as mechanisms for integrated weed management, *Weed Sci.* 30(1): 30-34.

Ridings, W.H., 1986, Biological control of stranglervine in citrus - a researcher's view, *Weed Sci,* 34(1): 31-32.

Ridings, W.H., Mitchell, D.J., Schoulties, C.L. and El-Gholl, N.E., 1976, Biological control of milkweed vine in Florida citrus groves with a pathotype of *Phytophthora citrophthora,* in: *Proc. Int. Symp. Biol. Contr. Weeds* (T.E. Freeman, ed.), pp. 224-240, University of Florida, Gainesville.

Ridings, W.H., Schoulties, C.L., El-Gholl, N.E. and Mitchell, D.J., 1977, The milkweed vine pathotype of *Phytophthora citrophthora* as a biological control agent of *Morrenia odorata* in citrus groves, *Prac. Int. Soc. Citric.* 3: 877-881.

Sands, D.C., Ford, E.J. and Miller, R.V., 1990, Genetic manipulation of broad host-range fungi for biological control of weeds, *Weed Tech,* 4: 471-474.

Sands, D.C. and Miller, R.V., 1993a, Evolving strategies for biological control of weeds with plant pathogens, *Pestic. Sci.* 37: 399-403.

Sands, D.C. and Miller, R.V., 1993b, Altering the host range of mycoherbicides by genetic manipulation, in: *Pest Control with Enhanced Environmental Safety* (S.O. Duke, J.J. Menn and J.R. Plimmer, eds), pp. 101-109, ACS Symp. Ser. 524, American Chemical Society, Washington, DC.

Scheepens, P.C. and Hoogerbrugge, A., 1990, Control of *Prunus serotina* in forests with the endemic fungus *Chondrostereum purpureum,* in: *Proc. VII Int. Symp. Biol. Contr. Weeds* (E.S. Delfosse, ed.), pp. 545-551, CSIRO Publications, Australia.

Scheepens, P.C. and Van Zon, J.C.J., 1982, Microbial herbicides, in: *Microbial and Viral-Pesticides* (E. Kurstak, ed.), pp. 623-641, Marcel Dekker, New York.

Schroeder, D., 1995, Biological control of weeds and the prospects in Europe, *Mitteilungen-der-Deutschen-Geselschaft-fur-Allgemeine-und-Angewandte-Entomologie* 10(1-6): 221-226.

Shabana, Y.M., Charudattan, R. and Elwakil, M.A., 1995a, Identification, pathogenicity and host specificity of *Alternaria eichhorniae* from Egypt, a bioherbicide agent for waterhyacinth, *Biol. Control* 5: 123-135.

Shabana, Y.M., Charudattan, R. and Elwakil, M.A., 1995b, Evaluation of *Alternaria eichhorniae* as a bioherbicide for waterhyacinth *(Eichhornia crassipes)* in greenhouse trials, *Biol. Control* 5: 136-144.

Smith, R.J., Jr., 1978, Field efficacy of a fungus for control of northern jointvetch in a 3-year pilot test, *Proc. Rice Tech. Working Group* 17: 71-72.

Smith, R.J., Jr., 1986, Biological control of northern jointvetch in rice and soybeans - a researcher's view, *Weed Sci.* 34 (suppl. l): 17-23.

Smith, R.J., Jr., Daniel, J.T., Fox, W.T. and Templeton, G.E., 1973a, Distribution in Arkansas of a fungus disease used for biocontrol of northern jointvetch in rice, *Plant Dis. Rep.* 57: 695-697.

Smith, R.J., Jr., Fox, W.T., Daniel, J.T. and Templeton, G.E., 1973b, Can plant diseases be used to control weeds?, *Arkansas Farm Research* 22(4): 12.

Supkoff, D.M., Joley, D.B. and Marois, J.J., 1988, Effect of introduced biological control organisms on the density of *Chondrilla juncea* in California, *J. Appl. Ecol,* 1089-1095.

Sutton, B.C., 1980, *The Coelomycetes,* Commonwealth Mycological Institute, Kew, England.

TeBeest, D.O., 1984, Induction of tolerance to Benomyl in *Colletotrichum gloeosporioides* f. sp. *aeschynomene* by ethyl methanesulfonate, *Phytopathology* 74: 864.

TeBeest, D.O., 1997, Status of DeVine and Collego, *IBG News* 6(1): 8.

TeBeest, D.O. and Templeton, G.E., 1985, Mycoherbicides: progress in the biological control of weeds, *Plant Dis.* 69(1): 6-10.

TeBeest, D.O., Templeton, G.E. and Smith, R.J., Jr., 1978, Temperature and moisture requirements for development of anthracnose on northern jointvetch, *Phytopathology* 68: 389-393.

TeBeest, D.O., Yang, X.B. and Cisar, C.R., 1992, The status of biological control of weeds with fungal pathogens, *Ann. Rev. Phytopathol. 30:* 637-657.

Templeton, G.E., 1982, Biological herbicides: discovery, development and deployment, *Weed Sci.* 30: 430-433.

Templeton, G.E., 1986, Mycoherbicide research at the University of Arkansas - past, present and future, *Weed Sci.* 34 (Suppl. I): 35-37.

Templeton, G.E., 1990, Weed control with pathogens: future needs and directions, in: *Microbes and Microbial Products as Herbicides* (R.E. Hoagland, ed.), American Chemical Society, Washington, D.C.

Templeton, G.E., 1992, Regulatory encouragement of biological weed control with plant pathogens, in: *Regulations and Guidelines: Critical Issues in Biological Control* (R. Charudattan and H.W. Browning, eds.), pp, 61-63, Proc. USDA/CSRS National Workshop, IFAS, University of Florida, Gainesville.

Templeton, G.E. and Heiny, D.K., 1989, Improvement of fungi to enhance mycoherbicide potential, in: *Biotechnology of Fungi for Improving Plant Growth* (J.M. Whipps and R.D. Lumsden, eds.), pp. 127-152, Cambridge University, Cambridge.

Templeton, G.E., Smith, R.J., Jr., and TeBeest, D.O., 1986, Progress and potential of weed control with mycoherbicides, *Rev. Weed Sci.* 2: 1-14.

Templeton, G.E., Smith, R.J., Jr., TeBeest, D.O., Beasley, J.N. and Klerk, R.A., 1981, Field evaluation of direct fungus spores for biocontrol of curly indigo in rice and soybeans, *Arkansas Farm Research* 30 (6): 8.

Templeton, G.E., Smith, R.J. Jr., TeBeest, D.O. and Beasley, J.N., 1988, Mycoherbicides, *Arkansas Farm Research,* March-April, 1988, p.7.

Templeton, G.E., TeBeest, D.O. and Smith, R.J., Jr., 1979, Biological weed control with mycoherbicides, *Ann. Rev. Phytopathol,* 17: 301-310.

Templeton, G.E., TeBeest, D.O. and Smith, R.J., Jr., 1984, Biological weed control in rice with a strain of *Colletotrichum gloeosporioides* (Penz.) Sacc. used as mycoherbicide, *Crop Prot.* 3(4): 409-422.

Theriot, E.A. 1980, Large-scale field application of *Cercospora rodmanii,* in: *Proc. 15th Ann. Meet. Aquatic Plant Contr. Res. Plan. Operations Rev., Misc. Pap.* A-81-3, pp. 491-492, U.S. Army Waterways Experiment Station, Vicksburg, MS.

Theriot, E.A., 1982, Large-scale operations management test with insects and pathogens for the control of waterhyacinth in Louisiana, in: *Proc. 16th Ann. Meet. Aquatic Plant Contr. Res. Plan. Operations Rev. Misc. Pap.* A-81-3, pp. 187-192, U. S. Army Waterways Experiment Station, Vicksburg, MS.

Thompson, J.A., 1993, Biological control of plant pests and pathogens: alternative approaches, in: *Biotechnology in Plant Disease Control* (I. Chet, ed.), pp. 275-285, Wiley-Liss, New York.

Tomley, A.J. and Evans, B.C., 1995, Some problem weeds in tropical and sub-tropical Australia and prospects for biological control using fungal pathogens, in: *Proc. of the VIII. Int. Symp. Biol. Contr. Weeds* (E.S. Delfosse and R.R. Scott, eds.), pp. 477-482, DSIR/CSIRO Melbourne.

Trujillo, E.E., 1976, Biological control of hamakua pamakani with plant pathogens, (Abstr.) *Proc. Amer. Phytopath. Soc.* 3: 298.

Trujillo, E.E., 1984, Biological control of hamakua pamakani with *Cercosporella* sp. in Hawaii, in: *Proc. VI Int. Symp. Biol. Contr. Weeds* (E.S. Delfosse, ed.), pp. 661-671, Agriculture Canada, Ottawa.

Trujillo, E.E., 1992, Bioherbicides, in: *Frontiers in Industrial Mycology* (G.F. Leatham, ed,), pp. 196-211, Chapman and Hall, New York.

Trujillo, E.E., Aragaki, M. and Shoemaker, R.A., 1988, Infection, disease development and axenic culture of *Entyloma compositarum,* the cause of hamakua pamakani blight in Hawaii, *Plant Dis.* 72: 355-357.

Turgeon, G. and Yoder, O.C., 1985, Genetically engineered fungi for weed control, in: *Biotechnology: Applications and Research* (P.N. Cheremisinoff and R.P. Ouellete, eds.), pp. 221-230, Technomic Publ. Co. Lancaster, Pennsylvania.

Walker, H.L., 1982, A seedling blight of sicklepod caused by *Alternaria cassiae, Plant Dis.* 66: 426-428.

Walker, H.L., 1983, Biocontrol of sicklepod with *Alternaria cassiae, Proc. South. Weed Sci. Soc.* 36: 139.

Walker, H.L. and Boyette, C.D., 1985, Biocontrol of sicklepod *(Cassia obtusifolia*) in soybeans *(Glycine max)* with *Alternaria cassiae, Weed Sci.* 33: 212-215.

Walker, H.L. and Riley, J.A., 1982, Evaluation of *Alternaria cassiae* for the biocontrol of sicklepod *(Cassia obtusifolia), Weed Sci.* 30: 651-654.

Wang, R., 1990, Biological control of weeds in China: A status report, in: *Proc. VII Int. Symp. Biol. Contr. Weeds* (E.S. Delfosse, ed.), pp. 689-693, CSIRO Publications, Australia.

Watson, A.K., 1991, The classical approach with plant pathogens, in: *Microbial Control of Weeds* (D.O. TeBeest, ed.), pp. 3-23, Chapman and Hall, New York.

Wilson, C.L., 1969, Use of plant pathogens in weed control, *Ann. Rev. Phytopathol.* 7: 411-434.

Woodhead, S., 1981, Field efficacy of *Phytophthora palmivora* for control of milkweed vine, *Phytopathology* 71: 913.

Wymore, L.A., Watson, A.K. and Gotlieb, A.R., 1987, Interaction between *Colletotrichum coccodes* and thidiazuron for control of velvetleaf *(Abutilon theophrasti), Weed Sci.* 35: 377-382.

Yang, X.B. and TeBeest, D.O., 1992, The stability of host-pathogen interactions of plant disease in relation to biological weed control, *Biol. Control* 2: 266-271.

Yoder, O.C., Turgeon, B.G., Ciuffetti, L.M. and Schafer, W., 1989, Genetic analysis of toxin production by fungi, in: *Phytotoxins and Plant Pathogenesis* (A. Graniti, R.D. Durbin and A. Ballio, eds.), pp. 43-60, Springer-Verlag, Inc. Berlin.

BIOTECHNOLOGICAL APPLICATIONS OF THERMOPHIILIC FUNGI IN MUSHROOM COMPOST PREPARATION

Raj Kumar and K.R. Aneja

Department of Botany
Kurukshetra University
Kurukshetra-136 119
India

INTRODUCTION

Fungi once thought of as killing or damaging organisms are utilised by man in many ways and their uses are expanding rapidly. The Egyptians and Romans prized mushrooms and there are records of their being eaten in China between 25 BC and AD 220 (Wang, 1985). The use of large fungi or mushrooms as food is quite commonplace, and has traditionally been associated with meats. In Malawi they are regarded as a meat analogue (Morris, 1984). This view was also held by Francis Bacon who, in his *Sylva Sylvarum* of 1927 described mushrooms as yielding 'so delicious a meat'. But more recently filamentous fungi have found their way in the industrial development of various kinds of mycoprotein and/or to supplement various substrates. The development of 'Quorn', a mycoprotein produced from *Fusarium graminearum,* is a success story and now festoons the shelves of British supermarkets (Trinci, 1992).

During the last fifteen years several researchers have tried to utilise the filamentous fungi, particularly thermophilic strains, for the development of environmentally-friendly compost that does not emit odours into the environment. With the expansion of suburban populations encroaching upon formerly rural areas, more people are being exposed to mushroom compost preparation facilities. Recently the general public has been unwilling to tolerate odours from substrate preparation and these odours are becoming a serious problem for composting operations (Beyer et al., 1997). Presently the mushroom composting process is mostly based on Sinden's work, who cut phase I down to about a week (exclusive of the collecting and wetting of horse manure) followed by a short phase II of about six days in trays or tunnels (Sinden and Hauser, 1953). This is well known as a short method of composting and is used by most of the mushroom growers in Europe, and developing countries like India and China are also nurturing their mushroom industry with the short method of composting.

Mushroom compost production is based on a mixture of biological, chemical, physical and ecological parameters which are not exactly defined. Traditionally compost is produced from wheat straw, straw-bedded horse manure, chicken manure and gypsum. After mixing and moistening these ingredients for about 2 to 3 days, the mixture is subjected to the phase I composting process. The mixed ingredients are stacked in windrows in the open air for an uncontrolled self-heating process for up to one week. During self-heating, temperatures in the windrows range from ambient to 80°C and ammonia and foul-smelling compounds are emitted, causing environmental problems. Phase II is essentially an aerobic process carried out by maintaining the compost temperature at 45°C for 6 days in shallow layers in mushroom houses or tunnels. This is preceded by pasteurisation for 8 hours at 70°C to prevent outbreak of pests and moulds during phase II (Straatsma et al., 1995b). The biotechnological application of thermophilic fungi in composting stems from (i) their ability to hydrolyse plant polymers, thereby hastening the process of decomposition, (ii) their role in the production of nutritionally rich compost to increase the quality and yield of mushrooms, and (iii) their suitability as agents of experimental systems for genetic manipulation for use in recombinant DNA technology.

In this paper, bioconversion, ecology of thermophilic fungi in mushroom compost, indoor phase II composting and growth promotion of *Agaricus bisporus* (= *A. brunnescens*) by thermophilic fungi are reviewed and the prospects for future study are identified.

BIOCONVERSION

The most important basic material for the production of compost is various kinds of straw, for example wheat, rye, barley and oat. However, wheat and rye are preferred because they are firmer and give better texture (Gerrits, 1985). The main function of straw is to provide a reservoir of cellulose, hemicellulose and lignin which is utilised by the mushroom mycelium as the carbon source during its growth. During composting a great deal of organic matter is broken down. Readily accessible compounds are degraded first and recalcitrant materials such as inorganic compounds, cellulose, lignin (Waksman and Nissen, 1932) and biomass (Derikx et al., 1990) accumulate. Cellulose is a linear molecule composed of repeating cellobiose units held together by ß-glycosidic linkages. Lignin is a complex phenylpropanoid polymer that surrounds and strengthens the cellulose-hemicellulose framework (Kringstad and Lindstrom, 1984). It is believed that hemicellulose acts as a molecular bonding agent between the cellulose and lignin fractions (Torrie, 1991). The optimal temperature of 50°C for the mineralization rate during phase I coincides with cellulase production by bacteria and fungi (Gilbert and Hazelwood, 1993). Cellulases are mixtures of several enzymes that act in concert to hydrolyse crystalline cellulose to its monomeric component glucose.

The exact mechanism of cellulose hydrolysis is not known (Duff and Murray, 1996), although a number of possible models have been proposed. The classic and perhaps simplest one is presented in Figure 1. The mechanism of cellulose hydrolysis is envisioned as an initial attack by endocellulases that cleave internal ß-1,4 linkages in amorphous sections of cellulose. Exocellulases (CBH I and CBH II in *Trichoderma*) then cleave cellobiose units from the non-reducing ends of the cellulose chain. Finally ß-glucosidase converts cellobiose to glucose monomers (Goyal, Ghosh and Eveleigh, 1991; Walker and Wilson, 1991). These enzymes act in a synergistic or co-operative manner.

In general, composting is considered to be an aerobic process (Finstein and Morris, 1975) suggesting high biological activity. The oxygen consumption rates of compost during phase I increase logarithmically with temperature from 20 to 70°C (Schulze, 1962). Derikx et al. (1990) observed a sharp decline in the oxygen consumption rates above 60°C, demonstrating the inhibitory effect of elevated temperatures on biological activity. They also observed that mineralization dropped to zero at 70°C and the oxygen consumption rate decreased to 25% of its original value. High biological activity is characterised by the production of stench, hydrogen sulfide, methane thiol, dimethyl sulfide and dimethyl trisulfide. Anaerobic conditions at elevated temperatures favour the formation of carbon disulfide and dimethyl disulfide indicating their non-biological formation which cause much of the environmental problem. Their formation can be decreased to 75% if phase I is performed in tunnels because of the aerobic environment provided by the forced ventilation (Derikx et al., 1991).

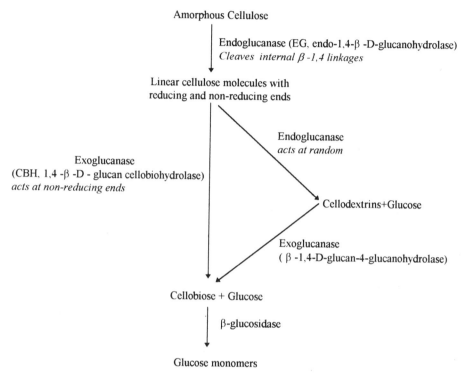

Figure 1. Scheme of breakdown of cellulose in lignocellulosic materials [Based on Chahal and Overend (1982) and Duff and Murray (1996)].

During the composting process, thermophilic/thermotolerant fungi show a whole spectrum of physiological behaviour with regard to their ability to decompose plant polymers. Evidently some quite complex interspecific relationships exist in this respect (Chang, 1967). Some can hydrolyse cellulose present in the straw viz. *Chaetomium thermophile, Humicola insolens, Aspergillus fumigatus* and *Torula thermophila* (Flannigan and Sellars, 1972). A few fungi such as *Talaromyces thermophilus* may be weakly lignolytic (Tansey et al., 1977). Sharma (1991) found no lignin degradation during phase I and phase II, however, he reported a decrease in the lignin content of the spent compost. Bonnen, Anton and Orth (1994) convincingly demonstrated the ability of *A. bisporus* to produce lignin-degrading peroxidases. In contrast, Iiyama, Stone and Macauley (1994) reported a relative increase in the lignin content of compost during composting and cultivation of mushrooms. They found that the structure of the lignin was altered rather than degraded. Till (1962) showed that good yields of mushrooms can be obtained on autoclaved straw supplemented with organic nitrogen. Extensive degradation of straw is not a prerequisite for a high yield of mushrooms. Tunnel phase I compost produces the same amount of mushrooms as traditional outdoor compost that has been degraded much more intensively (Straatsma et al., 1995b).

ECOLOGY OF THERMOPHILIC FUNGI IN MUSHROOM COMPOST

It has been clearly demonstrated that the rise in temperature and the decomposition of composting plant materials is brought about by thermophilic microorganisms including fungi. Successions of fungi in wheat straw compost have been studied by Chang and Hudson (1967). Their results are probably typical for any compost. Thermophilic microorganisms in compost have received extensive study (Fermor, Randle and Smith, 1985). An overview of the thermophilic fungi found in mushroom compost and their temperature optima are presented in Table 1. The course of fungal succession may be partially explained by the ecophysiological data available (Aneja, 1988; Chapman, 1974; Evans, 1971; Fergus and Amelung, 1971; Rosenberg, 1975, 1978; Satyanarayana et al., 1992).

It is well established that the ability to use complex carbon sources and the ability to thrive at high temperatures are the two most important characteristics of the successful colonisers of composts. On wheat straw composts *Rhizomucor pusillus* disappears early in the succession and seems to be a primary sugar fungus. *Thermomyces lanuginosus* persists as a secondary sugar fungus in mutualistic relationships with some of the true cellulose decomposers of composts (Chang, 1967; Deacon, 1985; Hedger and Hudson, 1974). *Chaetomium thermophile*, *Malbranchea sulfurea* and *Scytalidium thermophilum* (=*T. thermophila*) grow fast and are cellulolytic. The latter is reported to be the climax species in composts (Straatsma and Samson, 1993). The only time that fungi are not active in the compost is during the peak-heating phase during outdoor composting. The maximum temperature phase kills off all fungi in the centre and allows recolonization as the temperature falls to below each fungus's upper temperature limit for growth (Chang and Hudson, 1967). In general one might suspect that the higher the maximum temperature for growth the more rapid the recolonization. From the data of Cooney and Emerson (1964), Kumar (1996), Rosenberg (1975), Singh and Sandhu (1982) and Tansey and Brock (1972), there are only a few degrees difference in the upper temperature limits of the various thermophiles: all lie between 55 and 60°C.

Willenborg and Hindorf (1985) studied the fungal flora in mushroom culture substrate from the beginning of composting to the final stage of mushroom picking. They observed that 60.7% of the total flora were thermophilic fungi. Eighteen thermophilic fungi including 2 basidiomycetes and 1 sterile form from phase I and phase II composts were isolated and these represented most of the known thermophilic taxa (Table 1). Most species that colonised/appeared in the earlier stages disappeared after phase II composting. Fungi recovered from the compost at the end of phase II were exclusively *T. thermophila*, *H. insolens* and *C. thermophile*. However, the population density of *T. thermophila* was very high, with $10^{4.2}$ CFU g^{-1} (Kumar, 1996).

INDOOR PHASE II COMPOSTING

In practice two phases of composting are distinguished. Phase I is performed in the open or under a roof where the ingredients are stacked in long heaps. A transverse section of a compost pile is given in Figure 2. The material heats up and is turned several times. Processing of phase I in tunnels is presently under investigation and its acceptability is debated (Gerrits et al., 1995; Straatsma et al., 1995a). Phase II composting is essentially performed in tunnels or mushroom houses which allow satisfactory control of environmental conditions, viz. temperature and air supply. The tunnels are used for large scale operations. Before filling the 'young compost' (any substrate prior to phase II) in tunnels, it is subjected to phase I (see Figure 3) to soften the straw which allows an optimal amount of compost to be filled into the cropping rooms. This is also essential for high biological efficiency (BE) of the substrate

Table 1. Thermophilic and thermotolerant fungi found in mushroom compost[a] and their optimum temperature for growth.

Fungi	Optimum temperature (°C)	Reference(s)[b]
Thermophilic		
Chaetomium thermophile[1,6]	50	La Touche, 1950
Corynascus thermophilus[4]	45	Sigler et al., 1998
Hormographiella aspergillata[7]	NR	
Humicola insolens[6]	40-45	Cooney & Emerson, 1964; Rosenberg, 1975
Malbranchea sulfurea[1,6]	45	Cooney & Emerson, 1964; Rosenberg, 1975
Myriococcum albomyces[6]	37-42	Cooney & Emerson, 1964; Rosenberg, 1975
M. thermophilum[7]	NR	
Rhizomucor miehei[2,6]	35-45	Cooney & Emerson 1964; Rosenberg, 1975
R. pusillus[1,5,6]	35-45	Cooney & Emerson, 1964; Rosenberg, 1975
Scytalidium thermophilum[1,3,5,6] (= Torula thermophila)	45	Cooney & Emerson, 1964
Stilbella thermophila[3,5,6]	50	Rosenberg, 1975
Talaromyces emersonii[6,7]	40-45	Rosenberg, 1975
T. thermophilus[1,6]	45-50	Rosenberg, 1975
Thermoascus aurantiacus[6,7]	40-45	Cooney & Emerson, 1964
T. aurantiacus var. levispora[7]	NR	
T. crustaceus[7]	40	Stolk, 1965
Thermomyces lanuginosus[1,3,5,6]	45-50	Rosenberg, 1975
Thielavia terrestris[7]	45	Anonymous, 1992
Chaetomium sp.[7]	NR	
Thermotolerant		
Absidia corymbifera[1,6]	45	Kumar, 1996
Aspergillus fumigatus[1,3,5,6]	40	Cooney & Emerson 1964; Rosenberg, 1975
Coprinus cinereus[7]	26-37	Anonymous, 1992
Emericella nidulans[6,7]	35-37	Anonymous, 1992
Paecilomyces variotii[7]	23	Anonymous, 1992

[a]data from: 1. Chang, 1967; 2. Eicker, 1977; 3. Fergus, 1964; 4. Fergus and Sinden, 1969; 5. Hays, 1969; 6. Kumar, 1996; 7.Straatsma et al., 1994a.

[b]reference(s) for optimum temperature data; NR = Not reported.

(Gerrits et al., 1994). BE is the per cent fresh weight of mushrooms produced from a given dry weight of compost ingredients; it is an indication of the efficiency underlying the bioconversion process that transforms straw and supplements into mushrooms (Leonard and Volk, 1992).

Young compost is filled in tunnels and pasteurised at 70°C for 8 hours before processing at 45°C for 5 to 6 days. The dimension of a typical tunnel is 100 m² to fill compost at 2 m height or 1 tm^{-2} (Van Lier et al., 1994). For inoculating the compost with a thermophilic isolate, the compost is cooled to 40°C and inoculated at filling in the second tunnel. Ventilative heat management is applied to the compost in the tunnel to keep the temperature within the range of 45 to 50°C. Typically, the total circulation is 200 m³ t^{-1} h^{-1} and ventilation with fresh ambient air is about 20 m³ t^{-1} h^{-1} (Van Lier et al., 1994). For oxidation of the substrate only 2 m^3 t^{-1} h^{-1} would be required, therefore most of the ventilation air is used to remove excess water to control temperature and not to supply oxygen. The loss of water and high process temperature is important for a good quality compost. Air circulation rates below 100 m^3 t^{-1} h^{-1} result in lower water loss from the substrate, thus making it vulnerable to desiccation especially in the bottom section of the tunnel (Van Lier et al., 1994). Processing of the compost is continued until volatile NH_3 is undetectable (< 10 ppm). After processing the compost is cooled and inoculated with *A. bisporus* spawn.

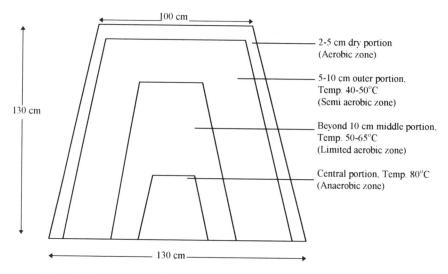

Figure 2. A transverse section of compost pile (Modified from Chahal et al., 1976).

The processing of compost has been rationalised during the last two decades. In general, tunnel processing is advantageous for bulk treatment of compost because it is cost effective; secondarily it provides the opportunity to clean the exhaust air of ammonia and stench. Ammonia can easily be removed by acid washing and stench caused by sulphur-containing organics may be removed by biofiltration or sodium hypochlorite washing (Op den Camp et al., 1992).

GROWTH PROMOTION OF *AGARICUS BISPORUS* BY THERMOPHIILIC FUNGI

In the last decade a new driving force has entered into the field of composting which is associated with the growth promotion of *A. bisporus* mycelium by thermophilic fungi, in particular *S. thermophilum* (Straatsma et al., 1995a). The inoculation of thermophilic fungi has shown that compost colonisation by selected isolates is successful and that microbial manipulation of phase II composting is feasible (Kumar, 1996; Straatsma et al., 1994b).

The radial growth rate of mushroom mycelium never exceeds 3 mm per day (Last, Hollings and Stone, 1974). Unfortunately a growth-promoting effect of *S. thermophilum* on the mycelium of *A. bisporus* is not found on agar media (Renard and Cailleux, 1973). The high hyphal extension rates of *A. bisporus* on compost in the presence of thermophilic fungi may have an ecological significance: it may be able to grow as fast as possible, thereby colonising as much substrate as possible. Once the substrate has been occupied, the mushroom mycelium seems to be able to prevent occupation by other microorganisms, either by consuming them (Fermor and Grant, 1985; Fermor and Wood, 1981) or by excretion of carbon monoxide (Stoller, 1978), which effectively inhibits growth of most competing organisms but inhibits the growth of the mushroom mycelium itself only partly (Derikx et al., 1990).

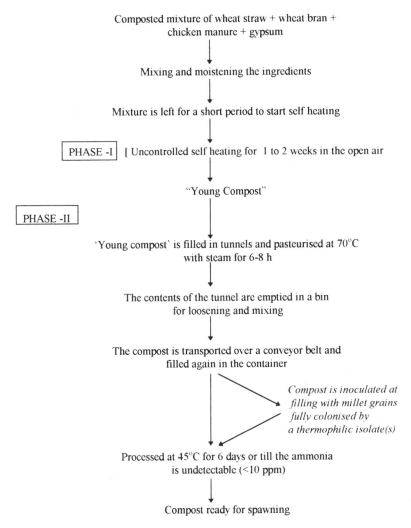

Figure 3. Protocol for the environmentally controlled production of compost inoculated with thermophilic isolate(s).

Our studies of the effect of thermophilic fungi on the mycelial extension rate of *A. bisporus* when inoculated in sterilised compost singly and in combinations (Table 2) have indicated that thermophilic fungi, in particular *T. thermophila* (Kurukshetra isolate, IMI No. 361370) and *M. sulfurea* (Kurukshetra isolate, IMI No. 361367), provide a trigger for enhanced growth of *A. bisporus* acting by an unknown mechanism. Therefore, combinations of fungal inocula are better than inoculating the compost with a single fungus. Our studies substantiated the results of Overstijns (1995) who reported that the addition of phase II compost, which contains all necessary thermophilic microorganisms, seemed to be better than an inoculum of a pure culture of *S. thermophilum.* Ross and Harris (1983) suspected that a viable but dormant biomass should be present to fill an otherwise biological vacuum. This prevents colonisation by unwanted competitors.

Table 2. Growth rates of *A. bisporus* on sterilised compost inoculated with different thermophilic fungi singly and in combinations.

Species	Mycelial extension rate (Kr, mm/day)[a]
Control	6. 1e
Chaetomium thermophile	6.3d
Malbranchea sulfurea	7.1b
Thermomyces lanuginosus	5.4g
Torula thermophila	6.6c
C. thermophile + M. sulfurea	7. 1b
C. thermophile + T. lanuginosus	6.0e
C. thermophile + T. thermophila	6.6c
M. sulfurea + T. lanuginosus	5.8f
M. sulfurea + T. thermophila	7.7a
T. lanuginosus + T. thermophila	6.0e
C. thermophile + M. sulfurea + T. lanuginosus + T. thermophila	6.5c

CV= 1.5%
[a]Treatments receiving the same letter are not significantly different (DMRT; P < 0.05).

Wiegant et al. (1992) suggested that CO_2 produced by *S. thermophilum* at 0.4 to 0.5% CO_2 (v/v) explained the growth-promoting effect. Heterotrophic CO_2 fixation in primary metabolism of *A. bisporus* is well known (Bachofen and Rast, 1968; Le Roux and Couvy, 1972) and established mycelia could maintain growth by the production of their own respiratory CO_2. Straatsma et al. (1995a) interpreted that CO_2 level influences the duration of the adaptation period rather than the extension rate. They showed that at optimal CO_2 level the mycelial extension rate of *A. bisporus* on compost fully grown with S. *thermophilum* was twice that of *A. bisporus* on compost without *S. thermophilum*. Therefore, the effects of thermophilic fungi and CO_2 seemed to be distinct.

CONCLUSIONS AND FUTURE PROSPECTS

Since conventional composting is coupled with emission of ammonia and annoying odours, the search for environmentally-friendly composting techniques is of great interest. Indoor composting research is now being targeted towards the microbial manipulation of phase II compost. Thermophilic fungi are believed to contribute significantly to the quality of compost. Thermophilic fungi in phase II compost are believed to contribute to a good crop of mushrooms in the following ways: (a) by decreasing the concentration of ammonia in the compost which would otherwise counteract the growth of the mushroom mycelium; (b) by immobilising nutrients in a form apparently available to the mushroom mycelium; and (c) possibly by having a growth-promoting effect on the mushroom mycelium, as has been demonstrated for *S. thermophilum* and several other thermophilic fungi (Wiegant, 1992). However, mechanisms behind the growth stimulation of *A. bisporus* by these fungi remain unresolved. It is advisable to perform phase I as a natural semi-controlled composting process for a good crop yield of mushrooms (Overstijns, 1995).

However, because of the growing concern about the environmental impact of production of compost and pressure of environmentalists, it will be necessary to reduce the length of phase I. By using tunnels it is now already possible to drop phase I outdoors completely (Gerrits et al., 1995). There are prospects of using spent compost for bioremediation purposes (Buswell, 1994) and as an organic soil stabiliser (Gerrits, 1994). Further research on indoor phase I composting and biodegradation of lignocellulosics is needed. Although *A. bisporus is* capable of degrading the lignin in compost, the specific components of the enzyme system utilised by this fungus during the biodegradation of lignin are not known with certainty. Bonnen et al. (1994) found that the activities of laccase and of manganese peroxidase of *A. bisporus* in compost are related to the degradation of compost lignin.

What is presently needed in the mushroom industry is to render the compost preparation procedure more rational and controllable by identifying the important factors that affect yield and quality. It is hoped that recombinant DNA technologies may provide an impetus to the study of enzymes involved in bioconversion and carbon metabolism. Recently, the cellulase and laccase genes of *A. bisporus* have been cloned (Perry et al., 1993; Wood and Thurston, 1991) and Schaap et al. (1994) cloned some housekeeping genes of carbon metabolism. Our knowledge of mechanisms underlying straw and compost degradation before cropping will be increased rapidly if *in situ* microscopic studies in straw and compost can be performed using gene probes for *S. thermophilum* including other thermophilic fungi and *A. bisporus*. Such work requires a multidisciplinary approach with input from disciplines such as mycology, mushroom science, molecular biology, biotechnology and biochemistry.

REFERENCES

Aneja, K.R., 1988, Biology of plant litter decomposition - fungi as decomposers of plant litter, in: *Perspectives in Mycology and Plant Pathology* (V.P. Agnihotri, A.K. Sarbhoy and D. Kumar, eds.), pp. 387-417, Malhotra Publishing House, New Delhi.

Anonymous, 1992, *Catalogue of the Culture Collection,* 10th ed., International Mycological Institute, Egham, Surrey, U.K.

Bachofen, R. and Rast, D., 1968, Carboxylierungsreaktionen in *Agaricus bisporus* III. Pyruvat und Phosphoenolpyruvat als CO_2-Acceptoren, *Arch. Mikrobiol.* 60: 217-234.

Beyer, D.M., Heinemann, P., Labance, S. and Rhoades, T., 1997, The effect of covering compost piles with microporous membrane on mushroom substrate preparation process and fresh mushroom yield, *Mushroom News* 45: 14-20.

Bonnen, A.M., Anton, L.H. and Orth, A.B., 1994, Lignin degrading enzymes of the commercial button mushroom, *Agaricus bisporus, Appl.Env. Microbiol.* 60: 960-965.

Buswell, J.A., 1994, Potential of spent mushroom substrate for bioremediation purposes, *Compost Sci. Util.* 2: 31-36.

Chahal, D.S. and Overend, R.P., 1982, Ethanol fuel from biomass, in: *Advances in Agricultural Microbiology* (N.S. Subba Rao, ed.), pp. 581-641, Oxford and IBH Publishing Co., New Delhi.

Chahal, D.S., Sekhon, A. & Dhaliwal, B.S. 1976, Degradation of wheat straw by the fungi isolated from synthetic mushroom compost, in: *Proc. 3rd International Biodegradation Symposium* (J.M. Shenpley and AM. Kaplan, eds.), pp. 665-671, Applied Science Publishers, London.

Chang, Y., 1967, Fungi of wheat straw compost II. Biochemical and physiological studies, *Trans. Br. Mycol. Soc.* 50: 667-677.

Chang, Y. and Hudson, H.J., 1967, The fungi of wheat straw compost I. Ecological studies, *Trans. Br. Mycol. Soc.* 50: 649-666.

Chapman, E.S., 1974, Effect of temperature on growth rate of seven thermophilic fungi, *Mycologia* 66: 542-546.

Cooney, D.G. and Emerson, R., 1964, *Thermophilic Fungi: An Account of their Biology, Activities and Classification,* W.H. Freeman and Co., San Francisco and London.

Deacon, J.W., 1985, Decomposition of filter paper cellulose by thermophilic fungi acting singly, in combination and in sequence, *Trans. Brit. Mycol. Soc.* 85: 663-669.

Derikx, P.J.L., Op den Camp, H.J.M, Van der Drift, C., Van Griensven L.J.L.D. and Vogels, G.D., 1990, Biomass and biological activity during the production of compost used as a substrate in mushroom cultivation, *Appl.*

Env. Microbiol. 56: 3029-3034.

Derikx, P.J.L., Simons, F.H.M., Op den Camp, H.J.M., Van der Drift, C., Van Griensven L.J.L.D. and Vogels, G.D., 1991, Evolution of volatile sulfur compounds during laboratory-scale incubations and indoor preparation of compost used as a substrate in mushroom cultivation, *Appl. Env. Microbiol.* 57: 563-567.

Duff, S.J.B. and Murray, W.D., 1996, Bioconversion of forest products industry waste cellulosics to fuel ethanol: A review, *Biores. Technol. 55:* 1-33.

Eicker, A., 1977, Thermophilic fungi associated with the cultivation of *Agaricus bisporus, J. S. Afr. Bot.* 43: 193-207.

Evans, H.C., 1971, Thermophilous fungi of coal spoil tips II. Occurrence, distribution and temperature relationships, *Trans. Brit. Mycol. Soc.* 57: 255-266.

Fergus, C.L., 1964, Thermophilic and thermotolerant molds and actinomycetes of mushroom compost during peak heating, *Mycologia* 56: 267-284.

Fergus, C.L. and Amelung, R.M., 1971, The heat resistance of some thermophilic fungi on mushroom compost, *Mycologia* 63: 675-679.

Fergus, C.L, and Sinden, J.W., 1969, A new thermophilic fungus from mushroom compost: *Thielavia thermophila* sp. nov., *Can. J. Bat.* 47: 1635-1637.

Fermor, T.R. and Wood, D.A.,1981, Degradation of bacteria by *Agaricus bisporus and* other fungi, *J. Gen. Microbiol.* 126: 377-387.

Fermor, T.R. and Grant, W.D., 1985, Degradation of fungal and actinomycete mycelia by *Agaricus bisporus, J. Gen. Microbiol.* 131: 1729-1734.

Fermor, T.R., Randle, P.E. and Smith, J.F., 1985, Compost as a substrate and its preparation, in: *The Biology and Technology of the Cultivated Mushroom* (P.B. Flegg, D.M. Spencer and D.A. Wood, eds.), pp. 81-109, John Wiley and Sons, Chichester, United Kingdom.

Finstein, M.S. and Morris, M.L., 1975, Microbiology of municipal solid waste composting, *Adv. Appl. Microbiol.* 19: 113-151.

Flannigan, B. and Sellars, P.N., 1972, Activities of thermophilous fungi from barley kernels against arabinoxylan and carboxymethylcellulose, *Trans. Brit. Mycol. Soc.*58: 338-341.

Gerrits, J.P.G., 1985, Developments in composting in the Netherlands, *Mushroom J.* 146: 45-53.

Gerrits, J.P.G., 1994, Composition, use and legislation of spent mushroom substrate in the Netherlands, *Compost Sci. Util,* 2: 24-30.

Gerrits, J.P.G., Amsing, J.G.M., Straatsma, G. and Van Greinsven, L.J.L.D., 1994, Indoor Compost: fase I processen van 5 dagen in tunnels, *Champignoncultuur* 38: 338-345.

Gerrits, J.P.G., Amsing, J.G.M., Straatsma, G. and Van Greinsven, L.J.L.D., 1995, Phase I process in tunnels for the production of *Agaricus bisporus* compost with special reference to the importance of water, in*: Science and Cultivation of Fungi* (T.J. Elliot, ed.), pp. 203-211, Balkema, Rotterdam.

Gilbert, H.J. and Hazelwood, G.P., 1993, Bacterial cellulases and xylanases, *J. Gen. Microbiol.* 139: 187-194.

Goyal, A., Ghosh, B. and Eveleigh, D., 1991, Characteristics of fungal cellulases, *Biores. Technol.* 36: 37-50.

Hays, W.A., 1969, Microbiological changes in composting wheat straw/horse manure mixtures, *Mushroom Sci.* 7: 173-186.

Hedger, J.N. and Hudson, H.J., 1974, Nutritional studies of *Thermomyces lanuginosus* from wheat straw compost, *Trans. Brit. Mycol. Soc.* 62: 129-143.

Iiyama, K., Stone, B.A. and Macauley, B.J., 1994, Compositional changes in compost during composting and growth of *Agaricus bisporus, Appl. Env. Microbiol.* 60: 1538-1546.

Kringstad, K.P. and Lindstrom, K., 1984, Spent liquors from pulp bleaching, *Env. Sci. Technol.* 18: 236A-248A.

Kumar, R., 1996, Taxophysiological studies on thermophilous fungi from north Indian soils, *Ph.D. Thesis,* Kurukshetra University.

Last, F.T., Hollings, M. and Stone, O.M., 1974, Effects of cultural conditions on the mycelial growth of healthy and virus-infected cultivated mushroom, *Agaricus bisporus, Annals Appl. Biol.* 76: 99-111.

La Touche, C.J., 1950, On thermophilic species of *Chaetomium, Trans. Brit. Mycol. Soc.* 33: 94-104.

Leonard, T.J. and Volk, T.J., 1992, Production of speciality mushrooms in north America: Shitake and Morels, in: *Frontiers in Industrial Mycology* (G.F. Leatham, ed.), pp. 1-23, Chapman and Hall, New York.

Le Roux, P. and Couvy, J., 1972, Fixation et metabolisme du gaz carbonique dans le mycelium et le cordon mycelien d'*Agaricus bisporus, Mushroom Sci.* 8: 641-646.

Morris, B., 1984, Macrofungi of Malawi: some ethno-botanical notes, *Bulletin of the British Mycological Society* 49-57.

Op den Camp, H.J.M., Pot, A., Van Griensven, L.J.L.D. and Gerrits, J.P.G., 1992, Stankproduktie tijdens 'Indoor Verse Compostbereiding' (IVC) en het effect van luchtbehandeling met een luchtwasser, *Champignoncultuur* 36: 319-325.

Overstijns, A., 1995, Indoor composting, *Mushroom News* 43: 16-26.

Perry, C.R., Matcham, S.E., Wood, D.A. and Thurston, C.F., 1993, The structure of laccase protein and its synthesis by the commercial mushroom *Agaricus bisporus, J. Gen. Microbiol.* 139: 171-178.

Renard, Y. and Cailleux, R., 1973, Contribution a l'etude des microorganismes du compost destiné a la culture

124

du champignon de couche, *Revue de Mycologie* 37: 36-47.

Rosenberg, S.L., 1975, Temperature and pH optima for 21 species of thermophilic and thermotolerant fungi, *Can. J. Microbiol.* 21: 1535-1540.

Rosenberg, S.L., 1978, Cellulose and lignocellulose degradation by thermophilic and thermotolerant fungi, *Mycologia* 70: 1-13.

Ross, R.C. and Harris, P.J., 1983, The significance of thermophilic fungi in mushroom compost preparation, *Sci. Hortic.* 20: 61-70.

Satyanarayana, T., Johri, B.N. and Klein, J., 1992, Biotechnological potential of thermophilic fungi, in: *Handbook of Applied Mycology Vol. 4* (D.K. Arora, R.P. Elander and K.G. Mukerji, eds.), pp. 729-761, Marcel Dekker, New York.

Schaap, P.J., Van der Vlugt, R.A.A., De Greet, P.W.J., Mueller, Y., Van Griensven, L.J.L.D. and Visser, J., 1994, Strategies for cloning of housekeeping genes of *Agaricus bisporus,* in: *Fifth International Mycological Congress* (August 14-21, 1994), Vancouver, B.C, p. 190 (Abstr.).

Schulze, K.L., 1962, Continuous thermophilic composting, *Appl. Microbiol.* 10: 108-122.

Sharma, H.S.S., 1991, Biochemical and thermal analyses of mushroom compost during preparation, *Mushroom Sci.* 13: 169-179.

Sigler, L., Aneja, K.R., Kumar, R., Maheshwari, R. and Shukla, R.V., 1998, New records from India and redescription of *Corynascus thermophilus* and its anamorph *Myceliophthora fergusii, Mycotaxon* (in press).

Sinden, J.W. and Hauser, E., 1953, The nature of composting process and its relation to short composting, *Mushroom Sci.* 2: 123-130.

Singh, S. and Sandhu, D.K., 1982, Growth response of some thermophilous fungi at different incubation temperatures, *Proc. Indian Acad. Sci. (Plant Sci.)* 91: 153-158.

Stolk, A.C., 1965, Thermophilic species of *Talaromyces* Benjamin and *Thermoascus* Miehe, *Antonie van Leeuwenhoek,* 31: 262-276.

Stoller, B.B., 1978, Detection and evaluation of carbon monoxide, ethylene, and oxidants in mushroom beds, *Mushroom Sci.* 10: 445-449.

Straatsma, G. and Samson, R.A., 1993, Taxonomy of *Scytalidium thermophilum,* an important thermophilic fungus in mushroom compost, *Mycol. Res.* 97: 321-328.

Straatsma, G., Samson, R.A., Olijnsma, T.W., Op den Camp, H.J.M., Gerrits, J.P.G. and Van Griensven, L.J.L.D., 1994a, Ecology of thermophilic fungi in mushroom compost, with emphasis on *Scytalidium thermophilum* and growth stimulation of *Agaricus bisporus* mycelium, *Appl. Env. Microbiol.* 60: 454-458.

Straatsma, G., Olijnsma, T.W., Gerrits, J.P.G., Amsing, J,G.M., Op den Camp, H.J.M. and Van Griensven, L.J.L.D., 1994b, Inoculation of *Scytalidium thermophilum* in Button Mushroom compost and its effect on yield, *Appl. Env. Microbiol.* 60: 3049-3054.

Straatsma, G., Olijnsma, T.W., Van Griensven, L.J.L.D. and Op den Camp, H.J.M., 1995a, Growth promotion of *Agaricus bisporus* mycelium by *Scytalidium thermophilum* and CO_2 in: *Science and Cultivation of Fungi* (T.J. Elliott, ed.), pp. 289-291, Balkema, Rotterdam.

Straatsma, G., Samson, R.A., Olijnsma, T.W., Gerrits, J.P.G., Op den Camp, H.J.M. and Van Griensven, L.J.L.D., 1995b, Bioconversion of cereal straw into mushroom compost, *Can. J. Bot.* 73 (Suppl. l): S1019-SIO24.

Tansey, M.R. and Brock, T.D., 1972, The upper temperature limits for eukaryotic organisms, *Proc. Nat. Acad. Sci.* (USA) 69: 2426-2428.

Tansey, M.R., Murrmann, D.N., Behnke, B.K. and Behnke, E., 1977, Enrichment, isolation, and assay of growth of thermophilic and thermotolerant fungi in lignin containing media, *Mycologia* 69: 463-476.

Till, O., 1962, Champignonkultur auf sterilisiertem Nahrsubstrat und die Wiederverwendung von abgetragenem Kompost, *Mushroom Sci.* 5: 127-133.

Torrie, J., 1991, Extracellular ß-D-mannanase activity from *Trichoderma harzianum* E58, *Ph.D. Thesis,* University of Ottawa.

Trinci, A.P.J., 1992, Myco-protein: A twenty-year overnight success story, *Mycol. Res.* 96: 1-13.

Van Lier, J.J.C., Van Ginkel, J.T., Straatsma, G., Gerrits, J.P.G. and Van Griensven, L.J.L.D., 1994, Composting of mushroom substrate in a fermentation tunnel: compost parameters and a mathematical model, *Neth. J. Agric. Sci.* 42: 271-292.

Waksman, S.A. and Nissen, W., 1932, On the nutrition of the cultivated mushroom, *Agaricus campestris,* and the chemical changes brought about by this organism in the manure compost, *Am. J. Bot.* 19: 514-537.

Walker, L.P. and Wilson, D.B., 1991, Enzymatic hydrolysis of cellulose: an overview, *Biores. Technol.* 36: 3-14.

Wang, Yun-Chang, 1985, Mycology in China with emphasis on review of the ancient literature, *Acta Mycologica Sinica* 4: 133-140.

Wiegant, W.M., 1992, Growth characteristics of the thermophilic fungus *Scytalidium thermophilum* in relation to production of mushroom compost, *Appl. Env. Microbiol.* 58: 1301-1307.

Wiegant, W.M., Wery, J., Buitenhuis, E.T. and de Bent, J.A.M., 1992, Growth-promoting effect of thermophilic fungi on the mycelium of the edible mushroom *Agaricus bisporus, Appl. Env. Microbiol.* 58: 2654-2659.

Willenborg, A. and Hindorf, H., 1985, Investigations on the fungal flora in mushroom culture substrate from various Rhineland installations, *Champignoncultuur* 284: 26-38.

Wood, D.A. and Thurston, C.F., 1991, Progress in the molecular analysis of *Agaricus* enzymes, in: *Genetics and Breeding of Agaricus* (L.J.L.D. Van Griensven, ed.) pp. 81-86, Pudoc, Wageningen, The Netherlands.

EFFECTIVENESS OF LYTIC ENZYMES IN ISOLATION OF PROTOPLASTS FROM *TRICHODERMA HARZIANUM*

S. Kaur and K.G. Mukerji

Applied Mycology Laboratory
Department of Botany
University of Delhi
Delhi-110 007
India

INTRODUCTION

Research into the biological control of plant diseases has arisen from a need to reduce dependence on chemical pesticides whose misuse in many crop systems has led to negative effects on the environment and health. Moreover, the use of chemical pesticides and fungicides is a relatively short term measure. On the other hand biological control of plant diseases offers an answer to many persistent problems in agriculture including problems of resource limitation, non-sustainable agricultural systems and over-reliance on pesticides (Cook and Baker, 1983). Bioprotectants provide unique opportunities for crop protection because they grow and proliferate and can colonize and protect newly formed plant parts to which they were not initially applied (Harman, 1990).

Despite these advantages, beneficial microbes are rarely used in the field to control plant diseases primarily because biocontrol agents have been both less effective and more variable than competitive chemical pesticides. A principal reason for this is the poor growth of bioprotectants. Competitive microflora that rapidly colonize planted seeds may inhibit the bioprotectant so that biocontrol fails (Harman, 1990; Harman et al., 1989; Hubbard et al., 1983). If biocontrol is to become an important component of integrated disease management (IDM) systems, it must be as effective and reliable as competitive chemical pesticides. Biocontrol agents must therefore be manipulated and improved for their effective use in the control of plant diseases (Mukerji and Garg, 1988a,b; Mukerji et al.,1992).

Among the wide variety of strain improvement techniques, protoplast fusion seems to be an efficient way of inducing genetic recombination in fungi which have no sexual cycle. The prerequisite of protoplast fusion is to obtain maximum yield which in totality is capable of retaining the ability to regenerate. There are a number of factors which influence the protoplast yield, of which the lytic enzyme system in the incubation mixture is of prime importance. The factors present in the lytic enzyme system may enhance or reduce the protoplast yield.

This chapter deals with the biocontrol of plant diseases and improvement of biocontrol agents. Emphasis has been given on protoplast fusion in the improvement of biocontrol agents.

BIOCONTROL OF PLANT DISEASES

Biocontrol can be defined as a reduction in the pathogen inoculum or its disease-producing capacity by the action of one or more organisms accomplished naturally or through manipulation of the environment, host or antagonist or by mass introduction of one or more antagonists (Cook and Baker, 1983). Basic aspects of strain selection, efficient production of biomass, formulation, storage ability and method of application are some of the main obstacles to the use of biocontrol agents (Kumar et al., 1997). Further, the potential for genetic manipulation of fungi to create genetically superior strains that can perform better than wild types has been extensively investigated (Lumsden and Lewis, 1989; Upadhyay et al.,1996, 1997, 1998 a,b,c).

Biocontrol Agents

Biocontrol agents are microorganisms with the potential to interfere with the growth or survival of plant pathogens and thereby contribute to biological control. Potential agents for biocontrol activity are rhizosphere competent fungi and bacteria, which in addition to their antagonistic activity are capable of inducing the growth response either by controlling minor pathogens or by producing growth stimulating factors (Chet et al., 1993). The environmental impact of biocontrol agents is assumed to be less significant than that of agrochemicals and their usage is presumed to reduce the consumption of chemical pesticides which causes serious ecological problems.

Among the large number of microorganisms used as biocontrol agents, *Trichoderma harzianum* is regarded as a particularly potent biocontrol agent. It shows antagonistic properties against a wide range of pathogens, for example species of *Rhizoctonia, Sclerotium, Fusarium, Colletotrichum, Verticillium, Pythium, Phytophthora* etc. *Trichoderma* spp. are also known for their ability to secrete a number of distinct enzymes capable of degrading the cell walls of other fungi.

Mechanism of Action

Biocontrol agents may utilize several modes of action, namely competition, mycoparasitism and antagonism. All of these mechanisms may operate independently or together and their activities can result in the suppression of plant pathogens (Singh and Faull, 1988).

Biocontrol agents compete well for nutrition and space with pathogens. Competition is an injurious effect of one organism on another because of the utilisation or removal of some resources of the environment, thereby determining the growth and infection of soil-borne plant pathogens in competition with other microorganisms. They parasitize plant pathogenic fungi which cover a multitude of different interactions including minor and major morphological disturbances, the overgrowth of hyphae of one fungus by another, penetration and direct parasitism by the formation of haustoria and lysis of one hypha by another (Singh and Faull, 1988). The production of volatile or non-volatile compounds by biocontrol agents has a direct impact on pathogenic microorganisms resulting in the denaturation of cell contents before they come in contact with the mycelium of the antagonist.

METHODS FOR IMPROVEMENT OF BIOCONTROL AGENT

Biocontrol agents must be manipulated and improved in order to be effectively used for control of plant diseases. Only some of the characteristics of the biocontrol agent may be considered to be beneficial in growth promotion and /or biocontrol, so that its performance has to be improved before it is commercially used in agriculture (Baker, 1989). Greater tolerance to environmental stress may increase the effectiveness of an organism and extend its useful range to new crop situations (Wilson and Pusey, 1985). The biocontrol ability of an organism can be improved either by altering the environment to make it more conducive to the particular biocontrol strain than to the competitive microflora, or by modifying the genetics of the organism to produce superior strains, or both.

The prime strategy for improvement is by genetic manipulation of biocontrol agents which can enhance their biocontrol activity and expand their spectrum. There are a number of methods available for this purpose including mutation and protoplast fusion. Ultraviolet radiation-induced mutants of *Trichoderma* spp. were found to be resistant to benomyl and were also effective as biocontrol agents against *Rhizoctonia solani, Sclerotium rolfsii* and *S.cepivorum* (Abd-El Moity et al., 1982; Papavizas, 1987). These mutants also helped to increase the level of rhizosphere competence (Baker, 1991) and yield of antibiotics (Howell and Sipanovic, 1983).

The biocontrol ability is under the control of a large number of genes, hence the use of mutagenesis treatment is not always desirable. Therefore, the introduction of superior characters may be best achieved by crossing of strains with appropriate characters (Hocart and Peberdy, 1989). Genetic recombination is a more powerful technique for developing effective biocontrol strains than mutation and selection. Protoplast fusion provides a means for recombination of whole genomes even between incompatible strains. The successful application of the technique of protoplast fusion has great potential in genetic recombination of species in which sexual and parasexual mechanisms are not present or are difficult to exploit.

PROTOPLAST FUSION IN IMPROVEMENT OF BIOCONTROL AGENTS

Although protoplast fusion has been reported in a large number of fungal species there are few studies related to the fungal biocontrol agents. Recently a few strains with improved biocontrol performance have been generated by protoplast fusion technology. Protoplast fusion has been used to improve the biocontrol ability of *T. harzianum* (Harman 1991; Harman et al., 1989; Lalithakumari et al.,1996; Pe'er and Chet, 1990; Sivan and Harman, 1991; Stasz et al., 1988). Recombinant progeny derived from protoplast fusion proved to be more efficient than either of the parents and a strain 1295-22, a fusant progeny, exhibited better biocontrol activity and more effectively colonized the rhizosphere of several crops (Harman et al., 1989; Sivan and Harman, 1991; Stasz et al., 1988). Much emphasis is being given to *Trichoderma* spp. as they possess useful traits for production of cellulolytic enzymes and biocontrol activity. It may be possible with protoplast fusion to combine these beneficial traits into a single superior strain.

Protoplast fusion in turn depends on a high yield of viable protoplasts in the shortest possible time. There are a number of factors which influence the protoplast yield, for example the nature of lytic enzyme, osmotic stabilizers, physiological status of an organism, cultural conditions, pretreatment method, pH, buffer, temperature and time of protoplasting. Of these, lytic enzyme is of prime importance because factors present in lytic enzyme may affect the protoplast yield.

EFFECT OF LYTIC ENZYME ON PROTOPLAST YIELD

The arrangement of cell wall components has a marked effect on the efficiency of lytic enzymes used for protoplast isolation from various species. In most cases the enzymes used for protoplast isolation are commercially available or are obtained from the microorganisms. A number of commercial enzymes, for example Cellulase CP, Driselase, β-glucuronidase, Glusulase, Novozym 234, Zymolyase and Sulphatase, have been used to isolate protoplasts from yeasts and filamentous fungi (Peberdy 1985). Most of them are crude or partially purified preparations with multiple activities, the main component being chitinase, α & β-glucanase and protease.

Although commercial enzymes are readily available and convenient to use, many workers have found that lytic enzymes produced from microorganisms are more effective in obtaining protoplasts. These mycolytic enzymes are prepared by growing a suitable microorganism on purified fungal cell walls as the sole carbon source, or they are derived from autolytic cultures of the specific fungus (Lynch et al., 1985; Peberdy, 1985; Reyes et al., 1984). A number of inducible substrates such as cellulose, chitin and laminarin have also been used to promote the synthesis of lytic enzyme complexes (Hocart and Peberdy, 1989). A wide range of microorganisms have been used to produce enzymes capable of totally or partially digesting the cell walls of fungi, including species of *Streptomyces, Micromonospora* (Villanueava and Garcia-Acha, 1971), *Penicillium purpurogenum* (Musilkova et al., 1969) and *Trichoderma* spp. (Ogawa et al., 1987). The lytic enzyme obtained from different species of *Trichoderma,* for example *T. harzianum* (Kitamato et al., 1988; Kumari and Panda, 1992), *T. viride* (Ogawa et al., 1989; Znidarsic et al., 1992) and *T. reesei* (Oberma et al., 1990) are used in protoplast isolation.

The analysis of various enzyme activities present in the lytic enzyme throws light on their contribution to wall lysis and protoplast formation. High levels of β-1,3-glucanase, endo β-1,3-glucanase, endo β-1,4-glucanase, chitinase and protease were present in lytic enzyme preparation from *T. harzianum* (Kumari and Panda, 1992) and *T. reesei* (Znidarsic et al., 1992).

ISOLATION OF PROTOPLASTS

The protoplasts of *T. harzianum* were isolated by incubating the young mycelium in the incubation mixture containing 50 µg/ml of crude extract of lytic enzyme, 0.15M Na-phosphate buffer, pH 5.8, 0.6M $(NH_4)_2SO_4$ (as osmotic stabilizer). It was incubated at 30°C with gentle shaking at 125 rpm. After 4 hours of contact with lytic enzyme, the protoplasts were obtained by centrifugation. Pure protoplasts were isolated from a mixture of protoplasts and cell wall debris by density-gradient centrifugation and were collected from the interphase of two solutions, i.e. 0.6M $(NH_4)_2SO_4$ and 0.6M sucrose. The pure protoplasts were washed and stored in 0.6M $(NH_4)_2SO_4$ at 0-4°C.

When the crude extract of lytic enzymes prepared from different isolates of *Trichoderma* and mixture of commercial enzymes were studied for maximum generation of protoplasts, it was observed that a maximum number of protoplasts was obtained with crude extract of lytic enzyme from 15 days old culture filtrate of *T. harzianum* DU/MS/1070 followed by *T. hamatum* and *T. harzianum* 2895, whereas the minimum number of protoplasts was obtained by using crude extract of lytic enzyme from *T. pseudokoningii*. The mixture of commercial enzymes also yielded a very low number of protoplasts (see Table 1). The application of crude extract of lytic enzyme from *Trichoderma* isolates alone are found to be suitable for effective protoplast isolation.

Table 1. Effect of crude extract of lytic enzymes prepared from 5, 10 and 15 days old culture filtrate of *Trichoderma* strains and mixture of commercial enzymes on protoplast yield.

Source of Lytic Enzyme		Protoplast Yield x 10^7 (mean ± S.D.)
T. harzianum 2895	A	3.322 ± 0.107
	B	4.203 ± 0.167
	C	4.667 ± 0.211
T. harzianum DU/MS/1070	A	4.138 ± 0.042
	B	4.614 ± 0.158
	C	5.125 ± 0.104
T. hamatum	A	4.139 ± 0.098
	B	4.628 ± 0.169
	C	4.718 ± 0.240
T. koningii	A	1.281 ± 0.138
	B	1.305 ± 0.099
	C	1.442 ± 0.119
T. pseudokoningii	A	1.142 ± 0.091
	B	1.218 ± 0.035
	C	1.339 ± 0.049
T. reesei	A	3.550 ± 0.233
	B	3.694 ± 0.059
	C	4.125 ± 0.086
T. viride	A	3.177 ± 0.125
	B	3.611 ± 0.151
	C	4.162 ± 0.113
Mixture of Commercial Enzymes*		1.946 ± 0.151

Crude extract of lytic enzyme prepared from 5 days old culture filtrate (A), 10 days old culture filtrate (B) and 15 days old culture filtrate (C).
* Mixture of commercial enzymes containing ß-glucuronidase (200 µg/ml), Driselase (1 mg/ml) and Chitinase (10 µg/ml).

The lytic enzymes from culture filtrates of different isolates of *Trichoderma* and commercial enzymes were assayed for different enzymes *viz.*, carboxymethyl cellulase (CMCase), filter paper activity (FPA), α-glucanase, xylanase, chitinase and proteinase and total protein. It is evident from data presented in Table 2 that the total protein increased with the age of culture. Similarly, high levels of CMCase, FPA and xylanase activities were associated with crude extract of lytic enzyme from different isolates of *Trichoderma*, whereas high levels of glucanase were detected in crude extract of lytic enzyme from 10 days old culture filtrate and maximum proteinase activity was detected in lytic enzymes prepared from 5 days old culture filtrate. The highest proteinase activity in 5 days old culture filtrate resulted in reduced protoplast yield, as it may have caused bursting of protoplasts and removal of proteinase from lytic enzyme resulting in increase in protoplast yield (Kitamato et al.,1988). The increase in cellulase (CMCase and FPA), xylanase activity and decrease in proteinase activity with the age of culture resulted in effective digestion of cell walls.

Table 2. Comparison of enzyme activities in crude extract of lytic enzymes from *Trichoderma* isolates and mixture of commercial enzymes.

Source of Lytic Enzyme		Total Protein mg/ml	FPA IU/ml	CMCase IU/ml	α - glucanase IU/ml	Xylanase IU/ml	Chitinase IU/ml	Proteinase IU/ml
T. harzianum	A	2.40	0.195	0.220	0.220	0.180	1.66	0.96
DU/MS/1070	B	2.56	0.235	0.270	0.230	0.320	1.33	0.69
	C	2.80	0.247	0.365	0.215	0.440	1.46	0.66
T. harzianum	A	2.73	0.189	0.315	0.175	0.255	2.33	0.87
2895	B	2.77	0.197	0.385	0.285	0.355	1.73	0.70
	C	3.27	0.226	0.470	0.270	0.390	1.93	0.55
T. hamatum	A	2.48	0.206	0.375	0.275	0.280	1.73	1.26
	B	2.56	0.252	0.480	0.305	0.355	2.19	1.25
	C	2.85	0.267	0.495	0.300	0.600	1.19	1.08
T. koningii	A	2.75	0.217	0.350	0.255	0.335	1.33	1.24
	B	3.17	0.242	0.470	0.295	0.350	2.53	0.85
	C	3.88	0.261	0.480	0.255	0.480	3.07	0.80
T. pseudokoningii	A	2.40	0.157	0.315	0.230	0.415	1.73	1.24
	B	2.53	0.199	0.455	0.485	0.605	1.79	1.08
	C	3.00	0.206	0.425	0.370	0.665	2.19	0.79
T. reesei	A	2.22	0.182	0.275	0.205	0.375	1.73	1.66
	B	2.48	0.215	0.330	0.315	0.565	2.06	0.87
	C	2.98	0.217	0.360	0.240	0.580	2.27	0.78
T. viride	A	3.50	0.150	0.315	0.185	0.500	3.13	1.09
	B	3.53	0.160	0.325	0.250	0.485	1.39	1.08
	C	3.80	0.185	0.330	0.185	0.550	1.93	0.63
Mixture of Commercial Enzymes*		4.75	0.550	0.450	0.395	0.900	3.20	2.77

A,B,C, * - Same as in Table 1.

CONCLUSIONS

Biological control of plant pathogens has been considered as a potential control strategy. However, because of practical constraints there are only a few successful field applications of antagonists used as biocontrol agents. Owing to the potential adverse effects of chemical pesticides and fungicides on the health of people and the environment there is a need for integrated control of plant pathogens. The use of biocontrol strategies offers several advantages over chemical control of plant diseases. In practical terms, biological control will not immediately nor totally replace chemicals but the judicious use of biocontrol agents with limited use of chemical pesticides can significantly enhance the quality of the environment and agricultural productivity. Genetic manipulation of biocontrol agents for improved ability to control various diseases and improved compatibility with chemical fungicides may reduce the use of chemical pesticides and fungicides.

Although protoplasts have been isolated from a large number of fungi there are very few reports of improved biocontrol strains with enhanced performance. Effective protoplast fusion depends on a high yield of protoplasts. The use of protoplast fusion to overcome incompatibility barriers allows characteristics from even distantly related isolates to be combined. Inter-species hybridization would seem to be a worthwhile approach to strain improvement in *Trichoderma*. Increased cellulolytic and mycolytic activities could enhance the effectiveness of the organism for biocontrol purposes.

FUTURE STRATEGIES

It has been visualised that the next decade will see widespread exploitation of genetic engineering, especially genetic manipulation and hybridization, in the improvement of biocontrol agents. Genetic alteration of the pathogen or potential microbial biocontrol agent may be possible in the near future. As the basis of biocontrol interactions becomes better understood, breeding programmes to improve strains will become more common. Experience with pesticides and fungicides has clearly shown how adaptable pest and pathogen populations can be. It is not unlikely, therefore, that with the increasing use of biocontrol organisms, even in an integrated approach, there will be a continual need to update the strains in use as the pathogen populations change to meet the challenge. Investigation of new technologies is required to develop biocontrol agents with improved performance. The biocontrol agents should be economically feasible and environmentally safe.

Protoplast fusion will undoubtedly play an important role in the exploitation of biocontrol fungi. Effective protoplast fusion depends on the high yield of protoplasts. Owing to the diversity of the cell walls of different organisms, the effective conditions of protoplasting should be standardized for each and every organism.

ACKNOWLEDGEMENTS

The senior author is thankful to the Council of Scientific and Industrial Research. The authors are also thankful to Prof. J.P. Tiwari, University of Alberta, Edmonton, Canada for providing commercial enzymes.

REFERENCES

Abd-El Moity, T.H., Papavizas, G.C. and Shatla, M.N., 1982, Induction of new isolates of *Trichoderma harzianum* tolerant to fungicides and their experimental use for control of white rot of onion, *Phytopath.* 72: 396-100.

Baker, R., 1989, Some perspectives on the application of molecular approaches to biocontrol problems, in: *Biotechnology of Fungi for Improving Plant Growth* (J.M. Whipps and R.D. Lumsden, eds.), Cambridge University Press, Cambridge.

Baker, R., 1991., Induction of rhizosphere competence in the biocontrol fungus *Trichoderma*, in: *The Rhizosphere and Plant Growth* (D.L. Keister and P.B. Cregan, eds.), Kluwer Academic Publishers, Netherlands.

Chet, I., Barak, Z. and Oppenheim, A., 1993, Genetic engineering of microorganisms for improved biocontrol activity, in: *Biotechnology in Plant Disease Control* (I. Chet, ed.), Willey-Liss Inc., USA.

Cook, R..J. and Baker, K.F., 1983, *The Nature and Practice of Biological Control of Plant Pathogens*, 2nd ed., American Phytopathological Society, St. Paul, Minnesota, USA.

Harman, G.E., 1990, Deployment tactics for biocontrol agents in plant pathology, in: *New Directions in Biological Control, Alternative for Suppressing Agricultural Pests and Diseases* (R.R. Baker and P.E.

Dunn, eds.), Alan R. Liss, New York.

Harman, G.E., 1991, Seed treatment for biological control of plant diseases, *Crop Prot.* 10:166-171.

Harman, G.E., Taylor, A.G. and Stasz, T.E., 1989, Combining effective strains of *Trichoderma harzianum* and solid matrix priming to improve biological seed treatment, *Plant Dis.* 73: 631-637.

Hocart, M.J. and Peberdy, J.F., 1989, Protoplast technology and strain selection, in: *Biotechnology of Fungi for Improving Plant Growth* (J.M. Whipps and R.D. Lumsden, eds.), Cambridge University Press, Cambridge.

Howell, C.R. and Sipanovic, R.D., 1983, Gliovirin, a new antibiotic from *Gliocladium virens* and its role in the biological control of *Phythium ultimum, Can. J. Microbiol.* 29: 321-324.

Hubbard, J.P., Harman, G.E. and Hadar, Y., 1983, Effect of soilborne *Pseudomonas* spp. on the biocontrol agent, *Trichoderma hamatum,* on the pea seeds, *Phytopath.* 73: 655-59.

Kitamato, Y., Mori, N., Yamamoto, M., Ohiwa, T. and Ichikawa, Y., 1988, A simple method for protoplast formation and improvement of protoplast regeneration from various fungi using an enzyme from *Trichoderma harzianum, Appl. Microbiol. Biotechnol.* 28: 445-450.

Kumar, R.N., Upadhyay, R.K. and K.G. Mukerji, 1997, Strategies in biological control of plant diseases, in: *IPM System in Agriculture, Vol. 2., Biocontrol in Emerging Biotechnology,* (R.K. Upadhyay, K.G. Mukerji and R.L. Rajak, eds.), Aditya Books, New Delhi, India.

Kumari, J.A. and Panda, T., 1992, Studies on critical analysis of factors influencing improved production of protoplasts from *Trichoderma reesei* mycelium, *Enz. Microbiol. Technol.* 14: 241-248.

Lalithakumari, D., Mrinalini, C., Chandra, A.B. and Annamalai, P., 1996, Strain improvement by protoplast fusion for enhancement of biocontrol potential integrated with fungicide tolerance in *Trichoderma* spp., *Zeitschrift für Pflanzenkrankeiten und Pflanzenschutz* 103: 206-212.

Lumsden, R.D. and Lewis, J.A., 1989, Problems and progress in the selection, production, formulation and commercial use of plant disease control fungi, in: *Biotechnology of Fungi for Improving Plant Growth* (J.M. Whipps and R.D.Lumsden, eds.), Cambridge University Press, Cambridge.

Lynch, P.T., Collin, H.A. and Issac, S., 1985, Use of autolytic enzyme for isolation of protoplasts from *Fusarium tricinctum* hyphae, *Trans. Br. Mycol. Soc.* 84: 473-478.

Mukerji, K.G. and Garg, K.L., 1988a, *Biocontrol of Plant Diseases, Vol. I,* CRC Press, Boca Raton, Florida.

Mukerji, K.G. and Garg, K.L., 1988b, *Biocontrol of Plant Diseases, Vol II,* CRC Press, Boca Raton, Florida.

Mukerji K.G., Tewari, J.P., Arora, D.K. and Saxena, G., 1992, *Recent Developments in Biocontrol of Plant Diseases,* Aditya Books, New Delhi, India.

Musilkova, M., Fencl Z. and Seichertova, O., 1969, Release of *Aspergillus niger* protoplasts by *Penicillium purpurogenum* enzyme, *Folia Microbiol.* 14: 47-50.

Oberma, H., Stobinska, H. and Kregiel, D., 1990, Obtaining and regeneration of *Trichosporon cutaneum* TrNu18 protoplasts, *Acta Aliment. Pol.* 16: 83-96.

Ogawa, K., Brown, J.A. and Wood, T.M., 1987, Intraspecific hybridization of *Trichoderma reesei* QM 9414 by protoplast fusion using colour mutants, *Enz. Microb. Technol.* 9: 229-232.

Ogawa, K., Tsuchimochi, M., Taniguchi, K. and Nakatsu, S., 1989, Interspecific hybridization of *Aspergillus usamii* mut *shirousamii* and *Aspergillus niger* by protoplast fusion, *Agric. Biol. Chem.* 53: 2873-2880.

Papavizas, G.C., 1987, Genetic manipulation to improve the effectiveness of biocontrol fungi for plant diseases control, in: *Innovative Approaches to Plant Disease Control* (I. Chet, ed.), John Wiley and Sons, New York.

Peberdy, J.F., 1985, Mycolytic enzymes, in: *Fungal Protoplasts: Application in Biochemistry and Genetics* (J.F. Peberdy and L. Ferenezy, eds.), Marcel Dekker Inc., New York.

Pe'er, S. and Chet, I., 1990, *Trichoderma* protoplast fusion: a tool for improving biocontrol agents, *Can. J. Microbiol.* 36: 6-9.

Reyes, F., Perez-Leblic, M.I., Martinez, M.J. and Lahoz, R., 1984, Protoplast production from filamentous fungi with their own autolytic enzymes., *FEMS Microbiol. Let.* 24: 281-283.

Singh , J. and Faull, J.L., 1988, Antagonism and biological control, in: *Biocontrol of Plant Diseases. Vol.II.* (K.G. Mukerji and K.L. Garg, eds.), CRC Press, Boca Raton, Florida.

Sivan, A. and Harman, G.E., 1991, Improved rhizosphere competence in a protoplast fusion progeny of *Trichoderma harzianum, J. Gen. Microbiol.* 137: 23-29.

Stasz, T.E., Harman, G.E. and Weeden, N.F., 1988, Protoplast preparation and fusion in two biocontrol strains of *Trichoderma harzianum, Mycologia* 80: 141-150.

Upadhyay, R.K., Mukerji, K.G. and Rajak, R.L., 1996, *IPM System in Agriculture, Vol. 1: Principles and Perspective,* Aditya Books, New Delhi, India.

Upadhyay, R.K., Mukerji, K.G. and Rajak, R.L, 1997, *IPM System in Agriculture, Vol. 2: Biocontrol in Emerging Biotechnology,* Aditya Books, New Delhi, India.

Upadhyay, R.K., Mukerji, K.G. and Rajak, R.L., 1998a, *IPM System in Agriculture, Vol. 3: Cereals,* Aditya Books, New Delhi, India.

Upadhyay, R.K., Mukerji, K.G. and Rajak, R.L., 1998b, *IPM System in Agriculture, Vol. 4: Pulses,* Aditya Books, New Delhi, India.

Upadhyay, R.K., Mukerji, K.G., Chamola, B.P. and Dubey, O.P., 1998c, *Integrated Pest and Disease Management,* APH Publishing Corporation, New Delhi, India.

Villanueava, J.R. and Garcia Acha, I., 1971, Production and use of fungal protoplasts, in: *Methods in Microbiology, Vol IV* (C.Booth, ed.), Academic Press, New York.

Wilson C.L. and Pusey, P.L., 1985, Potential for biological control of post harvest plant diseases, *Plant Dis.* 69: 375-378.

Znidarsic, P., Pavko, A. and Komel, R., 1992, Laboratory-scale biosynthesis of *Trichoderma* mycolytic enzymes for protoplast release from *Cochliobolus lunatus, J. Ind. Microbiol.* 9: 115-119.

TIMBER PRESERVATION: THE POTENTIAL USE OF NATURAL PRODUCTS AND PROCESSES

John W. Palfreyman[1], Nia A. White[1] and Jagjit Singh[2]

[1]Dry Rot Research Group
School of Molecular and Life Sciences
University of Abertay Dundee
Dundee DD1 1HG
U.K.

[2]Oscar Faber Applied Research
Marlborough House
Upper Marlborough Road
St. Albans AL1 3UT
U.K.

INTRODUCTION

Timber represents a rich source of potential nutrients for a wide variety of macro-organisms (e.g. insects and marine borers) and micro-organisms (notably fungi and bacteria). Timber is susceptible to these agents during both production and in-service use, though degradation by them is not always inevitable and is usually associated with misuse or inappropriate use. However when timber is to be used in some types of environment, notably where relatively high levels of moisture are available, protective systems are necessary to prevent decay. Currently such systems often rely on the use of chemical treatments but alternative strategies are under development at a number of centres and these come under the headings of environmental control, biological control and the use of natural products as wood preservatives. This chapter will explore each of these systems in the context of the control of the dry rot fungus *Serpula lacrymans*, a wood decay basidiomycete which causes severe damage to wooden elements of buildings in temperate areas and whose natural home appears to be the Himalayan foothills. Discussion of these natural processes and products will be preceded by a description of the current wood preservation scenario and of the pressures which are bringing about change in the associated industry.

BACKGROUND

The susceptibility of wood to decay has been recognised for millennia (e.g. biblical references to treating Noah's ark internally and externally with pitch). However the durability of wood, if kept in an appropriate environment, must also have been familiar to ancient peoples as, for example, Egyptian sycamore coffins have survived intact for over 4000 years. Early preservative systems depended upon surface coating or soaking to introduce chemicals into wood and the modern era of preservation may be considered to have developed after the invention, in 1838 by John Bethell, of the full cell impregnation system for wood treatment. This system, which uses a combination of pressure and vacuum to load up the wood cell, has been followed by various empty cell processes, oscillating and alternating pressure systems and sap displacement. Interestingly diffusion methods, relying on natural movement of chemicals into wood, are now becoming increasingly popular as they use water-borne active ingredients whose use may have environmental benefits. However despite many years of research the actual number of different chemical compounds used in wood preservation is fairly limited (for example, see Table 1) and there are various economic and social pressures demanding new ideas in this field.

Table 1. Chemicals licensed for use as wood preservatives in the UK in 1996 (see 'Pesticides 1996').

Acypetacs copper/ Acypetacs zinc	Creosote	Pentachlorophenol
Alkyl trimethyl ammonium chloride	Cypermethrin	Permethrin
	Dialkyldimethyl ammonium chloride	Propiconazole
Arsenic pentoxide	Dichlofluanid	Sodium 2-phenyphenoxide
Azaconazole	Disodium octaborate	Sodium dichromate/ fluoride/ PCP/ tetraborate
Benzalkonium chloride	Dodecylamine lactate/ salicylate	Tebuconazole
Benzothiazole	Furmecyclox	Tri(hexylene glycol) biborate
Boric acid	Gamma HCH	Tributyl tin naphthenate/ oxide/ phosphate
Carbendazim	Isothiazole Phenyphenol	
Chromium acetate/ trioxide	IPBC	Zinc naphthenate/ octoate/ versatate
Copper carbonate/ hydroxide/ naphthenate/ oxide/ versatate	Methylene (bis) thiocyanate Oxine copper	

The use of the term 'chemical' in discussions on wood preservatives can cause ambiguity since any substance used for preservation can be considered to be a chemical. For the sake of clarity in this document 'chemical' is taken to mean any substance produced or extracted in an industrial process and which is not a natural product, i.e. a substance produced biologically. This definition does not distinguish between toxic and non-toxic compounds, and many natural products are highly toxic, since it can be assumed that any compound used as a wood preservative must be either toxic to, or inhibit biological growth of, its target organism(s). Details of the actual production of treated wood is given in Table 2 and the European system for defining 'at-risk' timbers, i.e. timber usage 'Hazard Classes', is detailed in Table 3.

Changing trends in preservative usage have seen a decrease in the use of creosote in many countries with a corresponding increase in LOSP's and water soluble preservatives (e.g. Yeadon, 1995 for figures for the USA). Preservatives are used in three main ways, as anti-sapstain treatments to prevent disfigurement of wood prior to its use, as preventative systems to protect wood used in high hazard situations and as remedial treatments where failure of wooden components has occurred due to, for example, water ingress into normally dry environments or leaching of preservative from pretreated wood.

Table 2. Estimated wood preservative usage in selected European countries and the US for 1992 for three types of formulation: (i) light organic solvent preservatives (LOSP's), (ii) creosotes and (iii) water soluble preservatives (UNEP, 1994).

Country	Creosote	LOSP	Water soluble
UK	66 000	897 000	1 237 000
Denmark	0	29 900	205 000
Germany	150 000		600 000
Sweden	69 700	37 800	446 000
USA	2 452 000	885 000	12 000 000

Table 3. European Hazard Classes defining 'at risk' timbers.

Hazard Class	Situation	Risk	Moisture Content	Comments
Class 1	above ground	insects	<20%	not subject to wetting
Class 2	above ground	insects/fungi	>20%	occasional wetting
Class 3	above ground	insects/fungi	>20%	frequent wetting
Class 4	ground contact	insects/fungi	>20%	permanent wetting
Class 5	in salt water	insects/fungi marine borers	>20%	permanent wetting

For many years the search for new wood preservatives was driven by a desire to reduce cost and improve performance. In the last decade a new driving force has entered the field which is associated with the need to protect and sustain the environment, though it is probably true that economics still represents the bottom line (see also Table 4). For some time the perceived view was that chemical preservatives were not consistent with environmental protection and, though this view is no longer widely supported, some of the actual chemicals used in wood preservation are still seen in some areas as unacceptable. Current research aims to find new preservatives of low mammalian toxicity and with better fixation systems to ensure that preservatives stay in the wood. Alternative strategies which do not involve the use of chemical preservatives are also being considered more seriously in an effort to respond to consumer pressures and the increasing legislation involved with preservative use.

Table 4. The pressure for change.

Current wood preservatives are often chemicals of general toxicity
Desire for 'natural', 'sympathetic' systems/ consumer pressure
Availability of research funding, e.g. EU initiatives to developed new processes
Intellectual climate, the development of new ideas in biotechnology
Commercial and legislative pressures

THE DRY ROT SCENARIO

Decay of timber by the brown rot fungus *Serpula lacrymans* represents the most serious wood degradation problem found in buildings in temperate regions of Europe and Australasia though not, surprisingly, the Americas (Singh, 1994). As detailed elsewhere in this book, *S. lacrymans* is not native to Europe and its natural worldwide distribution appears to be limited to very specific areas of India and the U.S.A. Traditional methods for treating dry rot have centred around removal of infected wood, treatment of surrounding masonry with wood preservatives and, in many cases, replacement of removed timber with preservative-treated wood. Much of the strategy was developed in the U.K. at a time when there were large numbers of war-damaged buildings extant and funds for general maintenance were often scarce.

In the current economic climate alternative treatment strategies are being sought to reduce (i) the amount of both damaged and undamaged wood removed from buildings (notably when the buildings are of cultural significance) and (ii) the amount of additional chemicals, particularly those of low vapour pressure, introduced into the built environment. In order to develop effective alternative strategies, a full understanding of *S. lacrymans* is required. The biology, physiology and decay ability of the dry rot fungus has been reviewed in Jennings and Bravery (1991), Singh et al. (1993), Bech-Anderson (1995) and Palfreyman et al. (1995). New strategies being developed can be grouped under three distinct headings: (i) environmental control, (ii) biological control, and (iii) use of natural products from plants and other organisms.

Environmental Control

Environmental control essentially involves analysis of the decay processes, and the organisms involved, followed by the development of avoidance strategies. Within the context of environmental control comes the use of agents which render timber toxic, however for the purposes of this discussion only those processes which rely on control without the addition of extra agents will be considered.

Despite its ability to destroy wooden components in buildings the dry rot fungus is peculiarly sensitive to environmental conditions so, for example, it is unable to survive temperatures even mildly in excess of its optimal growth temperature. In addition the viability of *S. lacrymans* is rapidly compromised in the absence of accessible water or in relative humidities of less than around 90% (e.g. Low et al., 1997). However, the ability of the organism to transport water over relatively large distances, by means of its well developed strand structures, means that apparently non-viable mycelium can be reactivated once conditions improve if a suitable moisture sink is available.

In addition to a woody resource to provide nutrients, *S. lacrymans* also requires non-woody components of the built environment in order to grow optimally (Bech-Anderson, 1989; Palfreyman et al., 1996). Low (1998) studied a range of different building materials including sandstones, granite and lath/plaster (a plaster mix layered on to strips of wood, frequently used in historic buildings). Lath/plaster was particularly effective at supporting the decay capacity of *S. lacrymans*. Organisms grown on some sandstones demonstrate accumulation of metal ions and which no doubt changes the crystalline structure of the material. Both calcium and iron are probably utilised by *S. lacrymans* from such materials, iron to initiate the non-enzymatic depolymerisation of the cellulose substrate, and calcium to neutralise excessive oxalic acid production which the organism does not seem to be able to control itself. Many building components are rich in these two elements and their entrapment, in some way, within the building matrix would severely limit the ability of *S. lacrymans* to cause extensive damage.

From this, and other evidence, comes the conclusion that the first line of defence against dry rot is to ensure that water penetration into wood does not occur. Secondary lines of defence would relate to minimising the chances of the development of confined damp areas in buildings by introducing air channels into structures and restricting the contact between timber elements and masonry components (particularly those that contain calcium and/or iron). Many of the problems associated with dry rot can be avoided by the introduction of rigorous maintenance schedules. However, there will undoubtedly be limitations on the usefulness of the environmental control approach and current studies are designed to define these limits

(Palfreyman et al., 1995; Low et al., 1997).

Finally, recent studies have confirmed that the natural environment for *S. lacrymans* is certain restricted mountainous regions of the world (Bech-Anderson 1995; White et al., 1997). Additional analysis and definition of this biotope is likely to produce further clues for new environmental control methods.

Biological Control

Various definitions for biological control have been proposed. Possible definitions from agriculture which are of significance in wood preservation would be: 'the suppression of a pest by means of the introduction, propagation and dissemination of the predators, parasites and diseases which attack it' and 'the directed use of parasitoids, predators and pathogenic microbes to reduce and regulate pest populations to subeconomic levels' (see also Bruce, 1992). Systems designed to use biocontrol in pretreatment of timber have been evaluated for many years (Bruce, 1997) and continue to be actively researched. The use of biocontrol for the remedial treatment of dry rot, and in the prevention of re-occurrence of infections, is less well understood. Table 5 lists some of the advantages and limitations of biological control systems.

Table 5. Feature of biological control systems (modified from Bruce, 1992).

Advantages	Limitations
Environmentally acceptable Specificity - can be well targeted Responsive/reactive	Subject to environmental stress No history of use Insufficient research and development

Score and Palfreyman (1994) and Doi and Yamada (1991) have both demonstrated that a range of *Trichoderma* isolates are able, on preinoculation, to prevent subsequent colonisation of wood by *S. lacrymans*. The ability to arrest growth once established and effectively to kill the dry rot fungus in heavily infected wood is less easy to demonstrate (Score et al., 1998). It may well be that *Trichoderma* isolates are only able to operate effectively when used with an additional control method, for example, environmental control (see Table 6 for their properties).

Table 6. Properties of *Trichoderma* spp. used as biological control for the protection of timber (from Bruce, 1992).

Passive properties	Active properties
Nutritionally non-exacting Prolific spore production Fast growth rate Tolerance of environmental change High chemical tolerance	Mycoparasitism Soluble metabolites Volatile metabolites Siderophore production

It seems probable that the ideal control organism could be isolated from the same ecological niche as the organism controlled, so that it has maximum chance of surviving under conditions which favour the target. It was not possible to use such organisms for control of *S. lacrymans* until recently, since the natural environment of this organism had not been identified (Bech-Anderson, 1995; White et al., 1997). Now, though, the search for natural predators for *S. lacrymans* should be targeted towards the Himalayan foothills and specific regions of the Rocky Mountains in the USA where *S. lacrymans* has reliably been identified as being an indigenous organism.

Use of live fungi to control *S. lacrymans* may still present difficulties even if suitable agents can be found. For example *Trichoderma* spp. are normally prolific sporulators and spores could cause allergy problems if produced within buildings in confined spaces. However it may not be necessary to use live control systems since (i) the growth of many basidiomycetes and the germination of their spores can be inhibited by culture filtrates of specific *Trichoderma* species, (ii) volatile substances produced by *Trichoderma* species can inhibit basidiomycete growth, (iii) enzymes from control organisms may be able to inhibit growth of wood decay fungi, and (iv) the production of siderophores by biocontrol agents may be sufficient to inhibit decay (Srinivasan et al, 1995; Bruce, 1997; Score et al., 1998).

Use of Natural Products from Plants and Other Organisms

Throughout history mankind has utilised and manipulated natural extracts, notably from plants and fungi, for benefits ranging from food improvement to pharmaceutical development. Applications of such materials in the field of wood preservation are more limited due mainly to the relatively low value of the final product (treated wood) and the ready availability of a range of inexpensive chemicals. One area which has received considerable attention is the use of wood extractives, particularly tannins which are a by-product of timber pulping, as preservatives. Complexes of various flavonoid and hydrolysable tannins with copper ions can afford protection to timber but unfortunately they are easily leached out. An improvement on this technology involves the use of tannins to fix boron into timber. This produces a material which is non-toxic to humans, and other mammals, and is of use in hazard classes 1 to 3 (see Pizzi, 1997 for review).

The use of natural products to combat dry rot is a new field of research, although Mori et al. (1997) demonstrated that the bark of *Magnolia obovata* could inhibit the growth of a range of wood-destroying fungi including *S. lacrymans*. Extracts of a range of other plants, e.g *Miconia cannabina* (Miles et al., 1991), *Liriodendron tulipifera* (Hsu and Chen, 1991), *Cunninghamia lanceolata* (Shieh and Sumimoto, 1992) and *Acacia mearnsii* (Ohara et al., 1994) have been shown to be active against fungi including wood-destroying basidiomycetes. Many of these extracts are well know to indigenous populations as having medicinal properties resulting in the development, in recent years, of the subject of 'ethnobotany' and the related 'ethnomycology' (Cotton, 1996). Initial studies at the University of Abertay Dundee, sponsored by the Karl Meyer Foundation (Switzerland), on methanol extracts from a range of medicinal plants from northern India have not so far produced any useful preservatives. However, these studies emphasize the need for closer cooperation between those sourcing potential materials and those who wish to apply them to solve specific problems.

CONCLUSIONS

Wood preservation research is now being targeted towards the definition of novel solutions in a variety of different areas. With respect to the prevention of dry rot it is probable that these solutions will (i) be much more organism specific and (ii) relate much more closely to the microbial physiology of the causative organism. Environmental control will certainly have an important part to play in dry rot control in the future. The potential for biological control is less clear, unless more appropriate competitor fungi can be found, and even so its use is likely to be limited to specific situations. Natural products would seem ideally to complement the use of environmental control, giving additional protection where the continual monitoring needed for effective environmental control cannot be guaranteed.

Preservation of timber in use is a universal problem in a world where natural resources are limited. By contrast, dry rot is largely a problem of the developed world and indeed is only relevant in certain areas of the developed world. However, it seems probable that the solution to dry rot, or at least a new set of tools to combat *S. lacrymans*, will be found on the Indian sub-continent. Closer cooperation between mycologists and other scientists in India, with dry rot research organisations in Europe and beyond, should allow these solutions and tools to be identified.

ACKNOWLEDGEMENTS

The authors wish to thank (i) the Karl Meyer Foundation for a grant which allowed them to initiate studies on natural products and (ii) Historic Scotland for their participation in a partnership agreement with the University of Abertay Dundee which funded the environmental control studies outlined in this chapter.

REFERENCES

Bech-Anderson, J., 1995, *The Dry Rot Fungus and Other Fungi in Houses*, Hussvamp Laboratoriet, Copenhagen, Denmark.

Bruce, A., 1992, *Biological Control of Wood Decay*, IRG Wood Preservation, Doc. No. IRG/WP/1531-92.

Bruce, A., 1997, Biological control of wood decay, in: *Forest Products Biotechnology* (A. Bruce and J.W. Palfreyman, eds.), Taylor & Francis, London, U.K.

Cotton, C.M., 1996, *Ethnobotany: Principles and Applications*, John Wiley and Sons, Chichester, U.K.

Doi, S. and Yamada, A., 1991, *Antagonistic Effects of Three Isolates of Trichoderma spp. against Serpula lacrymans in the Soil Treatment Test*, IRG Wood Preservation, Doc. No. IRG/WP/1473.

Hsu, C.Y.H. and Chen, C.L., 1991, Antimicrobial activities of aporphine alkaloids from heartwood and discoloured sapwood of *Liriodendron tulipfera*, Holzforschung 45: 325-331.

Jennings, D.H. and Bravery, A.F. (eds), 1991, *Serpula lacrymans: Fundamental Biology and Control Strategies*, John Wiley and Sons, Chichester, U.K.

Low, G.A., Palfreyman, J.W., White, N.A., Staines, H.J. and Bruce, A., 1997, *Preliminary Studies to Assess the Effects of Aeration and Lowered Humidity on the Decay Capacity, Growth and Survival of the Dry Rot Fungus Serpula lacrymans*, IRG Wood Preservation, Doc. No. IRG/WP/97-10208.

Low, G.A., 1998, The environmental control of the dry rot organism, *Serpula lacrymans*, University of Abertay Dundee, Higher Degree Transfer Report.

Miles, D.H., Meideros, J., Chen, L., Cittawong, V., Swithenbank, C., Lidert, Z., Payne, A.M. and Hedin, P.A., 1991, A search for agrochemicals from Peruvian plants, *ACS Symposium Series* 449: 399-406.

Mori, M., Aoyama, M. and Doi, S., 1997, Antifungal constituents in the bark of *Magnolia obovata* Thunb., *Holz als Roh-und Wekstoff* 55: 275-278.

Ohara, S., Suzuki, K. and Ohira, T., 1994, Condensed tannins from *Acacia mearnsii* and their biological activities, *Mokuzai Gakkaishi* 40: 1363-1374.

Palfreyman, J.W., Phillips, A.M. and Staines, H.J., 1996, The effect of calcium ion concentration on the growth and decay capacity of *Serpula lacrymans* (Schumacher ex Fr.) Gray and *Coniophora puteana* (Schumacher ex Fr.) Karst., *Holzforschung* 50: 3-8.

Palfreyman, J.W., White, N.A., Buultjens, T.E.J. and Glancy, H., 1995, The impact of current research on the treatment of infestations by the dry rot fungus *Serpula lacrymans*, *Int. Biodeterioration & Biodegradation* 35: 369-395.

Pesticides, 1996, *Pesticides Approved under the Control of Pesticides Regulations*, 1986, HMSO, London.

Pizzi, A., 1997, Wood/ bark extractives as adhesives and preservatives, in: *Forest Products Biotechnology* (A. Bruce and J.W. Palfreyman (eds.), Taylor and Francis, London.

Score, A.J. and Palfreyman, J.W., 1994, Biological control of the dry rot fungus *Serpula lacrymans* by *Trichoderma* species: the effects of complex and synthetic media on interaction and hyphal extension rates, *Int. Biodeterioration & Biodegradation* 33: 115-128.

Score, A.J., Bruce, A. and Palfreyman, J.W., 1998, The biological control of *Serpula lacrymans*: use in the treatment of dry rot infected wood, *Holzforschung* (in press).

Shieh, J.C. and Sumimoto, M., 1992, Antifungal wood component of *Cunninghamia lanceolata*, *Mokuzai Gakkaishi* 38: 482-489.

Singh, J., 1994, Nature and extent of deterioration in buildings due to fungi, in: *Building Mycology* (J. Singh, ed.), E. & F.N. Spon, London, pp 34-53.

Singh, J., Bech-Andersen, J., Elborne, S.A., Singh, S., Walker, B. and Goldie, F., 1993, The search for wild dry rot fungus (*Serpula lacrymans*) in the Himalayas, *The Mycologist* 7(3): 124-131.

Srinivasan, U., Highley, T.L. and Bruce, A., 1995, The role of siderophore production in the biological control of wood decay fungi by *Trichoderma* spp., in: *Biodeterioration and Biodegradation 9* (A. Bousher, M. Chandra and R. Edyvean, eds.), Institute of Chemical Engineering, Rugby, pp. 226-230.

UNEP, 1994, *Environmental Aspects of Wood Preservation. A Technical Guide*, UNEP IE/PAC Technical Report Series, United Nations Publication.

White, N.A., Low, G.A., Singh, J., Staines, H. and Palfreyman, J.W., 1997, Isolation and environmental study of 'wild' *Serpula lacrymans* and *Serpula himantioides* from the Himalayan forests, *Mycol. Res.* 101: 580-584.

Yeadon, E., 1995, *Wood Preservation in the U.S.A.: Proceedings of the B.W.P.D.A. Convention*, B.W.P.D.A., U.K., pp. 28-32.

A BIOCONTROL FORMULATION FOR PROTECTION OF STORED BAMBOO FROM DECAY FUNGI

N.S.K. Harsh[1] and N.K. Kapse[2]

[1]Forest Pathology Division
Tropical Forest Research Institute
Jabalpur-482 021
Madhya Pradesh
India

[2]National Agricultural Research Project
Sindewahi-441 222
Chandrapur
M.S.
India

INTRODUCTION

Different species of bamboo constitute the major, and in most cases the only, raw material for pulp in India, China and other south-east Asian countries (Maheshwari and Satpathy, 1990). They also have a multitude of other uses, for example building material, construction of huts, fencing material, scaffolding, food, agricultural implements, fishing rods, parquet manufacture, furniture, oars, water conduits, baskets, mats, screens, toys, stabilizer of soil, and in plywood and particle board manufacturing.

The pulp and paper industry in India utilizes 53% bamboo, 40% hardwoods and 7% other materials as raw materials (Anon., 1986). Nearly 62% of bamboo production is used for pulp manufacture in India. Of a total 1.6 metric tonnes supplied to the pulp and paper industry annually in India, Madhya Pradesh alone contributes about 20% of the supply. According to the estimates of the National Commission on Agriculture, the pulp and paper requirement in India in 2000 A.D. will be approximately 9.369 metric tonnes, of which an estimated 3.645 metric tonnes will be derived from bamboo alone (Tewari, 1992).

We have estimated that a loss of about 26.5% of bamboo by weight occurs during one year of outside storage due to decay fungi alone. Guha et al. (1958) also reported the loss of 20 to 25% due to microbial decay during one year of storage. The damage is, therefore, enormous in terms of financial and material loss. Moreover, the decay of bamboo results in considerable reduction in the yield (reduced by 20%), quality and strength of the pulp (reduced by 50%), increase in consumption of bleaching chemicals, bulking problems in digestors, and loss of material during chipping and screening (Bakshi et al, 1968; Hatton et al, 1968). The protection of bamboo from decay fungi during storage is therefore essential. There are known preservatives for this purpose but oil-based preservatives have restricted use in pulp-making, and water-borne preservatives are harmful to users, are short-term, cause pollution hazards and are ineffective if the decay fungi have already become established.

The alternative, bioprotection, involves two critical points: (i) the biocontrol agent must be capable of colonizing the substrate (bamboo in this case); and (ii) the biocontrol agent should not damage the substrate. A successful bioprotectant must also have the following essential qualities: (i) long-term survival in the substrate, (ii) specific and direct contact with the target pathogens, (iii) simple and cheap multiplication, packing, distribution and application, (iv) no health hazards, (v) active under a wide range of environmental conditions, and (vi) efficient and economical.

FUNGI CAUSING DECAY IN STORED BAMBOO

During the surveys of various depots used for storing bamboo prior to sale, fifteen wood-decaying fungi were collected and identified, namely *Cyathus limbatus* Tul., *Datronia caperata, Earliella scabrosa* (Pers.) Gilb. & Ryv., *Flavodon flavus* (Klotz.) Ryv., *Ganoderma lucidum* (Leyss.) Karst., *Gloeophyllum striatum* (Fr.) Murr., *Lenzites acuta* Berk., *Microporus xanthopus* (Beac.: Fr.) Kunt., *Polyporus arcularius* (Batsch.) Fr., *P. grammocephalus* Berk., *P. tenuiculus* (Beauv.) Fr., *Poria rhizomorpha* Bagchee, *Schizophyllum commune* Fr., *Trametes cingulata* Berk. and *Trichaptum byssogenum* (Jungh.) Ryv. Of these only two, *Gloeophyllum striatum* and *Poria rhizomorpha,* are brown-rot fungi and rest are white-rot fungi. *Cyathus limbatus* is a Gasteromycete while others belong to the order Aphyllophorales of Hymenomycetes. *Schizophyllum commune* is a pioneer coloniser decay fungus and was found to occur frequently in every storage depot of bamboo, whereas *Polyporus grammocephalus* was found to cause maximum weight loss in bamboo.

SELECTION OF BIOCONTROL AGENTS

Keeping in view the essential prerequisites of a successful bioprotectant as mentioned above, *Trichoderma* species have particular potential for the biological control of basidiomycetous decay fungi. Species of *Trichoderma* Pers.: Fr. are ubiquitous, easy to isolate, grow rapidly on a variety of substrates, affect a wide range of plant pathogens, and are not pathogenic to host plants. They also act as mycoparasites, compete for nutrition and sites, produce antibiotics and possess an enzyme system capable of attacking plant pathogens. Various workers, for example Bergman and Nilsson (1966), Ricard (1976), Bruce and King (1983) and Bruce et al. (1984) have reported the effectiveness of *Trichoderma* spp. against the basidiomycetous decay fungi. In view of the above, thirteen isolates of seven species of *Trichoderma* were isolated from bamboo storage depots (stored chips, decayed bamboo, fruit bodies of decay fungi and soil under the stacks). Preliminary laboratory tests, i.e. presumptive tests, bioculture tests, group action tests, well in agar method, filter paper discs and cellophane paper method, revealed that four species, *Trichoderma atroviride* Rifai (isolate A 4), *T. harzianum* Rifai (A 8), *T. koningii* Oud. (A 2) and *T. pseudokoningii* Rifai (A 12), exhibited antagonistic activity against the test decay fungi. Among these, *T. pseudokoningii* exhibited the best antagonistic activity followed by *T. harzianum* (A 8) in suppressing the growth of test decay fungi.

146

LABORATORY TESTS WITH BAMBOO BLOCKS

Agar block and soil block tests conducted in the laboratory exhibited the effectiveness of *T. pseudokoningii* (A 12) and *T. harzianum* (A 8), evident from the comparative weight loss caused by test decay fungus, *Schizophyllum commune,* in treated and control bamboo blocks (Figure 1).

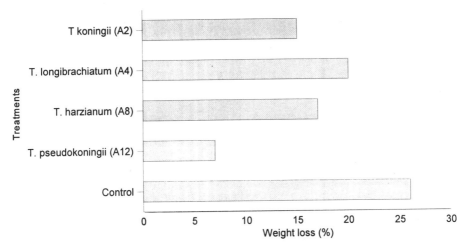

Figure 1. Effect of *Trichoderma* spp. on weight loss caused by *Schizophyllum commune* in bamboo blocks.

T. pseudokoningii (A 12) was found equally effective when tested against different decay fungi (see Table 1). The application time of *T. pseudokoningii* was also worked out and it was found most effective when introduced at the initial stages, that is soon after felling of the bamboo (Figure 2). *T. pseudokoningii* also caused a little weight loss in bamboo blocks which indicated that it was able to establish and grow in the bamboo blocks. However, unless the biocontrol agent is established in the bamboo it will not be effective against the decay fungi. *Trichoderma* species are known to be cellulolytic, but the comparison of weight loss caused by test decay fungus (32.78%) and that caused by T *pseudokoningii* (2.68%) was in favour of its application in the field.

Table 1. Effect of *Trichoderma pseudokoningii* (A 12) on different test fungi in causing decay in bamboo blocks.

Treatments	% weight loss
Inoculated with A 12 + *Schizophyllum commune*	5.41
Inoculated with A 12 + *Polyporus arcularius*	13.86
Inoculated with A 12 + *Datronia caperata*	7.59
Inoculated with *S. commune*	24.76
Inoculated with *P. arcularius*	33.33
Inoculated with *D. caperata*	28.70
Uninoculated control	0

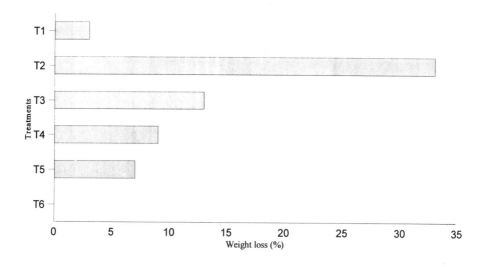

Key: T1=A12 at start
        T2=*P. arcularius* at start
        T3=A12+*P. arcularius* at start
        T4=*P. arcularius* after 2 weeks of A12
        T5=*P. arcularius* after 7 weeks of A12
        T6=Control, no inoculation

Figure 2. Effect of different period of inoculation of test fungus abd *Trichoderma* (A 12) on decay in bamboo blocks.

PRELIMINARY FIELD TRIALS

Freshly felled bamboo *(Dendrocalamus strictus* (Roxb.) Nees) sticks of similar diameter (3 cm) and 1 m length were used to make stacks. The sticks were treated prior to this (a) by dipping the ends into a solution of chemicals (boric acid and borax 1:2), (b) with biological treatment (spore suspension of A 12) and (c) with water for control. They were then stored outside for 12 months and observations of weight loss were recorded which revealed that biological treatment with A 12 provided better protection than the chemical treatment (Figure 3).

Another experiment was conducted when freshly cut bamboo sticks of almost equal diameter (3 cm) and 40 cm length were inoculated simultaneously with the test decay fungi and *T. pseudokoningii* at the cut ends and erected in the soil vertically for 12 months. The results (see Table 2) revealed the effectiveness of *T. pseudokoningii* once again.

PREPARATION OF THE FORMULATION

The formulation of *Trichoderma* spp. (A 12 and A 8) was prepared in powder form by growing them sugarcane bagasse, yielding 2×10^9 colony forming units per gram. The powder was packed in polythene bags and tight sealed for storage at room temperature. The formulation was found to be viable for more than 2 years at room temperature, which is as high as 39°C during the summer months. The cost of preparation of the formulation has been calculated at Rs 175 per kg.

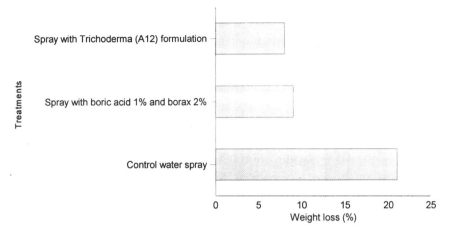

Figure 3. Effect of biological and chemical treatments on decay in bamboo during storage.

Table 2. Observations on field experiment with bamboo sticks.

Treatment	Observations at the end of experiment (mean of 3 replicates)
T1: *Trichoderma pseudokoningii* (A 12) alone at start	Sticks remained sound, a little discolouration in 2 sticks (out of 9), 2 sticks damaged by termites in ends touching the soil
T2: *Schizophyllum commune* at start	Weight loss noticed in all the sticks, fruit bodies observed in 3 sticks, discolouration in all sticks, 2 sticks damaged by termites
T3: *T. pseudokoningii* + *S. commune* simultaneously at start	Discolouration in 2 sticks, 2 sticks had fruit bodies near soil surface, 2 sticks damaged by termites, rest unaffected
T4: *T. pseudokoningii* + *S. commune* + *D. caperata* at start	Discolouration in 2 sticks, 2 sticks had fruit bodies near soil, 2 sticks had termite damage, rest unaffected
T5: Boric acid 1% + Borax 2% at start	Two sticks remained solid, 2 found damaged by termites, rest had fruit bodies of different fungi
T6: *T. pseudokoningii* alone after 3 weeks and again 5 weeks after start	All sticks found unaffected, only 2 sticks had termite damage at the ends
T7: Control untreated	All sticks found discoloured, weight loss in 3 sticks, end damaged by termites in 3 sticks

FIELD TRIALS

A large scale field trial was conducted at one of the largest bamboo depots of Madhya Pradesh at Garra depot of Balaghat (21°19'N and 22°24'N latitude, 79°39'E and 81°3'E longitude).

The biocontrol formulation (0.1 g/l), chemicals boric acid and borax (10 g + 20 g/l) and chemicals in combination with the biocontrol formulation were sprayed as three treatments on the stacks prepared from freshly felled bamboo. A suitable control sprayed with water was also maintained and treatments were replicated three times. Observations of decay to the bamboo were taken after one year which revealed the effectiveness of the formulation (Table 3).

Table 3. Observations on the field trial for the control of decay fungi in stored bamboo.

Treatment	Observations after one year of storage
T1: Treatment with *T. pseudokoningii* formulation	Discolouration in a few sticks, 4 sticks showed fruit bodies of *S. commune*, av. no. of fruit bodies per stick < 5
T2: Treatment with boric acid and borax (1:2)	Most of the sticks had fruit bodies of *S. commune*, av. no. of fruit bodies per stick 10-15
T3: Treatment with *T. pseudokoningii* formulation and boric acid and borax mixture	About 12-15 sticks showed fruit bodies of *S. commune*, av. no. of fruit bodies per stick 6-10.
T4: Control (untreated)	Nearly all sticks showed fruit bodies of *S. commune*, av. no. of fruit bodies per stick >20

CONCLUSION AND SCOPE

This technique of preparing a biocontrol formulation with *Trichoderma* species using bagasse is new. The development of formulation for biological control of decay caused by basidiomycetous fungi in stored bamboo has not been reported elsewhere. The cost of preparation of the formulation is very low and could be further reduced if calculated for large scale production. Chemical preservatives such as boric acid and borax cost around Rs 180 per kg.

Commercial preparations of *Trichoderma* species are available in India for the biological control of various diseases of agricultural crops such as root rot, seedling rot, damping-off, collar rot, *Fusarium* wilt, etc. This technique could therefore also be used for the commercial preparation of other biocontrol agents (fungi) for wider application. Prepared in this way, the formulation has the additional advantage of having a sugar base in the form of bagasse powder which, on spraying with water, provides the initial nutrients for the growth and establishment of the biocontrol agent in the substrate. Besides, the formulation remains viable for a longer period in a natural medium at room temperature and provides a greater number of propagules as compared to those derived from fermenter biomass technology.

REFERENCES

Anon., 1986, *Report of the inter-ministerial group on wood substitution*, Ministry of Environment and Forests, Govt. of India, New Delhi.

Bakshi, B.K., Guha, S.R.D., Singh, Sujan, Panth, P. and Taneja, K., 1968, Decay in flowered bamboo and its effect on pulp, *Pulp and Paper* 22, 9:1-4.

Bergman, O. and Nilsson, T, 1966, On outside storage of pine chips at Lovholmeh's paper mill, Skogshogskolan, Inst. for Virueslara Uppsatser Nr.R. 53 p. 60.

Bruce, A., Austin, W.J. and King, B., 1984, Control of growth of *Lentinus lepideus* by volatiles from *Trichoderma, Trans. Brit. Mycol. Soc.* 82:423.

Bruce, A. and King, B., 1983, Biological control of wood decay by *Lentinus lepideus* Fr. produced by *Scytalidium* and *Trichoderma* residues, *Material und Organismen* 18: 171-181.

Guha, S.R.D., Bakshi, B.K. and Thapar, H.S., 1958, The effect of fungus attack on bamboo on the preparation and properties of pulp, *J. Sci. Indust. Res.* 17C: 72-74.

Hatton, J.V., Smith, R.S. and Rogers, J.H., 1968, Outside chip storage - its effect on pulp quality, *Pulp and Paper*

    Mag. Can. 69(15): 33-36.

Maheshwari, S. and Satpathy, K.S., 1990, The efficient utilization of bamboo for pulp and paper marketing, in: *Bamboo: Current Research* (I.V.R. Rao, R. Gnanaharan and C.B. Shastry, eds.), Kerala Forest Research Institute, India.

Ricard, J.L., 1976, Biological control of decay in standing creosote treated poles, *J. Inst. Wood Sci.* 7(4):6-9.

Tewari, D.N., 1992, *A Monograph on Bamboo,* International Book Distributors, Book Sellers and Publisher, Dehradun, India.

USE OF FUNGI IN THE CONTROL OF PLANT PATHOGENS

Brajesh Kumar Singh[1], Sapna Arora[2], R.C. Kuhad[1] and K.G. Mukerji[3]

[1]Department of Microbiology
University of Delhi
South Campus
New Delhi-110 021
India

[2]Enbee Chemicals Ltd.
Bhopal
India

[3]Department of Botany
University of Delhi
Delhi-110 007
India

INTRODUCTION

Indiscriminate and excessive use of chemical pesticides to control plant diseases has recently received considerable criticism. The past thirty years of chemicalization of agriculture have led to severe environmental threats to plant, animal and human life around the world. According to World Health Organisation estimates, approximately one million people are taken ill every year with pesticide poisoning and up to 20,000 of those die in agony (Mukhopadhyay, 1997).

Pesticides are necessary at present but are not a long term solution to crop, human and animal health. Besides their non-target effects and hazardous nature, they are becoming more expensive and some are losing their effectiveness because of the development of resistant strains. In recent years, the need to develop disease control measures as alternatives to chemicals has become the priority of scientists worldwide. Biological control, especially using fungi against plant pathogens, has gained considerable attention and appears to be promising as a viable supplement or alternative to chemical control (Cook and Baker, 1983; Lumsden and Lewis, 1988; Papavizas, 1981, 1985).

According to Baker and Cook (1974) 'biological control is the reduction of inoculum density or disease-producing activities of a pathogen or parasite in its active or dormant state, by one or more organisms, accomplished naturally or through manipulation of the environment, host or antagonist or by mass introduction of one or more antagonists'. However, the greatest potential for an increased role of biological control in crop disease management lies not only in commercial biological control agents, but also in exploiting our ever increasing understanding of the role of various cultural practices on general biological activities in crop ecosystems (Nigam and Mukerji, 1992). Moreover, the main purpose of biological control of a plant disease is to suppress the inoculum load of the target pathogen below the level that potentially causes an economically significant outbreak of disease. A number of reviews have been carried out on reports dealing with various aspects of biocontrol, for example Baker and Cook, 1974; Baker and Dickman, 1992; Chet, 1989; Mukerji and Garg, 1988a, b; Mukerji et al., 1992; Papavizas, 1992; Whipps, 1993. In this chapter we mainly discuss control of plant pathogens by antagonistic fungi.

FUNGAL BIOCONTROL AGENTS

Fungi represent an excellent reservoir for isolating and demonstrating microbes of biocontrol ability. Many diverse physiological groups are found in the soil and rhizosphere; the type and amounts depend largely upon soil type, nature and age of the plant root and method used for isolation (Rovira, 1965). Some fungal genera routinely isolated from the rhizosphere and phylloplane of various plants which have been implicated in biocontrol are listed in Table 1. Most developmental research to date has been done with isolates of *Trichoderma* and *Gliocladium* which are effective against various diseases of many crops in the greenhouse and field (Chet, 1987; Papavizas, 1985).

MECHANISMS OF BIOCONTROL

The microorganisms used in biological control of plant diseases are termed 'antagonists'. An antagonist is a microorganism that adversely affects the growth of other microbes growing in association with it. The microbial antagonism that is seen in biological control of plant pathogens is broadly based on competition (for nutrients and space), parasitism and hyperparasitism. These mechanisms may operate together or independently and their activities can result in suppression of microbial plant pathogens.

Competition

The term 'competition' was defined by Clarke (1965) as the active demand in excess of the immediate supply of materials or conditions on the part of two or more organisms.

Competition operates in many biological control systems, but it is the least understood, most complex and perhaps least amenable to quantification of any physical or chemical characteristic (Baker and Dickman, 1992). Dormant propagules of pathogen in soil need some exogenous nutrients to germinate and to infect the host. Different components of microflora in soil compete for such requirements.

Competition of the biocontrol agents with other microorganisms can be triggered by abiotic factors. Ammonium sulphate solution as suggested by Seaby (1977) improves the biocontrol of *Heterobasidion annosum* by *Trichoderma viride* perhaps by providing supplemental nitrogen to *T. viride*, inhibiting antagonistic bacteria and killing some host cells enabling *T. viride* to colonise faster.

Table 1. Commonly studied antagonist-fungal pathogen systems.

Biocontrol agents evaluated	Diseases	Causal organism	References
Alternaria	Leaf spots and blights	*Cercospora, Venturia, Alternaria,*	Austin et al. (1977) and Windels & Lindow (1985)
Ampelomyces	Powdery mildew	*Sphaerotheca, Podosphaera*	Blackman & Fokkemma (1982)
Chaetomium	Leaf spots and blights	*Venturia*	Boudreau & Andrew (1987) Cullen et al. (1984)
Coniothryium	Root and crown rots	*Sclerotium*	Ahmed et al. (1977)
		Sclerotinia	Huang (1977), Turner & Tribe (1976)
Acremonium	Grey mould	*Botrytis*	Janisiewicz (1988)
Athelia	Leaf spots and blights	*Venturia*	Windels & Lindow (1985)
	Scab	*Venturia*	Heya & Andrew (1983)
Aspergillus	Anthracnose	*Colletotrichum*	Kanapathipillai (1988)
Cladosporium	Grey mould	*Botrytis*	Bhatt & Vaughan (1960)
Debaryomyces	Blue mould Grey mould	*Penicillium italicum P. digitatum*	Chalutz & Wilson (1989)
Fusarium	Tree decays	*Heterobasidion, Cryphonectria, Chondrostereum*	Mercer (1988)
Gliocladium	Root rots and crown rots Fruit rots	*Rhizoctonia Sclerotium Sclerotinia*	Janisiewicz (1987) Howell (1987) Tu (1980) Tu & Vaartaja (1981)
Laestisaria	Damping off	*Pythium*	Martin et al. (1983)
Penicillium (non-pathogenic)	Damping off Fruitlet core rot	*Pythium Penicillium* (pathogenic)	Lim & Rohrbach (1798) Windels & Kommedahl (1978)
Myrothecium	Green mould	*Penicillium*	Appel et al. (1988) Gees et al. (1988)
Pythium	Damping off	*Pythium*	Martin & Hancock (1987)
Trichoderma	Damping off	*Pythium*	Blackman & Fokkemma (1982)
	Root rots Crown rots Vascular wilts Stem blights / cankers Fruit rots	*Rhizoctonia Sclerotium Sclerotinia Fusarium Botrytis, Penicillium*	Elad et al. (1984) Hadar et al. (1979) Harman et al. (1980) Lifeshitz et al. (1986) Sivan & Chet (1986)
Verticillium	Tree decays Root rots Rust	*Rhizopus etc. Rhizoctonia Uromyces, Puccinia*	Tronsmo Ystaas (1980) Blackman & Fokkemma (1982) Jager & Velvis (1986)

The nature of competition is fungistatic (inhibitory) and there will be little change in the inoculum density of the rhizosphere adjacent to foci of infection before and after a competitive biological control strategy has been applied. Therefore the prospects for efficient application of fungal competition in biological control are very limited.

Mycoparasitism

Mycoparasitism occurs when one fungus exists in intimate association with another from which it derives some or all of its nutrients while conferring no benefit in return (Lewis et al., 1989). Based on their mode of parasitism, mycoparasites are separated into two major groups, biotrophs and necrotrophs (Barnett and Binder, 1973). Biotrophs are usually regarded as those organisms that are able to obtain nutrients from living cells and are recognised by the fact that invaded tissues do not die. This type of parasitic relationship is physiologically balanced and parasites seem to be highly adapted to this mode of life. Biotrophic mycoparasites tend to have a more restricted host range and produce specialised structures to absorb nutrition.

The necrotrophic or destructive mycoparasites kill the host cell slightly before or after the invasion and absorb nutrients from the dying or dead host cells. Necrotrophic mycoparasites, unlike the biotrophic parasites, tend to be more aggressive, have a broad host range and are relatively unspecialised in their mode of parasitism. The antagonistic activity of necrotrophic mycoparasites is attributed to the production of antibiotics, toxins or hydrolytic enzymes (Chet et al., 1997). Mycoparasitism is a commonly observed phenomenon *in vitro* and *in vivo* and its involvement in biological disease control has been reviewed extensively (Ayers and Adams, 1981; Lumsden, 1981; Manocha, 1991; Nigam and Mukerji, 1992; Sharma et al., 1998; Wells, 1988).

The hyphae of *Trichoderma* have been shown to parasitise many pathogenic fungi, viz. *Rhizoctonia, Sclerotium, Sclerotinia, Helminthosporium, Fusarium, Verticillium, Venturia, Pythium, Phytophthora, Rhizopus, Botrytis,* etc. (Beagle-Restaino and Papavizas, 1985; Chet et al., 1981; Kumar, 1993; Wolfehechel and Jensen, 1992). When a mycoparasite is grown with its host in dual culture, it grows towards the host and a typical branching pattern occurs. The recognition phenomenon is accomplished by binding of agglutinin (lectin) of the host to the carbohydrate residues on the cell walls of *Trichoderma* spp. Hydrolytic enzymes such as ß-1,3-glucanase, cellulase, chitinase and other different combinations of enzymes aid biocontrol agents in the penetration of the host hypha cell (Baker and Dickman, 1992).

To achieve the desired result from mycoparasitic biocontrol, the physical and nutritional environment must be favourable for the agents to be active. Many researchers have reported on the successful biocontrol of important diseases by using mycoparasitic agents in glasshouse trials (Ayers and Adams, 1981; Sharma et al., 1998). Whipps et al. (1989) reported self maintenance by *Coniothyrium minitans* against *Sclerotinia sclerotiorum* disease of lettuce and celery in glasshouse trials. Although success in the open environment is rare, mycoparasitism has the potential for eradication of pathogens and remains an attractive strategy in biological control.

Antibiosis

Baker and Griffin (1995) defined antibiosis as 'inhibition or destruction of an organism by the metabolic production of another'. The production of inhibitory metabolites such as toxic molecules, volatiles and lytic enzymes by fungal biocontrol agents has been reported for many years (Bryan and McGowan, 1945; Dennis and Webster, 1971; Joe, 1998). However, insufficient evidence exists for their contribution to pathogen suppression and disease reduction *in situ*. *Gliocladium virens* is a common example of the role of antibiotics in biological control by fungal antagonists. Gliovirin is a diketopiperazine antibiotic which appears to kill the fungus *Pythium ultimum* by causing coagulation of its protoplasm (Chet et al., 1997). Mycelium of *P. ultimum* that has been exposed to gliovirin does not grow even after washing and transferring to fresh medium (Howell, 1982; Howell et al., 1993). Howell (1991) observed that a combination of *G. virens* treatment of cotton seed with a reduced level of the fungicide metalaxyl provided disease suppression equal to that of a full fungicidal treatment. Antifungal, volatile alkyl pyrones produced by *Trichoderma harzianum* were inhibitory to a number of fungi *in vitro* and, when added to a peat soil mixture, they reduced the incidence of damping off in lettuce caused by *Rhizoctonia solani* (Chet et al., 1997). Ordentlich et al. (1992) isolated a novel inhibitory substance, 3-2 (hydroxypropyl),4-(2,4-hexadienyl)2(5H)-furanone, produced by one isolate of *T. harzianum* that was found to suppress the growth of *Fusarium oxysporum* and may be involved in biocontrol of *Fusarium* wilt. However, the real role of antibiosis in biological control is not fully understood. Even in cases where antifungal metabolite production by an agent reduces disease, other mechanisms may also be operating (Baker and Griffin, 1995).

ACHIEVEMENTS

Despite the lack of spectacular achievements in biological control of plant disease on a commercial scale, initial results from greenhouse trials and laboratory based screening are encouraging and worthwhile. Many research and review articles reported successful control of various plant diseases using different antagonists. Fungi in the genus *Trichoderma*, a filamentous deuteromycetes, have recently gained considerable recognition as effective biocontrol agents. *Trichoderma* spp. are most promising antagonists against a wide range of plant pathogenic fungi which attack several economically important plants. Some of the important species established as biocontrol agents are *T. harzianum, T. viride, T. polysporum* and *T. hamatum*. They are ubiquitous in distribution, occur in all soils and habitats and sporulate readily on natural and artificial substances. They are capable of surviving in diverse climatic conditions and it is advantageous to select the species according to the soil type and climate. For instance, *T. hamatum* and *T. pseudokonigii* are adapted to excessive soil moisture, *T. viride* and *T. polysporum* are adapted to low temperature and *T. harzianum* to both cool and warm climatic regions (Joe, 1998).

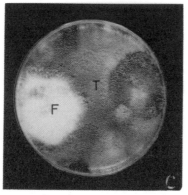

Figure 1. Colony of *Fusarium oxysporum* f. sp. *lycopersici* overgrown by *Trichoderma viride* DU/MS/1069.

The first practical use of *Trichoderma* species was achieved for the control of *Armillaria mellea* fungus which causes root-rot of citrus (Nigam, 1997). *T. harzianum* has been successfully used as a potential agent for controlling several diseases caused by *R. solani* on bean, tomato, peanut and cucumber (Hadar et al., 1979; Nigam, 1997; Wells et al., 1972). *Trichoderma* can also control wood-rotting, wood infecting fungi (Smith et al., 1981) and a variety of soil-borne pathogens of seedlings, mature plants and mushrooms. Weindling (1932) was the first to report *Trichoderma* as a mycoparasite of *R. solani, S. rolfsii, Phytophthora* spp. and *Pythium* spp.

These fungi combat plant pathogens also by antagonism in the form of antibiosis (anti-fungal metabolites such as trichodermin and viridin) and competition for nutrition and/or exploitation (Chet et al., 1997). Some of the commercial products of *Trichoderma* spp. available are Antagon, Biocure-F, Bioderma, Dermapak, Funginil, Trichofit and Trichosan in India and Binab-T, F-stop, Trichodermin and Trichodex in other countries. These species exhibit several advantages in their use as potent biofungicides. They involve neither plant parasitism nor post-harvest crop loss. They are non-toxic to animals and humans and can be applied to soil along with biofertilizers (Joe, 1998).

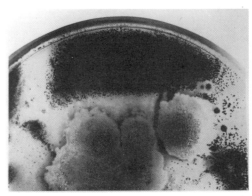

Figure 2. Colonies of *Sporotrichum carthusio-viride* inhibiting growth of *Aspergillus niger.*

Gliocladium roseum parasitises *Phomopsis sclerotioides*, the causal organism of black root rot of cucumber. The pathogen is killed by enzymatic activity of *G. roseum* (Moody and Gindrant, 1977). *G. roseum* has successfully controlled *Pythium ultimum* and *R. solani* by producing antibiotics. Some species of *Fusarium* are mycoparasites to rusts and smuts (Nigam, 1997). *Puccinia graminis* f. sp. *tritici, Claviceps purpurea* and *R. solani* were controlled by *Fusarium* spp. (Arora and Dwivedi, 1980; Hornok and Weloz, 1983; Prasad and Sharma, 1964).

LIMITS OF FUNGAL BIOCONTROL AGENTS

During the early years of biological control, following some of the initial exciting discoveries, it was thought that biological control of plant pathogens might herald the end of chemical fungicide use. However, this optimism was quickly followed by disappointment as control failures eventually outnumbered successes (Baker and Cook, 1974). A number of *in vitro* experimental successes of biocontrol have failed to be reproduced in natural systems.

A number of criteria must be met for the successful development of a commercially viable biocontrol agent (Scher and Castagno 1986). For example, the product should be (i) in demand in the market place; (ii) technically feasible; (iii) economically viable; (iv) competitively attractive; (v) acceptable to environmentalists and regulatory agencies and (vi) compatible with the company's activities and interests.

Currently, limits to the commercialization of biological control imposed by criteria (i), (ii) and (vi) have apparently been overcome in many instances (Nelson, 1991). Because most of the biocontrol products are relatively new to the market place, it is not yet clear whether they will compete well with chemical fungicides and be acceptable to environmentalists and regulatory agencies.

Another limitation is acceptance to growers and public awareness of biopesticides.

Growers have become so accustomed to chemical pesticides that slow control of disease by biological agents has not been successful in matching users' high expectations. Education efforts are needed to inform growers and the general public of the virtues and limitations of biological control approaches.

FUNGAL BIOCONTROL AGENTS IN INTEGRATED DISEASE MANAGEMENT PROGRAMMES

Integrated disease management (IDM) achieves ideal control of plant diseases by harmonizing techniques in an organized way, by making control practices compatible and by blending them into a multi-faceted, flexible, evolving system (Kumar and Mukerji, 1997). Simmon and Sivasithamparam (1989) have reviewed the literature on integration of antagonists with other biological and chemical pesticides. They suggested "the combination of fungal bio-control agents is needed for the control of take-all diseases of wheat". It has been pointed out that since *Trichoderma* spp. are ubiquitous and dominant in soil, then if treated with a sublethal dosage of fungicides, they can proliferate easily to produce antibiotics, compete for nutrients and also sometimes act as mycoparasites. Their role in integrated disease management may therefore be of practical significance (Kumar and Mukerji, 1997).

Many important pathogens have been successfully controlled by an integrated approach using biocontrol agents and fungicides. These include control of *Rhizoctonia solani* on radish, egg plant and cucumber by *T. harzianum* in combination with PCNB and chlorothalonil (Dennis and Webster, 1971; Lewis and Papavizas, 1980), *Sclerotium rolfsi* on bean by *T. harzianum* with PCNB (Chet et al., 1979), *Phytophthora capsici* on pepper by *T. harzianum* with ridonil (Papavizas and Lewis, 1981).

CONCLUSIONS

Although biological control of plant diseases with fungal antagonists cannot dramatically solve agricultural problems at present, laboratory and greenhouse results are exciting and encouraging. The initial response of growers towards a commercially available formulation of *Trichoderma* spp. has been good, however it may take several years before microbial biocontrol products are fully accepted by farmers.

To expedite the process for acceptance of biofungicides by growers, several important considerations must be implemented. Foremost among these is greater co-operation between industrial and non-industrial scientists. This is essential for the interchange of basic scientific information on research and development activities. Development of models and methods to integrate biocontrol agents in IDM programmes is urgently needed. There is also a need to incorporate new biotechnological tools into existing biological control programmes to improve mechanism of action and induced host response. These in turn will help in improving microbial action against the pathogens through the manipulation of both their biology and genetic constitution.

It is very important to launch awareness programmes among farmers and other users of pesticides about the benefits and potential of biocontrol agents. Despite the practical constraints, once the potential of microbial biocontrol agents is realised it should eventually become possible to overcome the factors impeding its successful development and commercialization.

REFERENCES

Ahmed, A.H.M. and Tribe, H.T., 1977, Biological control of white rot of onion *(Sclerotium cepivorum)* by *Coniothyrium minitans, Plant Pathol.* 26: 75-78.

Appel, D.J., Gees, R. and Coffey, M.D., 1988, Biological control of the post harvest pathogen *Penicillium digitatum* on Eureka lemons, *Phytopathol.* 78: 1593 (Abstr.).

Arora, D.K. and Dwivedi, R.S., 1980, Mycoparasitism of *Fusarium* spp. of *Rhizoctonia solani* Kunn, *Plant and Soil*, 55: pp. 43-53.

Austin, B., Dickinson, C.H. and Goodfellow, M., 1977, Antagonistic interaction of phylloplane bacteria with *Dreschlera dictoides* (Dreschler) Shoemaker, *Can. J. Microbiol.* 23: 710-715.

Ayers, W.A., and Adams P.B., 1981, Mycoparasitism and its application of biological control of plant diseases, in: *Biological Control in Crop Production* (G.C. Papavizas, ed.), Allanueld, Granada, pp. 91-103.

Baker, K.F. and Cook, R.J., 1974, Biological Control of Plant Pathogens, Freeman Press, San Francisco, 1-433.

Baker, R. and Dickman, M.B., 1992, Biological control with fungi, in: *Soil Microbial Ecology: Application in Agricultural and Environmental Management* (F. Blain, ed.) Metting, Jr. pp. 275-305.

Baker, R. and Griffin, G.J., 1995, Molecular strategy for biological control of fungal plant pathogens, in: *Novel Approaches to Integrated Pest Management* (R. Reuveni, ed.), CRC Press Boca Raton, 369-381.

Barnett, H.L. and Binder, F.L., 1973, The fungal host-parasite relationship, *Ann. Rev. Biochem.* 50: 207-292.

Beagle-Restaino, J.E. and Papavizas G.C., 1985, Biological control of *Rhizoctonia* stem canker and black scurf of potato, *Phytopathol.* 75: 560-564.

Bhatt, D.D. and Vaughan, E.K., 1962, Preliminary investigations on biological control of gray-mold *(Botrytis cinera)* of strawberries, *Plant Dis. Repter.* 46: 342-3 45.

Blackman, J.P. and Fokkemma,.N.J., 1982, Potential for biological control of plant diseases on the phylloplane, *Ann. Rev. Phytopathol.* 20: 167-192.

Boudreau, M.A. and Andrew. J.H., 1987, Factors influencing antagonism of *Chaetomium globosum* to *Venturia inaequalis:* A case study in failed biocontrol, *Phytopathol.,* 77: 1470-1475.

Bryan, P.W. and McGowan, J.C., 1945, Viridin: A highly fungistatic substance produced by *Trichoderma viride, Nature* 151: 144-145.

Chalutz, E. and Wilson, C.L., 1989, Postharvest biocontrol of green and blue mold and sour rot of citrus fruit by *Debaryomyces havsenil, Plant Dis.* 74: 134-137.

Chet, I., 1987, *Innovative Approaches to Plant Disease Control,* Wiley, New York.

Chet, I., 1989, Mycoparsitism - recognition physiology and ecology, in: *New Direction in Biological Control* (R. Baker and P.C. Dunn, eds.) R. Liss, New York, pp. 725-733.

Chet, I., Harman G.E. and Baker, R.J., 1981, *Trichoderma hamatum*: its hyphal interaction with *Rhizoctonia solani and Pythium* spp., *Microb. Ecol.* 7: 29-38.

Chet, I., Inbar, J. and Hadar, Y., 1997, Fungal antagonists and mycoparasites, in: *The Mycota* (K. Eser and P.A. Lemke, eds.), Springer-Verlag, New York, pp.163-184.

Chet, I., Hadar, Y., Elad, Y., Kalan, J. and Henis, Y., 1997, Biological control of soil borne plant pathogens by *Trichoderma harzianum,* in: *Soil-borne Plant Pathogens* (B. Schippers and W. Grams, eds.) Academic Press, London, pp. 585-590.

Clarke, F. E., 1965, The concept of competition in microbial ecology, in: *Ecology of Soil borne Plant Pathogens* (K.F. Barker and W.C. Snydari, eds.), California Univ. Press, Berkeley, pp. 339-345.

Claydon, N., Allan, M., Hanson, J.R. and Avent, A.G., 1987, Antifungal alkyl pyrones of *Trichoderma harzianum, Trans. Br. Mycol. Soc.* 88: 503-513.

Cook, R.J. and Baker, K.F., 1983, *The Nature and Practice of Biological Control of Plant Pathogens,* Am. Phytopathol. Soc., St. Paul, Minnesota, pp.1-539.

Cullen. D., Berbee, F.M. and Andrew, J.H., 1984, *Chelonium globosum* antagonizes the apple scab pathogen, *Venturia inaequalis,* under field conditions, *Can. J. Bot.* 62: 1814-1818.

Dennis, C. and Webster, J., 1971, Antagonistic properties of species group of Trichoderma II. Production of volatile antibiotics, *Trans. Br. Mycol. Soc.* 57: 4-48.

Elad, Y., Barak, R. and Chet, I., 1984, Parasitism of *Sclerotium rolfsii* sclerotia by *Trichoderma harzianum, Soil Biol. Biochem.* 16: 381-386.

Gees, R., Droby, S. and Cotley, M.D., 1988, Post harvest disease control of green mold on orange by a mutant strain of *Myrothecium roridum. Phytopathol..* 78: 1593-1594.

Hadar, Y., Chet, I. and Henis, Y., 1979, Biological control of *Rhizoctonia solani* damping off with wheat bran cultures of *Trichoderma harzianum, Phytopathol.* 69: 64-68.

Harman, G.E., Chet, I. and Baker, R., 1980, *Trichoderma hamatum* effects on sad and seedling disease induced in radish and pea by *Pythium* spp. or *Rhizoctonia solani, Phytopathol.* 70: 1167-1172.

Heye, C.C. and Andrew, J.H., 1983, Antagonism of *Athelia bombacina* and *Chaetomium globosum* to the apple scab pathogen, *Venturia inaequalis, Phytopathol.* 73: 650-654.

Hornok, L. and Welooz, T., 1983, *Fusarium heterospermum* a highly specialised hyperparasite of *Claviceps purpurea, Trans. Brit. Mycol. Soc.* 80: 377-380.

Howell, C.R., 1982, Effect of *Gliocladium virens* on *Pythium ultimum, Rhizoctonia solani* and damping off of cotton seedlings, *Phytopathol.* 72: 496-498.

Howell, C.R., 1987, Relevance of mycoparasitism in biological control *of Rhizoctonia solani* by *Gliocladium virens, Phytopathol.* 77: 992-994.

Howell, C.R., 1991, Biological control of *Pythium* damping off of cotton with seed coating preparation of *Gliocladium virens, Phytopathol.,* 81: 738-741.

Howell, C.R., Stipanovic, R.D. and Lumsden, R.D., 1993, Antibiotic production by strains of *Gliocladium virens* and its relations to biocontrol of cotton seedling diseases, *Biocontrol Sci. Technol.,* 3: 435-441.

Huang, H.C., 1977, Importance of *Coniothyrium minitans* in survival of sclerotia of *Sclerotinia sclerotivorum* in wilted sunflower, *Can. J. Bot.,* 55: 289-295.

Jager, G. and Velvis, H., 1986, Biological control of *Rhizoctonia solani* on potatoes with antagonists: The effectiveness of three isolates of *Verticillium biguttatum, Neth. J. Pl. Pathol.* 92: 231-238.

Janisiewicz, W.J., 1987, Post harvest biological control of blue mold on apples, *Phytopathol.* 77: 481-485.

Janisiewicz, W.J., 1988, Biological control of disease of fruits, in: *Biological Control of Plant Diseases. Vol. 2* (K.G. Mukerji and K.L. Garg, eds), CRC Press.

Joe, Y., 1998, *Trichoderma*, a potential biofungicide, *The Hindu,* Feb 12, p24.

Kanapathipillai, V.S., 1988, Non chemical control of *Colletotrichum musae* the anthracnose pathogen of banana, *Proc. 5th Int. Congr. of Plant Pathol, Koyoto,* p-422-425.

Kumar R.N., 1993, Biological control of damping off disease of Chir Pine, *Ind. J. Mycol. Plant Pathol.* 23 (3): 35-37.

Kumar R.N. and Mukerji, K.G., 1997, Integrated disease management - future perspectives, in: *Advances in Botany*, Adity Book Pvt. Ltd., New Delhi, pp. 91-118.

Lewis, J.A. and Papavizas, G.C., 1980, Integrated control of *Rhizoctonia* fruit rot of cucumber, *Phytopathol.* 70: 85-89.

Lewis, K., Whipps, J.M. and Cook, R.C., 1989, Mechanism of biological disease control with special reference to the case study of *Pythium oligandrum* as an antagonist, in: *Biotechnology of Fungi for Improving Plant Growth* (J.M. Whipps and R.D. Lumsden, eds.), Cambridge University Press, Cambridge, pp.191-209.

Lifeshitz, R., Windham, M.T. and Baker, R., 1986, Mechanism of biological control of preemergence damping off of pea by seed treatment with *Trichoderma* spp., *Phytopathol.* 26: 720-725.

Lim, T.K. and Rohrbach, K.G., 1980, Role of *Penicillium funiculosum* strains in the development of pineapple fitiseaes, *Phytopathol.* 70: 663-665.

Lumsden, R.D., 1981, Ecology of mycoparasitism, in: *The Fungal Community: Its Organisation and Role in the Ecosystem* (D.T. Wicklow and G.C. Carrol, eds.), Marcel Dekker, New York, pp. 295-318.

Lumsden, R.D. and Lewis, J.A., 1988, Selection, production, formulation and commercial use of plant disease biocontrol fungi: problems and progress, in: *Biotechnology of Fungi for Improving Plant Growth* (J.M. Whipps and R.D. Lumsden, eds.), Cambridge University Press, pp. 171-232.

Manocha, M.S., 1991, Physiology and biochemistry of biotrophic mycoparasitism, in: *Handbook of Applied Mycology Vol I* (D.K. Arora, B. Rai, K.G. Mukerji, G.R. Knudsen, eds.), Marcel Dekker, New York, pp. 273-300.

Martin, F.N. and Hancock, J.G., 1987, The use of *Pythium oligandrum* for biological control of preemergence damping off caused by *P. ultimum, Phytopathol.* 77: 1013-1020.

Martin, S.B., Hoch, H.C. and Abawi, G.S., 1983, Population dynamics of *Laelisaria arvalis* and low temperature *Pythium* spp. in untreated and pasteurized but field soils, *Phytopathol.,* 73: 1445-1449.

Mercer, P.C., 1988, Biological control of decay fungi in wood, in: *Biological Control of Plant Disease Vol I,* (K.G. Mukerji and K.L. Garg, eds.), CRC Press, Boca Raton, F.L., pp. 177-198.

Moody, A.R. and Gindrant, D., 1977, Biological control of cucumber black root rot by Gliocladium roseum, *Phytopathol.* 67: 1159-1196.

Mukerji, K.G. and Garg, K.L. (eds.), 1988a, *Biocontrol of Plant Diseases, Vol. I,* CRC Press, Florida, USA.

Mukerji, K.G. and Garg, K.L. (eds.), 1988b, *Biocontrol of Plant Diseases, Vol. II,* CRC Press, Florida, USA.

Mukerji, K.G.,Tewari, J.P., Arora, D.K. and Saxena, G.(eds.), 1992, *Recent Development in Biocontrol of Plant Diseases*, Aditya Books Pvt. Ltd., New Delhi.

Mukhopadhyay, A.N., 1997, Biopesticides for plant disease management, *Proc. Nat. Conf. Biopesticides,* pp. 75-77.

Nelson, E.B., 1991, Current limits to biological control of fungal pathogens, in: *Handbook of Applied Mycology Vol. I* (D.K. Arora, B. Rai, K.G. Mukerji and G.R. Khudsen, eds.), Marcel Dekker, New York, pp. 327-335.

Nigam, N., 1997, Fungi in biocontrol, in: *New Approaches in Microbial Ecology* (J.P. Tiwary, G. Saxena, N. Mittal, I. Tewari, B.P. Chamola, eds.) Aditya Books Pvt. Ltd., New Delhi, pp. 427-457.

Nigam, N. and Mukerji, K.G., 1992, Biocontrol of powdery mildew, in: *Recent Developments in Biocontrol of Plant Diseases* (K.G. Mukerji, J.P. Tiwari, D.K. Arora and G. Saxena eds.), Aditya Books Pvt. Ltd., New Delhi, pp. 17-22.

Ordentlich, A., Weisman, Z., Gottieb, H.E., Cojocaru, M. and Chit, I., 1992, New inhibitory natural product produced by the biocontrol agent *Trichoderma harzianum, Phytochem.* 31: 485-486.

Papavizas, G.C., 1981, *Biological Control in Crop Protection*, Allanheld Osmun, London.

Papavizas G.C., 1985, *Trichoderma* and *Gliocladium:* Biology, ecology and potential for biocontrol, *Ann. Rev. Phytopathol.* 23: 23-54.

Papavizas, G.C., 1992, *Biological control of selected soil-borne plant pathogens with Gliocladium and Trichoderma*, ISBN, 0-306-442223-230.

Papavizas G.C. and Lewis, J.A., 1981, Introduction and augmentation of microbial antagonists for the control of soil-borne plant pathogens, in: *Biological Control in Crop Production* (G.C. Papavizas, ed.) Allanheld Osmum and Co., Totawa, N.J., pp. 305-322.

Prasad, R. and Sharma S.K., 1964, *Fusarium* spp. mycoparasitic on stem rust of wheat *Puccinia graminis* var. *tritici* (Pers.) Eriks and Henn., *Ind. Phytopathol.* 17: 258-261.

Rovira, A.D., 1965, Interactions between plant roots and soil microorganisms, *Ann. Rev. Microbial,* 19: 241-266.

Scher, F.M. and Castagno, J.R., 1986, Biocontrol: A view from industry. *Can. J. Plant Pathol.* 8: 222-224.

Seaby, D.A., 1977, The possibility of using *Trichoderma viride* for the control of *Heterobasidium annosum* on conifer stumps, in: *Proc. Semin. Biol. Control*, Royal Irish Academy, Dublin, pp. 57-59.

Sharma, M., Mittal, N., Kumar R.N. and Mukerji, K.G., 1998, Fungi: Tool for plant disease management, in: *Microbes: For Health, Wealth and Sustainable Environment* (A. Varma, ed.), Malhotra Publishing House, New Delhi.

Simmon, A. and Sivasithamparam, K., 1989, Pathogen suppression. A case study in biological suppression of *Gaeumannnomyces graminis* var. *tritici* in soil, *Soil Biol. Biochem.* 21: 311-355.

Sivan, A. and Chet, I., 1986, Biological control of Fusarium crown rot of tomato by *Trichoderma harzianum* under field condition, *Plant Dis.* 71: 587-592.

Smith, K.T., Blanchard, R.O., and Schortle, W.C., 1981, Postulated mechanism of biocontrol of decay fungi in red maple wounds treated *with Trichoderma harzianum, Phytopathol.* 71: 496-498.

Tronsmo, A. and Ystaas, 1980, Biological control of *Botrytis cinera on* apple, *Plant Dis.* 64: 1009-1011.

Tu, J.C., 1980, *Gliocladium virens*: a destructive mycoparasite of *Sclerotinia sclerotiorum, Phytopathol.* 70: 670-674.

Tu, J.C. and Vaartaja, O., 1981, The effect of the hyperparasite *(Gliocladium virens)* on *Rhizoctonia solani* and on *Rhizoctonia* root-rot of white beans, *Can. J. Bot.* 59: 22-27.

Turner, G.J. and Tribe, H.T., 1976, On *Coniothyrium minitans* and its parasitism of *Sclerotinia* spp., *Trans. Br. Mycol. Soc.* 66: 97-105.

Weindling, R., 1932, *Trichoderma lingnorum* as a parasite of other fungi, *Phytopathol.* 22: 837-841.

Wells, H.D., 1988, *Trichoderma* as biocontrol agent in: *Biocontrol of Plant Disease, Vol. I.* (K.G. Mukerji & K.L. Garg, eds.) CRC Press, Florida, p. 7182.

Wells, H.D., Bell, D.K. and Jaworski, C., 1972, Efficiency of *Trichoderma harzianum* as a biocontrol for *Sclerotium rolfsii, Phytopathol.* 62: 442-447.

Whipps, T.M., 1993, A review of white rust *(Puccinia horiana)* disease *on Chrysanthemum* and the potential for its biological control with *Verticillium lecanii* Viegel, *Ann. Appl. Biol.* 122: 173-187.

Whipps, J.M., Budge, S.P. and Ebben, M.H., 1989, Effect of *Coniothyrium minitans* and *Trichoderma harzianum* on *Sclerotinia* disease of celery and lettuce in the glass house at a range of humidities, in: *Proc. CEC Joint Experts*, Meeting Cabrils, Spain.

Windels, C.E. and Kommedahl, T., 1978, Factors affecting *Penicillium oxalicum* as a seed protectant against seedling blight of pea, *Phytopathol.* 68: 1656-1661.

Windels, C.E. and Lindow, S.E., 1985, *Biological Control on Phylloplane*, Amer. Phytopathol. Soc., St. Paul, Minnessota pp. 169-178.

Wolfehechel, H. and Jensen, D.F., 1992, Use of *Trichoderma harzianum* and *Gliocladium virens* for biological control of post emergence damping off and root rot of cucumbers caused by *Pythium ultimum, J. Phytopathol.* 136 (3): 221-230.

BIOLOGICAL CONTROL OF ROOT-KNOT NEMATODE BY NEMATODE-DESTROYING FUNGI

Neelima Mittal, Geeta Saxena and K.G. Mukerji

Applied Mycology Laboratory
Department of Botany
University of Delhi
Delhi-110 007
India

INTRODUCTION

Biological control is a broad concept which encompasses a range of control strategies including cultural practices, host plant resistance and the introduction or encouragement of antagonistic organisms. Baker and Cook (1974) defined biological control of nematodes as reduction in nematode damage by organisms antagonistic to nematodes through the regulation of nematode populations and/or a reduction in the capacity of nematodes to cause damage, which occurs naturally or is accomplished through the manipulation of the environment or by the mass introduction of antagonists.

In the past, the use of pesticides has increased rapidly. In some countries, the fumigants dibromochloropropane and ethylene dibromide, which were the basis of nematode control, have been removed from the market because of health and environmental problems. With the increasing cost of testing and registering pesticides, the development of new nematicides has almost ground to a halt, so that additional non-chemical means of nematode control will have to be found. Biological control will play an increasing role in practical nematode control in the future.

All plant-parasitic nematodes spend at least part of their lives in soil, one of the most complex of habitats. A wide and diverse range of fungi which feed on nematodes also occur in the soil. These fungi are collectively known as nematode-destroying or nematophagous fungi. These fungi attack living nematodes or their eggs and utilize them as a source of nutrients. Depending upon the mode of attack, they have been categorised in three groups: (i) predators which develop special nematode-trapping structures (adhesive or non-adhesive) on the mycelium by which the nematodes are efficiently captured; (ii) endoparasites which produce adhesive or non-adhesive spores either to adhere to the surface of the nematodes or to become ingested by them; and (iii) parasites of root-knot and cyst-nematodes which attack the eggs or females of these nematodes.

PREDATORS AS BIOCONTROL AGENTS

Nematode-trapping fungi have long been considered as promising biological agents for the control of nematodes. Some of the early tests in greenhouse conditions tended to exaggerate their potential (Linford, 1937; Linford et al., 1938; Linford and Oliveira, 1938; Linford and Yap, 1938, 1939). Such experiments usually involved adding one fungus to soil which had been amended with organic matter. Some commercial preparations of nematode-trapping fungi have been marketed but the products have never been widely used because of quality control problems and inconsistent performance. *Arthrobotryis robusta* Duddington, commercially formulated as Royal 300, reduced populations of *Rotylenchus myceliophagus* Goodey and increased yields of the cultivated mushroom *Agaricus bisporus* (Lange) Sing (Cayrol et al., 1978) whilst Royal 350, a similar product containing *Arthrobotrys superba* Corda, gave adequate control of root-knot nematode on tomato provided it was used in situations where nematode populations were high (Cayrol, 1983; Cayrol and Frankowski, 1979).

Monacrosporium ellipsosporium was frequently observed associated with *Meloidogyne* egg masses in some field soils and in potted cultures of the nematode (Mankau and Wu, 1985). In a field test using two levels of fungus on wheat grain substrate mixed into transplant holes for tomato seedlings, the improved plant growth and *M. incognita* reduction obtained at harvest were in direct relationship to the amounts of fungus used.

ENDOPARASITES AS BIOCONTROL AGENTS

The host ranges of only a few endoparasitic fungi have been determined but in general these fungi were little more specific than those that form traps (Birchfield, 1960; Esser, 1976; Esser and Ridings, 1973).

Meria coniospora was used in biocontrol experiments because of its strong ability to attract nematodes (Jansson, 1982) and its known specificity of conidial adhesion to nematodes (Jansson et al., 1984; Jansson and Nordbrind-Hertz, 1983). *M. coniospora* significantly reduced root-knot nematode, *Meloidogyne* spp., galling on tomatoes in greenhouse pot trials (Jansson et al., 1985). Another endoparasitic fungus *Hirsutella* sp. has been suggested as a promising microbe for use in biological control of plant parasitic nematodes (Stürhan and Schneider, 1980). Jaffee and Zehr (1982) found *Hirsutella rhossiliensis* attacking the plant parasitic nematode *Circonemella xenoplax*.

The endoparasitic fungi are largely dependent on nematodes for their nutrition and produce only limited mycelial growth outside the host. Species such as *Nematophthora gynophila* and *Catenaria auxiliaris* (Kühn) Tribe, which are aggressive parasites of some cyst nematodes in moist environments, are obligate parasites, while many other species grow only slowly on standard mycological media. Because of the difficulties involved in culturing the endoparasitic fungi, there have been few attempts to utilise them for nematode control. Nematode numbers in sterile sand were reduced when conidia of *Nematoctonus concurrens* Dreschler and *N. haptocladus* were added (Giuma and Cooke, 1974).

FUNGI PARASITIC IN EGGS AS BIOCONTROL AGENTS

The efficiency and adaptability of *Paecilomyces lilacinus* in effectively controlling different pathogenic nematodes has been studied under different climatic and soil environmental conditions (Cabanillas and Barker, 1989; Candanedo et al., 1983; Davide and

Zorilla, 1983; de Sisler et al., 1985; Noe and Sasser, 1984; Roman and Rodriguez-Marcano, 1985).

As a potential biocontrol agent, *P. lilacinus* appears to have a number of advantages. Most reports indicate that it is a good competitor in most soils, particularly in warmer regions (Domsch et al., 1980). Its optimum growth temperature is 26 to 30°C. It readily produces abundant inoculum in the form of long chains of conidia. It has been reported to degrade chitin (Okafor, 1967) and it is also strongly proteolytic (Barout, 1960; Domsch, 1960; Endreeva et al., 1972).

Paecilomyces lilacinus and *P. variotii* have been reported to occur in cysts of *Globodera* and *Heterodera* in various geographical locations (Dowsett and Reid, 1977, 1979; Friman et al., 1985) and the former is known from eggs of *Meloidogyne arenaria* and *M. incognita* in North America and Peru (Tzean and Estey, 1979; Veenhuis et al., 1985). *Paecilomyces lilacinus* is typically soil-borne and is considered to be an important egg parasite, at least in some instances. It is common in the rhizosphere of a number of plants and produces the antibiotics leucinostatin and lilacin (Samson, 1974). Murao et al. (1976) have reported biosynthesis by *P. lilacinus* of a cell wall lytic enzyme capable of lysing the chitin-containing walls of *Rhodotorula* and *Sporobolomyces*. This enzyme may be involved in enabling *P. lilacinus* to colonise nematode eggs whose shells contain chitin (Persson et al., 1985).

P. lilacinus has proved to be an efficient biocontrol agent in its field application in controlling *M. incognita* and *Tylenchulus semipenetrans*. Jatala et al. (1979) reported parasitism of *M. incognita acrita* egg masses by *P. lilacinus* in Peru. The fungus consistently infected eggs and occasionally infected females of *M. incognita acrita*. Penetration of mature *Meloidogyne* females is generally through the anus or vulva. 80 to 90% of the eggs of the fungus-inoculated nematodes were found to be destroyed by the fungus. Jatala et al. (1980) reported that the artificial inoculation by *P. lilacinus* parasitised the egg masses of *M. incognita acrita* concluding the potentiality of *P. lilacinus* to control *M. incognita* on potato under field conditions. Noe and Sasser (1984) noted significant control of *M. incognita* infected with *P. lilacinus* in tomato and okra. Morgan-Jones et al. (1984) found that *P. lilacinus* occurred in significant numbers in eggs of *M. arenaria*.

Cabanillas et al. (1988) observed absence of root galling and giant cell formation in tomato roots inoculated with nematode eggs infected with *P. lilacinus*. Ekanayake and Jayasundara (1994) studied the effect of two nematophagous fungi *Paecilomyces lilacinus* and *Beauveria bassiana* for use as biocontrol agents against *Meloidogyne incognita* on tomato in Sri Lanka and compared with carbofuran. Carbofuran and *P. lilacinus* controlled the root-knot nematode and increased the growth of plants and *B. bassiana* was less effective.

ROLE OF SOIL AMENDMENTS IN BIOCONTROL

Organic amendment is the attempt to develop management strategies which do not rely on nematicides and which promote naturally occurring microorganisms capable of destroying plant parasitic nematodes at several stages of their life cycle. Addition of organic matter to the soil stimulates microbial activity as evidenced by increased populations of algae, fungi, bacteria and other organisms. Proliferation of microorganisms results in increased enzymatic activities of the amended soil and accumulation of specific end products which may be nematicidal (Alam et al, 1979; Badra et al., 1979; Johnson, 1974; Mian and Rodriguez-Kabana, 1982a,b; Mian et al., 1982; Rodriguez-Kabana and Hollis, 1965; Sitaramaiah and Singh, 1978; Walker, 1971). The magnitude of microbial stimulation and the qualitative nature of the responding microflora depend on the nature of the organic matter added.

Chitin as a Soil Amendment

Chitin amendments to soil are effective for the control of nematodes (Godoy et al., 1983; Mankau and Das, 1969; Mian et al., 1982; Rodriguez-Kabana et al., 1984; Spiegel et al., 1986, 1987, 1988). Chitin studies have suggested the possibility of selecting materials for soil amendment that will stimulate development of a specialized microflora antagonistic to nematodes. A possible amendment is to use nematode components, for example cuticle, egg shell, gelatinous matrix etc. Addition of these specific compounds to soil would be expected to stimulate development of microbial species capable of degrading similar compounds present in the nematode. Use of chitin, a specialised type of organic amendment, offers the advantage of determining the microbial species that will be present in the amended soil for some time after the application of the amendments.

Chitin, a polymer of unbranched chains of ß(1→4) linked 2-acetamino-2-deoxy-D-glucose residues, is the most common polysaccharide in nature whose basic unit is an amino sugar (Muzzarelli, 1977). Chitin is considered a permanent component of egg shells of plant parasitic nematodes (Bird and McClure, 1976) and has been detected also in the gelatinous matrix of the root-knot nematode, *Meloidogyne javanica* (Spiegel and Cohn, 1985).

Use of chitin for the control of soil-borne pathogens was first suggested by Mitchell and Alexander (1961). They found that the addition of small quantities of chitin to soil markedly reduced the severity of root-rot of beans caused by *Fusarium solani* f. *phaseoli* and vascular wilt of radishes caused by *F. oxysporum* f. *conglutinans* (Mitchell and Alexander, 1961).

Godoy et al. (1983), Mankau and Das (1974), Mian et al. (1982) and Rodriguez-Kabana et al. (1984) investigated several aspects of chitin treatment against plant-parasitic nematodes such as nematicidal activity range, soil enzymatic activity, microbial populations developed in chitin-treated soil and phytotoxicity phenomena. These workers found a reduction in the number of citrus nematode (*Tylenchulus semipenetrans*), root-knot nematode (*Meloidogyne incognita*) and soybean cyst nematode (*Heterodera glycines*) in soils treated with chitin. The action of chitin against nematodes was interpreted by Mian et al (1982) as resulting from the release of ammoniacal nitrogen into the soil through decomposition of the polymer and by the simulation of chitinolytic microflora, capable of parasitizing the eggs.

During the microbial breakdown of chitin, several substances are liberated. Characterization of these products has revealed the presence of N-acetyl glucosamine, glucosamine, acetic acid and ammonia (Muzzarelli, 1977). Accordingly, a mechanism for the degradation was postulated. The polymer is probably hydrolysed to yield N-acetyl glucosamine, which is then converted to acetic acid and glucosamine and the ammonia is liberated from the latter compound or one of its derivatives (Muzzarelli, 1977).

The effect of chitin amendments on nematodes may last for several months. After addition of chitin, sufficient time must be given to develop specialized organisms to levels adequate for effective nematode control. Good control often occurs in the second crop following harvest of the first (Rodriguez-Kabana et al., 1987). Numbers of chitinolytic microorganisms, especially actinomycetes and bacteria, were higher in chitin-amended soil compared with the control. The workers added an isolate of *Paecilomyces lilacinus*, which was chitinolytic, together with chitin to soil infested with *M. arenaria*. The study demonstrated that the fungus inoculum could increase the effectiveness of the chitin amendment against the nematode.

BIOCONTROL OF ROOT-KNOT NEMATODE *MELOIDOGYNE INCOGNITA* USING *PAECILOMYCES LILACINUS* IN *LYCOPERSICON ESCULENTUM*

The efficacy of *P. lilacinus* alone and in combination with chitin in controlling *M.*

incognita in tomato plants was studied. The culture of *M. incognita* was maintained and multiplied on the roots of Pusa Purple long cultivar of egg-plant plants. A suspension containing 1000 nematodes per ml was added to each pot. *P. lilacinus* was grown on potato dextrose agar for 7 days at $28°\pm1°C$ and a spore suspension was prepared by adding sterile water to the culture. Ten millilitres of this suspension containing 1.7×10^7 spores per ml were added to each pot. The experiment was conducted in fumigated soil. In each pot sand and soil were mixed in the ratio of 2:1. Urea and superphosphate were also added.

Treatments in sterilized soil consisted of:

1.   Control (sterilized soil only)
2.   Soil + Nematodes (N)
3.   Soil + Nematodes + 0.5g Chitin (N+C)
4.   Soil + Nematodes + Fungal spore suspension (N+F)
5.   Soil + Nematodes + 0.5g Chitin + Fungal spore suspension (N+C+F).

For each treatment there were 12 replicates. Chitin was used in a powdered form and was properly mixed in the soil before transplanting the plants. A week after planting, the nematode suspension was added near the roots of the plants by removing the upper layer of soil. After 7 days the spore suspension of the fungus was added near the roots of the plants by removing rhizospheric soil which was then replaced.

Growth parameters were measured after 30, 60 and 90 days. The growth was assessed in terms of shoot/root length, fresh weight and dry weight and number of galls per gramme fresh root weight.

The results show that N+C+F treated plants showed improved plant growth in terms of height and dry weight (Fig. 1a-f). The galling was lowest in this treatment. There was maximum galling in plants inoculated with nematodes only and the plant growth was also suppressed. Out of the two treatments N+F and N+C, the former was less effective (Fig. 2a-e).

It can be said that in the case of N+C+F treatment, either *P. lilacinus* could have increased the effectiveness of chitin against the nematodes or vice-versa. Chitin serves as the substrate or food base for selective development of the biocontrol agent in soil (Rodriguez-Kabana et al., 1987).

The approach to increase the effectiveness of organic amendments by using a fungus or several fungi antagonistic to nematodes will be of great benefit.

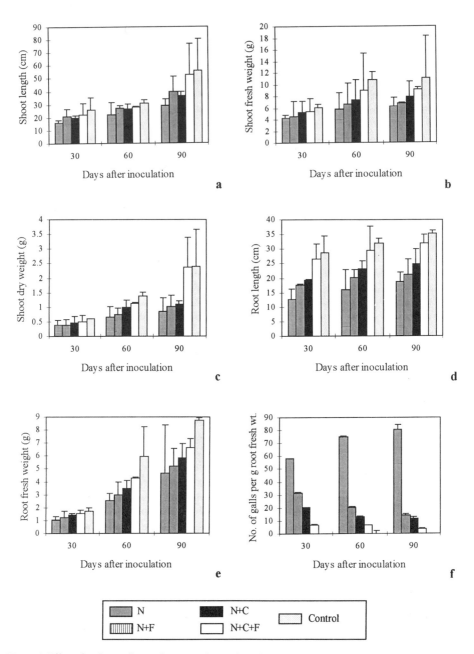

Figure 1. Effect of various soil amendments on the (a) shoot length, (b) shoot fresh weight, (c) shoot dry weight, (d) root length, (e) root fresh weight, (f) number of galls per gram root fresh weight in *Lyccpersicon esculentum.*

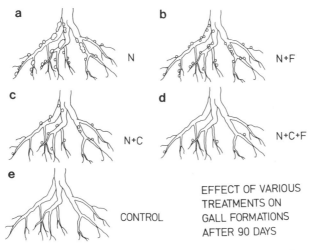

Figure 2. Effect of various treatments on gall formation after 90 days.

CONCLUSIONS

P. lilacinus has proved to be a very effective biocontrol agent in combination with chitin in the control of the root-knot nematode *Meloidogyne incognita*.

ACKNOWLEDGEMENTS

N. Mittal is grateful to the Council of Scientific and Industrial Research for providing financial assistance.

REFERENCES

Alam, M.M., Khan, A.M. and Saxena, S.K., 1979, Mechanism of control of plant-parasitic nematodes as a result of the application of organic amendments to the soil. V. Role of phenolic compounds, *Indian J. Nematol.* 9: 136-142.

Badra, T., Saleh, M.A. and Oteifa, B.A., 1979, Nematicidal activity and composition of some organic fertilizers and amendments, *Revue Nematol.* 2: 29-36.

Baker, K.F. and Cook, R.J., 1974, *Biological Control of Plant Pathogens,* W.H. Freeman and Co., San Francisco, California.

Barout, S., 1960, An ecological and physiological study on soil fungi of the northern Neger (Israel), *Bull. Res. Coun. Israel* 80: 65-80.

Birchfield, W., 1960, A new species of *Catenaria* parasitic on nematodes of sugarcane, *Mycopathologia* 13: 331-338.

Cabanillas, E. and Barker, K.R., 1989, Impact of *Paecilomyces lilacinus* inoculum level and application time on control of *Meloidogyne incognita* on Tomato, *J. Nematol.* 21: 115-120.

Cabanillas, E., Barker, K.R. and Daykin, M.E., 1988, Histopathology of the interactions of *Paecilomyces lilacinus* with *Meloidogyne incognita* on Tomato, *J. Nematol.* 20: 362-364.

Candanedo, E., Lara, J., Jatala, P. and Gonzales, F., 1983, Control biologico del nematode *Meloidogyne incognita* con el hongo *Paecilomyces lilacinus* (Abstr.), XXIII Meeting APS Caribbean Div.

Cayrol, J.C., 1983, Lutte biologique contre les *Meloidogyne* au moyen d'*Arthrobotrys irregularis*, *Rev. Nematol.* 6: 265.

Cayrol, J.C. and Frankowski, J.P., 1979, Une methode de lutte biologique contre les nematodes a galles des racines appartenant an genre *Meloidogyne* Pepinieristes Horticultures Maraichers, *Revue Horticole* 193: 15-23.

Cayrol, J.C., Frankowski, J.P., Laniece, A., d'Hardemare, G. and Talon, J.P., 1978, Contre les nematodes en champignonniere. Mise an point d'une method de lutte biologique a l'aide d'un hyphomycete predateur: *Arthrobotrys robusta* souche *antipolis* (Royal 300), *Rev. Hortic.*, 184: 23-30.

Davide, R.G. and Zorilla, R.A., 1983, Evaluation of a fungus *Paecilomyces lilacinus* (Thom.) Samson. for the biological control of the potato cyst nematode *Globodera rostochiensis* Woll. as compared with some nematicides, *Phil. Agric.* 66: 397-404.

de Sisler, G.M., Silvestri, L. and Acita, J.O., 1985, Utilizacion de *Paecilomyces lilacinus* para el control de *Nacobbus aberrans* (Nematoda, Nacobbidae) en campo, *Fitopathologia* 20: 17-20.

Domsch, K.H., 1960, Das Pilzspectrun einer Benedprobe 3. Nachweis der Einzelpilze, *Arch. Mikrobiol.*, 35: 310-339.

Domsch, K.H., Gams, W. and Anderson, T.H., 1980, *Compendium of Soil Fungi, Volume I*, Academic Press, New York, pp. 859.

Dowsett, J.A. and Reid, J., 1977a, Light microscope observations on the trapping of nematodes by *Dactylaria candida, Can. J. Bot.* 55: 2963-2970.

Dowsett, J.A. and Reid, J., 1977b, Transmission and scanning electron microscope observations on the trapping of nematodes by *Dactylaria candida, Can. J. Bot.* 55: 2963-2970.

Dowsett, J.A. and Reid, J., 1979, Observations on the trapping of nematodes by *Dactylaria scaphoides* using optical, transmission and scanning electron microscope techniques, *Mycologia* 71: 379-391.

Ekanayake H.M.R.K. and Jayasundara, N.J., 1994, Effect of *Paecilomyces lilacinus* and *Beauveria bassiana* in controlling *Meloidogyne incognita* on tomato in Sri Lanka, *Nematologia Mediteranea* 22 (1): 87-88.

Endreeva, N.A., Ushakova, V.I. and Egorov, N.S., 1972, Study of proteolytic enzymes of different strains of *Penicillium lilacinum* Thom. in connection with their fibrinolytic activity, *Mikrobiologiya* 41: 364-368.

Esser, R.P., 1976, *Haptoglossa heterospora* Drechsler, a fungus parasite of Florida nematodes, *Florida Dept. of Agriculture and Consumer Service*, No. 23, 2 pp.

Esser, R.P. and Ridings, W.H., 1973, Pathogenicity of selected nematodes by *Catenaria anguillulae, Soil and Crop Science, Florida* 33: 60-64.

Friman, E., Olsson, S. and Nordbring-Hertz, B., 1985, Heavy trap formation by *Arthrobotrys oligospora* in liquid culture, *FEMS Microbiol. Ecol.* 31: 17-21.

Giuma, A.Y. and Cooke, R.C., 1974, Potential of *Nematoctonus* conidia for biological control of soil borne phytonematodes, *Soil Biol. Biochem.* 6: 217-220.

Godoy, G., Rodriguez-Kabana, R. and Morgan-Jones, G., 1983, Fungal parasites of *Meloidogyne arenaria* eggs in an Alabama soil. A mycological survey and greenhouse studies, *Nematropica* 13: 201-213.

Jaffee, B.A. and Zehr, E.I., 1982, Parasitism of the nematode *Criconemella xenoplax* by the fungus *Hirsutella rhossiliensis, Phytopathology* 72: 1378-1381.

Jansson, H.B., 1982a, Attraction of nematodes to endoparasitic nematophagous fungi, *Trans. Br. Mycol. Soc.* 79: 25-29.

Jansson, H.B. and Nordbring-Hertz, B., 1983, The endoparasitic fungus *Meria coniospora* infects nematodes specifically at the chemosensory organs, *J. Gen. Microbiol.* 129: 1121-1126.

Jansson, H.B., Von Hofsten, A. and Von Mecklenburg, C., 1984, Life cycle of the endoparasitic nematophagous fungus *Meria coniospora*: A light and electron microscopic study, *Antonie Leeuwenhoek. J. Microbiol.* 50: 321-327.

Jansson, H.B., Jeyaprakash, A. and Zuckerman, B.M., 1985, Control of root-knot nematodes on tomato by the endoparasitic fungus *Meria coniospora, J. Nematol.* 17: 327-329.

Jatala, P., Kaltenbach, R. and Bocangel, M., 1979, Biological control of *Meloidogyne incognita acrita* and *Globodera pallida* on potatoes, *J. Nematol.* 11: 303.

Jatala, P., Kaltenbach, R., Bocangel, M., Devax, A.J. and Compos, R., 1980, Field application of *Paecilomyces lilacinus* for controlling *Meloidogyne incognita* on potatoes, *J. Nematol.* 12: 226-227.

Johnson, L.F., 1974, Extraction of oat straw, flax and amended soil to detect substances toxic to root-knot nematode, *Phytopathol.* 64: 1471-1473.

Linford, M.B., 1937, Stimulated activity of natural enemies of nematodes, *Science* 85: 123-124.

Linford, M.B. and Oliveira, J.M., 1938, Potential agents of biological control of plant parasitic nematodes, *Phytopathol.* 28: 14.

Linford, M.B. and Yap, F., 1938, Root-knot injury restricted by a nematode-trapping fungus, *Phytopathol.* 28: 14-15.

Linford, M.B. and Yap, F., 1939, Root-knot nematode injury restricted by a fungus, *Phytopathol.* 29: 596-609.

Linford, M.B., Yap, F. and Oliveira, J.M., 1938, Reduction of soil populations of the root-knot nematode during decomposition of organic matter, *Soil Sci.* 45: 127-141.

Mankau, R. and Das, S., 1969, The influence of chitin amendments on *Meloidogyne incognita, J. Nematol.* 1: 15-16.

Mankau, R. and Das, S., 1974, Effect of organic materials on nematode bionomics in citrus and root-knot nematode infested soil, *Indian J. Nematol.* 4: 138-151.

Mankau, R. and Wu, 1984, Biological control of *Meloidogyne incognita* on tomato with the predaceous fungus

Monacrosporium ellipsosporum, First Int. Cong. Nematol. Canada (Abstr.), 116.

Mian, I.H. and Rodriguez-Kabana, R., 1982a, Survey of the nematicidal properties of some organic materials available in Alabama as amendments to soil for control of *Meloidogyne arenaria, Nematropica* 12: 205-220.

Mian, I.H. and Rodriguez-Kabana, R., 1982b, Organic amendments with high tannin and phenolic contents for control of *Meloidogyne arenaria* in infested soil, *Nematropica* 12: 221-234.

Mian, I.H., Godoy, G., Shelby, R.A., Rodriguez-Kabana, R. and Morgan-Jones, G., 1982, Chitin amendments for control of *Meloidogyne arenaria* in infested soil, *Nematropica* 12: 71-84.

Mitchell, R. and Alexander, M., 1961, Chitin and biological control of *Fusarium* disease, *Pl. Dis. Reptr.* 45: 487-490.

Morgan-Jones, G., White, J.F. and Rodriguez-Kabana, R., 1984, Phytonematode pathology: Ultrastructural studies II. Parasitism of *Meloidogyne arenaris* egg and larvae by *Paecilomyces lilacinus, Nematropica* 14: 57-71.

Murao, S., Yamamoto, R. and Arai, M., 1976, Isolation and identification of red yeast cell wall lytic enzyme producing microorganisms, *Agric. Biol. Chem.* 40: 23.

Muzzarelli, R.A., 1977, *Chitin,* Pergamon Press, New York, pp. 309.

Noe, J.P. and Sasser, J.N., 1984, Efficacy of *Paecilomyces lilacinus* in reducing yield losses due to *Meloidogyne incognita, First Int. Cong. Nematol., Canada (Abstr.)* 116.

Okafor, N., 1967, Decomposition of chitin by microorganism isolated from a temperate and a tropical soil, *Nova Hedwigia* 13: 209-226.

Persson, Y., Veenhues, M. and Nordbring-Hertz, B., 1985, Morphogenesis and significance of hyphal coiling by nematode-trapping fungi in mycoparasitic relationships, *FEMS Microbiol. Ecol.* 31: 283.

Rodriguez-Kabana, R. and Hollis, J.P. 1965, Biological control of nematodes in rice fields: role of hydrogen sulfide, *Science* 148: 524-526.

Rodriguez-Kabana, R., Jones, G.M., Godoy, G. and Gintis, B.O., 1984, Effectiveness of species of *Gliocladium, Paecilomyces* and *Verticillium* for control of *Meloidogyne arenaris* in field soil, *Nematropica* 14(2): 155-170.

Rodriguez-Kabana, R., Morgan-Jones, G. and Chet, I., 1987, Biological control of nematodes. Soil amendments and microbial antagonists, *Plant and Soil* 100: 237-247.

Roman, J. and Rodriguez-Marcano, A., 1985, Effect of the fungus *Paecilomyces lilacinus* on the larval population and root-knot formation of *Meloidogyne incognita* in tomato, *J.Agric. Univ. Puerto Rico* 69: 159-167.

Samson, R.A., 1974, *Paecilomyces* and some allied hyphomycetes, *Stud. Mycol.* 6: 1.

Sitaramaiah, K. and Singh, R.S., 1978, Role of fatty acids in margora cake applied as soil amendment in the control of nematodes, *Indian J. Agric. Sci.* 48: 266-270.

Spiegel, Y. and Cohn, E., 1985, Chitin is present in gelatinous matrix of *Meloidogyne, Revue. Nematol.* 8: 184-186.

Spiegel, Y., Cohn, E. and Chet, I., 1986, Use of chitin for controlling plant-parasitic nematodes. I. Direct effects on nematode reproduction and plant performance, *Plant and Soil* 95: 87-95.

Spiegel, Y., Chet, I. and Cohn, E., 1987, Use of chitin for controlling plant-parasitic nematodes. II. Mode of action, *Plant and Soil* 98 (3): 337.

Spiegel, Y., Chet, I., Cohn, E, Galper, S. and Sharon, E., 1988, Use of chitin for controlling plant-parasitic nematodes. III. Influence of temperature on nematicidal effect, mineralisation and microbial population build up, *Plant and Soil* 109(2): 251-256.

Stürhan, D. and Schneider, R., 1980, *Hirsutella heteroderae,* a new nematode-parasitic fungus, *Phytopatholische Zeitschrift* 99: 105-115.

Tzean, S.S. and Estey, R.H., 1979, Transmission electron microscopy of fungal nematode-trapping devices, *Can. J. Plant Science* 59: 785-795.

Veenhuis, M., Nordbring-Hertz, B. and Harder, W., 1985, Development and fate of electron dense microbodiesin trap cells of the nematophagous fungus *Arthrobotrys oligospora, Antonie van Leeuwenhoek* 51: 399.

Walker, J.T., 1971, Populations of *Pratylenchus penetrans* relative to decomposing nitrogenous soil amendments, *J. Nematol.* 3: 43-49.

RHIZOSPHERE BIOLOGY OF ROOT-KNOT DISEASED *ABELMOSCHUS ESCULENTUS* IN RELATION TO ITS BIOCONOTROL

Renuka Rawat, Anjula Pandey, Geeta Saxena and K.G. Mukerji

Applied Mycology Laboratory
Department of Botany
University of Delhi
Delhi-110 007
India

INTRODUCTION

Soil, as a medium for the growth of plants, supports an invisible but nevertheless vital population of microorganisms. The roots of living plants create a special habitat by virtue of their exudations and materials sloughed off during root growth (Katznelson et al., 1948; Lynch and Whipps, 1990; Rovira, 1969). Root exudates play an important role in the establishment and maintenance of rhizosphere microflora and microfauna. In exchange, microbial metabolites also contribute to this unique niche by stimulating or inhibiting associated microorganisms and plant root growth. The qualitative and quantitative variation in the microflora of the rhizosphere is directly related to root exudations which in turn depend upon plant type, age and developmental factors, foliar application and microbial interactions (Bansal and Mukerji, 1994; Bowen and Rovira, 1992; Curl and Truelove, 1986).

The microbial population of the rhizosphere consists of pathogenic, symbiotic or saprophytic microbes which affect the growth and development of plants. A multiplicity of associative and antagonistic interactions amongst these microflora and fauna adds an extra dimension to the rhizosphere effect (Boosalis and Mankau, 1965; Bowen, 1991; Fravel and Keinath, 1991). Thus, the rhizosphere as an interface of soil and roots is a hub of microbial activity and a zone of special interest to plant pathologists. Since Hiltner (1904), extensive work has been done to elucidate the dynamics of the rhizosphere (Bowen and Rovira, 1992; Katznelson, 1965; Rovira and Davey, 1974; Starkey, 1958).

A complete knowledge of the ecology of soil-borne pathogens, their survival and interactions with other microbes in the rhizosphere is imperative to achieving biological control; it is a fundamental means of enhancing the role of beneficial or antagonistic microbes in order to reduce the threshold levels of pathogenic microbes, thereby improving plant growth.

From Ethnomycology to Fungal Biotechnology
Edited by Singh and Aneja, Plenum Press, New York, 1999

ROOT-KNOT NEMATODES IN THE RHIZOSPHERE: PATHOGENESIS AND INTERACTIONS

Root-knot nematodes, *Meloidogyne* spp., cause enormous damage to crops, especially vegetables. Of the fifty species of *Meloidogyne* known, four are very pathogenic, namely *Meloidogyne incognita, M. javanica, M. arenaria* and *M. hapla* which are highly adaptable to various agro-ecosystems (Roberts, 1995). Root-knot nematodes, being endoparasites, complete their entire life cycles in roots or soil. They are therefore influenced by rhizosphere effects and root exudates play an important role in the activation of quiescent nematode stages and their chemotactic movement towards the infection site (Bird, 1959; Prot, 1980; Sheperd, 1970).

Pathogenesis

The most preferred sites of nematodes in the root are the elongation zone, young lateral apices or injured tissue, out of which the former two are major exudation zones (Barker and Davis, 1996; Rovira, 1973). After entering the roots, *Meloidogyne* juveniles move through the cortex both intra- and inter-cellularly and take up position near the vascular tissue or in the meristematic cells of the root tips inducing specialized nurse systems through hypertrophy and hyperplasia (Wyss, 1988). Roots and underground parts are the worst affected due to the formation of galls, the principal symptoms reported for the first time by Berkeley (1855). Nematode-damaged roots are unable to utilize water and fertilizer and so they waste these increasingly expensive commodities. This, along with stunted growth, low vigour and reduced crop yield, is a major factor in limiting agricultural productivity (Sasser and Carter, 1985).

In general, vegetable crops such as okra, tomato, brinjal, etc. are highly susceptible to root-knot nematodes and suffer maximum damage often with complete loss in yield. According to Taylor and Sasser (1978) the average total annual yield loss in the world is approximately 5%, with developing countries of the tropics and subtropics experiencing the maximum loss. In India, an attempt to assess the yield losses in various vegetables has been made by different workers. The percent yield loss in okra ranged from 6.0 to 90.9% (Bhatti and Singh, 1981; Gill and Jain, 1995).

Interactions with Wilt and Root Rot Fungi

Fawcett (1931) stated that nature does not work in pure cultures; similarly root-knot nematodes along with their hosts are not isolated in the ecological system but are strongly influenced by antagonists and pathogens. Experimental evidence also indicates that root-knot nematodes share biological interactions with other soil microflora like bacteria, fungi and viruses.

Under natural field conditions, plants may develop multiple parasitic associations with soil microflora including fungal-nematode associations. Research on the interaction between soil fungi and plant parasitic nematodes has progressed greatly from the time that Atkinson (1892) first observed the interaction between *Meloidogyne* and *Fusarium* on cotton leading to wilting. There are approximately seventy described *F. oxysporum* forma speciales which are known to interact with nematodes (Armstrong and Armstrong, 1981). The *Meloidogyne* and *Fusarium* interaction has been extensively studied because of its significance on world crops such as cotton and tobacco (Milne, 1972; Sasser, 1972). Similar studies have also been conducted on vegetables such as beans, tomato, eggplant, squash, cabbage and cucumber by different workers (Bergeson et al., 1970; Caperton et al., 1986; Jenkins and Coursen, 1957; Ribeiro and Ferraz, 1983; Singh et al., 1981; Smith and Noguera, 1982).

Early observations on nematode-fungal interactions suggested a 'wound facilitation concept', however the relationship between root-knot nematodes and wilt and root rot fungi is more interesting. The exudates of root-knot infected roots in the rhizosphere is probably the first stage of synergistic interaction between the two (Mai and Abawi, 1987; Taylor, 1990). According to an hypothesis put forward by some workers, the interaction between *Meloidogyne* and *Fusarium* is more biological and physiological than physical in nature. It was further substantiated by studying changes brought about by nematodes in the host. *Meloidogyne* juveniles form giant cells and synctia in the host roots which bring about significant changes in the morphology, anatomy and biochemistry of plants. These giant cells act as a reservoir of carbohydrates, proteins, amino acids, and lipids for the development of the fungus.

Another kind of fungal-nematode relationship which is important but less clearly defined is the disease complex formed by root-rot fungi and nematodes. There is enhanced development of root rot fungi, for example *Pythium* sp., *Rhizoctonia* sp., *Fusarium solani*, *Phytophthora parasitica*, *Sclerotium rolfsii* and *Colletotrichum* sp., in roots attacked by nematodes (Khan, 1993). Golden and VanGundy (1975) reported that the galled regions of tomato roots infected with *M. incognita* were heavily colonized by *R. solani* indicating its preference for the latter due to the presence of rich nutrient medium in the infected region. A similar association was observed by Khan and Müller (1982) on radish roots infected with *M. hapla*, thereby indicating that modifications in root exudates and host substrate by nematode activity enhance susceptibility of host plants and infectiveness of pathogenic fungi such as wilt and root-rot fungi in the rhizosphere (Pitcher, 1965; Powell, 1971a, 1979; Sidhu and Webster, 1977).

Interactions with Antagonists

On the other hand, under natural field conditions nematodes are attacked by a wide variety of predators and parasites. Predators include fungi, nematodes, insects and mites while parasites comprise of viruses, protozoa, bacteria and fungi (Mankau, 1981). Soil fungi which are antagonistic towards nematodes broadly consist of predacious parasites of eggs and cysts and fungi that produce toxins. Common genera of such fungi are *Catenaria*, *Arthrobotrys*, *Dactyllela*, *Monacrosporium*, *Dactylaria*, *Myzocytium*, *Harposporium*, *Cylindrocarpon*, *Exophiala*, *Fusarium*, *Gliocladium*, *Paecilomyces*, *Phoma* and *Torula* (Alam, 1990; Barron, 1977). Such associations result in a biological balance manifested by attachment and penetration by one or more antagonistic fungi in the eggs, juveniles and adult nematodes (Jansson and Nordbring Hertz, 1980; Stirling, 1989).

The effective antagonists which bring about natural disease control are likely to be found in the rhizosphere and rhizoplane of the plants growing in disease-suppressive soil and in the root zones of resistant or escape plants in pathogen-infested areas. Harnessing of these natural enemies and their successful reports has increased research efforts toward biological control of plant parasitic nematodes (Davide, 1995; Esser and Sobers, 1964; Jatala, 1986; Kerry, 1987; Sikora, 1992; Tribe, 1980).

The impetus behind biological control is largely due to recent advances in the use of toxic pesticides: their harmful effects on humans and beneficial microflora; their residual nature; contamination of the water table; the time required for development of resistant cultivars and the economic pressure on land use which limits the use of rotation and other cultural methods.

RHIZOSPHERE BIOLOGY OF *A. ESCULENTUS*: A CASE STUDY

The present investigation was prompted by an awareness of disease-suppressive soil as a probable source of antagonists and the need to study the relationship between host, rhizosphere

(RS) and non-rhizosphere (NRS) microflora and root-knot nematodes. The study was conducted in experimental plots located in the Botanical Garden, University of Delhi. Seeds of okra cultivars Pusa Makhmali (PM) and New Pusa-4 (NP-4) used were procured from the National Bureau of Plant Genetic Resources, New Delhi.

The rhizosphere microflora was studied along two broad lines: quantitatively, to determine the abundance of major groups of soil microflora and qualitatively, to examine the composition of RS and NRS microflora. Soil dilution and plating method (Timonin, 1941) using selective and non-selective media was employed. The interaction between cultivars and root-knot nematodes was also studied by recording number of galls and fruit yield at regular intervals of growth period.

An earlier report on rhizosphere microflora of *Abelmoschus esculentus* was given by Ranga Rao and Mukerji (1972). They reported on quantitative and qualitative analysis of soil microflora in the RS and NRS of cv Pusa Sawani. Consistent enhancement in the rhizosphere microflora from the seedling stage to maturity was observed in comparison to that of NRS which showed periodic fluctuations.

In the present work, the rhizosphere and non-rhizosphere microflora in both the cultivars followed a nearly parallel trend (Figures 1A and 1B).

Qualitative analysis of microflora showed the presence of egg-parasitizing and antagonistic fungi such as *Paecilomyces lilacinus, P. variotii, P. fusisporus, Acremonium* sp., *Gliocladium* sp., *Trichoderma* sp. and *Verticillium* sp. in the rhizosphere of cv PM during all stages of plant growth, whereas *Paecilomyces* spp. appeared in the rhizosphere of cv NP-4 only during 75-90 days after germination. The presence of antagonists and egg-parasitizing fungi in the rhizosphere of PM (which is tolerant to root-knot nematode) and their absence in the rhizosphere of NP-4 (which is highly susceptible to root-knot nematode attack) is probably due to the role played by the root exudation of galled and non-galled roots (Mousa, 1991; Srivastava and Dayal, 1986; Stephan et al., 1996).

The exudation of galled roots of NP-4 also stimulated the wilt and rot fungi viz., *Fusarium oxysporum, Rhizoctonia* sp., *Phoma* sp., *and Macrophomina* sp. Similar results of suppression of antagonists and stimulation of wilt and rot fungi have been reported by many workers (Bergeson et al., 1970; Chalal and Chhabra, 1984; Hirano et al., 1979). The appearance of wilt fungi in the rhizosphere of NP-4 coincides with the formation of giant cells by nematodes ensuring better nutrient availability to the penetrated fungal pathogens, which is in accordance with the view given by previous workers (Bird, 1972; Wang and Bergeson, 1974; Webster, 1985).

The number of galls observed in NP-4 was negatively correlated to the fruit yield (Figure 2). Cultivar PM showed higher fruit yield as compared to that of cv NP-4. These results showed that apart from the fact that cv PM has an inherent property to be more tolerant than NP-4 against root-knot nematode, the rhizosphere microflora played a significant role in imparting protection to the former cultivar against the pest (Figure 3). The severity of infection was lowered to an appreciable level as antagonistic microflora and biocontrol agents such as *Paecilomyces lilacinus* reduced the population of *Meloidogyne* (juveniles and eggs) to a non-threatening level.

The present study also conforms to the already established biocontrol potential of *P. lilacinus*. This highly competitive saprophytic fungus, an effective parasite of *Meloidogyne* eggs, has been tested the world over with some very positive results (Alam, 1990; Frieire and Kerry, 1985; Gomes-Carneiro and Cayrol, 1991; Jatala et al., 1981; Mittal et al., 1995; Sasser and Neo, 1995).

Another important aspect of biological control is the bio-environmental management of indigenous antagonists in the field which involves organic amendments, crop rotation, and chemical and physical soil treatments designed to increase the population of the antagonists and reduce the severity of the disease (Abid and Maqbool, 1990; Akhtar and Mahmood, 1994;

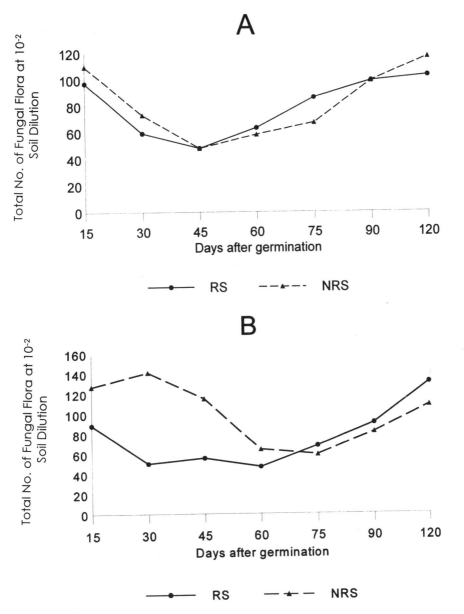

Figure 1. Rhizosphere and non-rhizosphere microflora of *A. esculentus.* A: Cultivar Pusa Makhmali (PM); B: Cultivar New Pusa-4 (NP-4).

Mankau, 1980, 1981; Nordbring Hertz, 1988; Owino and Mousa-E, 1995; Vincete and Acosta, 1992).

Organic amendments play an important role in biological control, as they activate or develop microflora whose antagonistic activity augments the effects of chemicals produced during the decomposition process. The use of oil cakes, green manure, organic manure, crop residue, plant extracts, chicken litter and chitinous material has been studied for the management of root knot disease (Müller and Gooch, 1982).

In okra, organic amendments such as azolla, water soluble extracts of oil cakes and Neem derivatives have been used in controlling root-knot with positive results (Abid et al., 1995; Khan et al., 1991; Thakar et al., 1987).

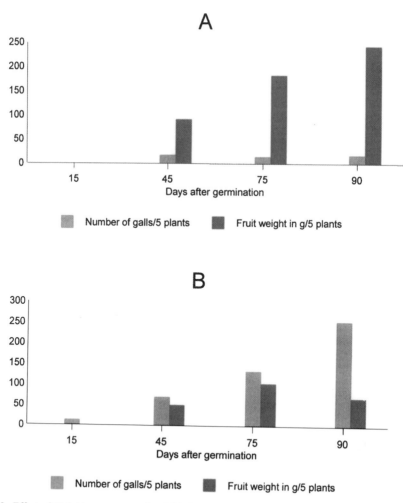

Figure 2. Effect of *Meloidogyne* spp. on the yield of *A. esculentus.* A: Cultivar Pusa Makhmali (PM). B: Cultivar New Pusa-4 (NP-4).

The observed differences in the qualitative and quantitative rhizosphere microflora of the two cultivars studied attest to the hypothesis that different genotypes in species (susceptible and tolerant) differentially influence root exudates which govern the dynamics of the microbial population in the rhizosphere (Atkinson et al., 1975; Krigsvold et al., 1982; Rengel, 1991; Roberts, 1995). Therefore, genetic management of the rhizosphere by introducing resistant lines shows promising potential which shall have a critical role in controlling the pathogen's activity and disease induction.

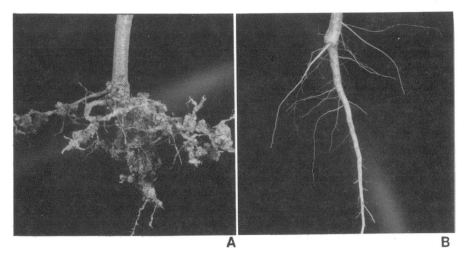

Figure 3. A: Profuse galling in roots of *A. esculentus* cv NP-4 infected with *Meloidogyne* spp. B: Non-galled roots of cv PM.

CONCLUSIONS

In the light of environmental, economic and practical concerns, ensuring restrictions on nematicide use is essential. Ensuing this, there is increased emphasis on integrated nematode management (INM) which is the paradigm adopted by all pest control disciplines. INM seeks to stabilize the population of target nematodes to non-threatening levels by employing direct, non-chemical cultural and physical methods; encouraging naturally-occurring biocontrol agents; enhancing the biodiversity inherent in multiple cropping and multiple cultivar traditional farming systems to increase the available resistance or tolerance to nematodes and by introducing new resistant lines. INM does not necessarily imply the complete exclusion of nematicides, which are still quick action control measures, but it signifies a cohesive approach where a new generation of safe and effective nematicides are used at low rates along with environmentally-friendly practices in sustainable agricultural systems.

FUTURE STRATEGIES FOR THE SUCCESSFUL BIOLOGICAL CONTROL OF ROOT-KNOT NEMATODES

Some future strategies for implementing a more integrated approach to nematode management are as follows: (i) detailed study of the rhizosphere biology of other commonly used cultivars of *A. esculentus* with relation to root exudation, rhizosphere microflora and root-knot nematode; (ii) *in vitro* tests to evaluate the potential of antagonists isolated from naturally suppressive soils against eggs and juveniles of root-knot nematodes; and (iii) assessment of the ability of introduced antagonists in competition and establishment in the rhizosphere and rhizoplane of plants and their efficacy in disease control through greenhouse and micro-plots experiments.

ACKNOWLEDGEMENTS

The authors acknowledge the University Grants Commission, New Delhi for financial support.

REFERENCES

Abid, M. and Maqbool, M.A., 1990, Effects of inter cropping of *Tagetes erecta* on root-knot disease and growth of tomato, *Int. Nematol. Net. Newsl.* 7(3): 41-42.

Abid, M., Zaki, M.J. and Maqbool, M.A., 1995, Neem derivatives for the control of root-knot nematode (*Meloidogyne javanica*) on okra, *Paki. J. Phytopathol.* 7(2): 212-214.

Akhtar, M. and Mahmood, I., 1994, Prophylactic and therapeutic use of oil cakes and leaves of neem and castor extracts for the control of root-knot nematode on chilli, *Nematologia-Mediterranea* 22(2): 127-129.

Alam, M.M., 1990, *Paecilomyces lilacinus* - A nematode biocontrol agent, in: *Nematode Biocontrol: Aspects and Prospects* (M.S. Jairajpuri, M.M. Alam and I. Ahmad, eds.), pp.72-82, CBS Publishers & Distributors, Delhi, India.

Armstrong, C.M. and Armstrong, J.K., 1981, Formae speciales and races of *Fusarium oxysporum* causing wilt diseases, in: *Fusarium: Diseases, Biology and Taxonomy* (P.E. Nelson, T.A. Toussoun and R.J. Cook, eds.), pp. 391-399, Pennsylvania State University Press.

Atkinson, G.F., 1892, Some diseases of cotton, *Ala. Agric. Exp. Stn. Bull.* 41: 1-65.

Atkinson, T.G., Neal, J.L.Jr. and Larson, R.I., 1975, Genetic control of the rhizosphere microflora of wheat, in: *Biology and Control of Soil-borne Plant Pathogens* (G.W. Bruehl, ed.), pp. 116-122, Am. Phytopathol. Soc. St. Paul, Minn.

Bansal, M. and Mukerji, K.G., 1994, Positive correlation between VAM induced changes in root exudation and mycorrhizosphere mycoflora, *Mycorrhiza* 5: 39-44.

Barker, K.R. and Davis, E.L., 1996, Assessing plant nematode infestation and infections, *Advn. Bot. Res.* 23: 103-127.

Barron, G.L., 1977, *The Nematode Destroying Fungi*, Canadian Biological Publication Ltd., Guleph, Ontario.

Bergeson, G.N., Van Gundy, S.D. and Thomason, I.J., 1970, Effect of *Meloidogyne javanica* on rhizosphere microflora and *Fusarium* wilt of tomato, *Phytopathol.* 60: 1245-1249.

Berkeley, M.J., 1855, Vibrio forming excrescences on the roots of cucumber plants, *Gard. Chron.*, April, p. 220.

Bhatti, D.S. and Jain, R.K., 1977, Estimation of loss in okra, tomato and brinjal yield due to *M. incognita, Indian J. Nematol.* 7: 37-41.

Bird, A.F., 1959, The attractiveness of roots to the plant parasitic nematodes, *Meloidogyne javanica* and *M. hapla, Nematologica*, 4: 322-335.

Bird, A.F., 1972, Quantitative studies on the growth of synctia induced in plants by root knot nematodes, *Int. J. Parasitol.* 2: 157-170.

Boosalis, M.G. and Mankau, R., 1965, Parasitism and predation of soil microorganisms, in: *Ecology of Soil-Borne Plant Pathogens* (K.F. Baker and W.C. Snyder, eds.), pp. 374-389, Univ. of California Press, Berkeley.

Bowen, G.D., 1991, Microbial dynamics in the rhizosphere: Possible strategies in managing rhizosphere population, in: *The Rhizosphere and Plant Growth* (D.L. Keister and P.B. Cregan, eds.), pp. 25-32, Kluwer Academic Publishers.

Bowen, G.D. and Rovira, A.D., 1992, The rhizosphere: the hidden half of the hidden half, in: *Roots: The Hidden Half* (Y. Waisel, A. Eshel and U. Kafkafi, eds.), pp. 641-669, Marcel Dekker Inc., New York.

Caperton, C.M., Martyn, R.D. and Starr, J.L., 1986, Effects of *Fusarium* inoculum density and root-knot nematodes on wilt disease in summer squash, *Pl. Dis.* 70: 207-209.

Chalal, P.P.K. and Chhabra, H.K., 1984, Interaction of *M. incognita* with *Rhizoctonia solani* on tomato, *Indian J. Nematol.* 14: 56-57.

Curl, E.A. and Truelove, B., 1986, *The Rhizosphere*, pp. 1-281, Springer-Verlag, Berlin, Germany.

Davide, R.G., 1995, Biological control of plant parasitic nematodes in the Philippines, *Biocontrol* 1(4): 21-23.

Esser, R.P. and Sobers, E.K., 1964, Natural enemies of nematodes, *Soil Crop Sci. Soc. Fla. Proc.* 24: 326-352.

Fawcett, H.S., 1931, The importance of investigations on the effect of known mixtures of organisms, *Phytopathol.* 60: 100-103.

Fravel, D.R. and Keinath, A.P., 1991, Biocontrol of soil borne pathogens with fungi, in: *The Rhizosphere and Plant Growth* (D.L. Keister and P.B. Cregan, eds.), pp. 237-243, Kluwer Academic Publishers.

Freire, F.C.O. and Kerry, B.R., 1985, Parasitism of eggs, females and juveniles of *Meloidogyne incognita* by *Paecilomyces lilacinus* and *Verticillium chlamydosporium*, *Fitopathologia Brasileira* 10: 577-596.

Gill, J.S. and Jain, R.K., 1995, Nematode problems of vegetable crops in India, in: *Nematode Pest Management: An Appraisal of Ecofriendly Approaches* (G. Swarup, D.R. Dasgupta and J.S. Gill, eds.), pp. 166-178, NSI, New Delhi.

Golden, J.K. and VanGundy, S.D., 1975, A disease complex of okra and tomato involving the nematode, *Meloidogyne incognita* and the soil inhabiting fungus, *Rhizoctonia solani*, *Phytopathol.* 65: 265-273.

Gomes-Carneiro, R.M.D. and Cayrol, J.C., 1991, Relationship between inoculum density of the nematophagous *Paecilomyces lilacinus* and control of *M. arenaria* on tomato, *Revue-de-Nematologie* 14(4): 629-634.

Hirano, K., Sugiyama, S. and Iida, W., 1979, Relation of the rhizosphere microflora to the occurrence of *Fusarium* wilt of tomato under presence of the root-knot, *Meloidogyne incognita*, *Jap. J. Nematol.* 9: 60-68.

Jansson, H.B. and Nordbring-Hertz, B., 1980, Interactions between nematophagous fungi and plant parasitic nematodes: Attraction, induction of trap formation and capture, *Nematologica* 26: 383-389.

Jatala, P., 1986, Biological control of plant parasitic nematodes, *Ann. Rev. Phytopathol.* 24: 453-489.

Jatala, P., Salas, R., Kaltenbach, R. and Bocangel, M., 1981, Multiple application and long term effect of *Paecilomyces lilacinus* in controlling *Meloidogyne incognita* under field applications, *J. Nematol.* 13: 445.

Jenkins, W.R. and Coursen, B.W., 1957, The effect of root-knot nematodes, *Meloidogyne incognita acrita* and *M. hapla*, on *Fusarium* wilt of tomato, *Plant Dis. Reporter* 41: 182-186.

Katznelson, H., 1965, Nature and importance of rhizosphere, in: *Ecology of Soil-Borne Plant Pathogens* (K.F. Baker and W.C. Snyder, eds.), pp. 187-207, Univ. of California Press, Berkeley.

Katznelson, H., Lochhead, A.G. and Timonin, M.I., 1948, Soil microorganisms and the rhizosphere, *Bot. Rev.* 14: 543-587.

Kerry, B.R., 1987, Biological control, in: *Principles and Practice of Nematode Control in Crops* (R.H. Brown and B.R. Kerry, eds.), pp. 232-263, Academic Press, Sydney.

Khan, M.W. and Müller, J., 1982, Interactions between *Rhizoctonia solani* and *Meloidogyne hapla* on radish in gnotobiotic culture, *Libyan J. Agric.* 11: 133-140.

Khan, M.W. (ed.), 1993, *Nematode Interactions*, Chapman and Hall.

Khan T.A., Nisar, S. and Hussain, S.I., 1991, Effect of water soluble extracts of certain oil cakes on the development of root-knot disease of okra (*Abelmoschus esculentus* var. Sevendhari), *Curr. Nematol.* 2(2): 167-170.

Krigsvold, D.T., Griffith, G.J. and Hale, M.G., 1982, Microsclerotial germination of *Cylindrocladium crotalariae* in the rhizosphere of susceptible and resistant peanut plants, *Phytopathol.* 72: 859-864.

Lynch, J.M. and Whipps, J.M., 1990, Substrate flow in the rhizosphere, *Pl. Soil* 129: 1-10.

Mai, W.F. and Abawi, G.S., 1987, Interaction among root knot nematodes and *Fusarium* wilt fungi on host plants, *Ann. Rev. Phytopathol.* 25: 317-338.

Mankau, R., 1980, Biological control of nematode pests by natural enemies, *Ann. Rev. Phytopathol.* 18: 415-440.

Mankau, R., 1981, Microbial control of nematodes, in: *Plant Parasitic Nematodes, Volume 3* (B.M. Zuckerman and R.A. Rohde, eds.), pp. 475-494, Academic Press, New York.

Milne, D.L., 1972, Nematodes of tobacco, in: *Economic Nematology* (I.M. Webster, ed.) pp. 159-186, Academic Press, London.

Mittal, N., Saxena, G. and Mukerji, K.G., 1995, Integrated control of root-knot disease in three crop plants using chitin and *Paecilomyces lilacinus*, *Crop Protection* 14(8): 647-651.

Mousa-E, S.M., 1991, Biological management of soil-borne pathogens and root-knot nematode complexes on soybean, in: *Proc. Second Afro-Asian Nematol. Symp. at Menoufiya, Egypt, 18-22 Dec.* pp. 110-114.

Müller, R. and Gooch, P.S., 1982, Organic amendments in nematode control: An examination of literature, *Nematoropica* 12: 319.

Nordbring-Hertz, B., 1988, Nematophagous fungi: strategies for nematode exploitation and for survival, *Microbiol. Sci.* 5(4): 108-116.

Owino, P.O. and Mousa-E, S.M., 1995, Effects of time of harvest, agro-chemicals and antagonistic plants on the biological control and fungal parasitism of *Meloidogyne javanica* eggs, in: *Proc. Second Afro-Asian Nematol. Symp. at Menoufiya, Egypt, 18-22 Dec.*, pp. 125-130.

Pitcher, R.S., 1965, Interrelationships of nematodes and other pathogens of plants, *Helminthological Abstr.* 34: 1-17.

Powell, N.T., 1971a, Interactions between nematodes and fungi in disease complexes, *Ann. Rev. Phytopathol.* 9: 253-274.

Powell, N.T., 1979, Internal synergisms among organism inducing disease, in: *Plant Disease: An Advanced Treatise, Volume 4* (J.G. Horsfall and E.B. Cowling, eds.), pp. 113-133, Academic Press, New York.

Prot, J.C., 1980, Migration of plant parasitic nematodes toward plant roots, *Rev. Nematol.* 3: 305-318.

Ranga Rao, V. and Mukerji, K.G., 1972, Fungal flora in the root zone of healthy and infected plants, *Ann. Inst. Pasteur* 122: 81-90.

Rengel, Z., 1997, Root exudation and microflora populations in rhizosphere of crop genotypes differing in tolerance to micronutrient deficiency, *Pl. Soil* 196: 255-260.

Ribeiro, C.A.G. and Ferraz, S., 1983, Studies on the interaction between *M. javanica* and *F. oxysporum* f.sp. *phaseoli* on bean, *Fitopathologia Brasileira* 8: 439-446.

Roberts, P.A., 1995, Conceptual and practical aspects of variability in root knot nematodes related to host plant resistance, *Ann. Rev. Phytopathol.* 33: 199-221.

Rovira, A.D., 1969, Plant root exudates, *Bot. Rev.* 35: 35-57.

Rovira, A.D., 1973, Zones of exudations along plant roots and spatial distribution of microorganisms in the rhizosphere, *Pestic. Sci.* 4: 361-366.

Rovira, A.D. and Davey, C.B., 1974, Biology of rhizosphere, in: *The Plant Root and its Environment* (E.W. Carson, ed.), pp. 153-204, Univ. Virginia Press, Charlottesville.

Sasser, J.N., 1972, Nematode diseases of cotton, in: *Economic Nematology* (J.M. Webster, ed.), pp. 187-214, Academic Press, London.

Sasser, J.N. and Carter, C.C., (eds.) 1985, *An Advanced Treatise on Meloidogyne, Volume 1: Biology and Control*, North Carolina State Univ. Graphics, Raleigh.

Sasser, J.N. and Noe, J.P., 1995, Evalaution of *Paecilomyces lilacinus* as an agent for reducing yield losses due to *Meloidogyne incognita*, *Biocontrol* 1(3): 57-67.

Sheperd, A.M., 1970, The influence of root exudates on the activity of some plant parasitic nematodes, in: *Root Disease and Soil Borne Pathogens* (T.A. Toussoum, R.V. Bega and P.E. Nelson, eds.), pp. 134-137, Univ. California Press, Berkeley.

Sidhu, G.S. and Webster, J.M., 1977, Predisposition of tomato to the wilt fungus (*Fusarium oxysporum lycopersici*) by the root-knot nematode (*Meloidogyne incognita*), *Nematologica* 23: 433-442.

Sikora, R.A., 1992, Management of the antagonistic potential in agricultural ecosystems for the biological control of plant parasitic nematodes, *Ann. Rev. Phytopathol.* 30: 245-270.

Singh, D.B. Reddy, P.P. and Sharma, S.R., 1981, Effect of root-knot nematode *Meloidogyne incognita* on *Fusarium* wilt of french beans, *Indian J. Nematol.* 11: 84-85.

Smith, B.G. and Noguera, R., 1982, Effect of *Meloidogyne incognita* on the pathogenicity of different isolates of *Fusarium oxysporum* on brinjal (*Solanum melongena* L.), *Agronomia Tropical* 32: 284-290.

Srivastava, L.S. and Dayal, R.A., 1986, Studies on rhizosphere mycoflora of *Abelmoschus esculentus* XIV. - Antibiosis by soil microorganism. *Indian Phytopathol.* 39: 104-106.

Starkey, R.L., 1958, Interactions between microorganisms and plant roots in the rhizosphere, *Bacteriol. Rev.* 22: 154-172.

Stephan, Z.A., El-Behadli, A.H, Al-Zahroon, H.H. and Georgees, M.S., 1996, Control of root-knot wilt disease complex on tomato plants, *Dirasat-Series B, Pure and Applied Sciences*, 23(1): 13-16.

Sterling, G.R., 1989, Biological control of plant-parasitic nematodes, in: *Diseases of Nematodes* (G.O. Poinar and H.R. Jansson, eds.), pp. 94-139, CRC Press.

Taylor, A.L. and Sasser, J.N., 1978, *Biology, Identification and Control of Root-knot Nematodes (Meloidogyne spp.)*, Coop. Publ. Dept. Plant Pathol., North Carolina State Univ. and US Agency Int. Deve., Raleigh, USA.

Taylor, C.E., 1990, Nematode interactions with other pathogens, *Annals Appl. Biol.* 116: 405-416.

Thakar, N.A., Patel, H.R. and Patel, C.C., 1987, Azolla in management of root-knot disease in okra, *Indian J. Nematol.* 17(1): 136-137.

Timonin, M.I., 1941, The interactions of higher plants and soil microorganisms III: Effects of byproducts of plant growth on activity of fungi and actinomycetes, *Soil Sci.* 52: 395-413.

Tribe, H.T., 1980, Prospects for the biological control of plant-parasitic nematodes, *Parasitology* 81: 619-639.

Vincete, N.E. and Acosta, N., 1992, Biological control and chemical control of nematodes in *Capsicum annum* L., *Jr. Agric. Univ. Puerto-Rico* 76: 171-176.

Wang, E.L.H. and Bergeson, G.B., 1974, Biochemical changes in root exudate and xylem sap of tomato plants infected with *Meloidogyne incognita*, *J. Nematol.* 6: 192-194.

Webster, J.M., 1985, Interactions of *Meloidogyne* with fungi on crop plants, in: *Advanced Treatise on Meloidogyne Volume 1: Biology and Control* (J.N. Sasser and C.C. Carter, eds.), pp. 183-192, North Carolina State Univ. Graphics, Raleigh.

Wyss, U., 1998, Pathogenesis and host-parasite specificity in Nematodes, in: *Experimental and Conceptual Plant Pathology, Volume 2* (W.M. Hess, U.S. Singh and D.J. Weber, eds.), pp. 417-432, Oxford and IBH Publishing Pvt. Ltd., New Delhi, India.

VESICULAR ARBUSCULAR MYCORRHIZAE IN THE CONTROL OF FUNGAL PATHOGENS

Mamta Sharma and K.G. Mukerji

Department of Botany
University of Delhi
Delhi-110 007
India

INTRODUCTION

Roots of higher plants are known to support the growth of complexes of microbes that in turn can have a profound effect on the growth and survival of the plant. Considerable amounts of dry matter produced by the plants may be released into the soil in the form of root exudates and cell sloughage (Bansal and Mukerji, 1994; Barber and Martin, 1976). Symbiotic mycorrhizal fungi are ubiquitous in occurrence and are known to colonize roots of almost all plants. Concomitant colonization and infection of roots by mycorrhizal fungi and by pathogens and other microbes inevitably lead to modification of each other's activity (Linderman, 1985;1989). These interactions are of great importance. Potentially beneficial interactions, if maintained or enhanced, could potentially result in biological control of pathogens.

Mycorrhizal colonization occurs after seed germination. At this time the zone of elongation is most extensive and the root has spent the phosphorus reserves of the seed. Root exudation is greatest in the zone of elongation, where vesicular arbuscular mycorrhizal (VAM) colonization is initiated. Carbon losses from the roots are sufficient to sustain the growth activity of the fungi, that is germination of spores, growth of hyphae, penetration of hyphae into the root cortex and development of external hyphae that take up phosphorus beyond the zone of depletion around the root.

Before phosphorus sufficiency is attained, however, the root is at risk to pathogen invasion because cellular permeability is increased due to phospholipid depletion in membranes and root exudation is at its maximum (Graham et al., 1981; Ratnayake et al., 1978). As hyphal uptake of phosphorus occurs, root phosphorus content increases, membrane permeability is reduced and more carbon is translocated to the VAM fungi in the roots resulting in less exudation from the roots. Mycorrhizal-induced decreases in root exudations have been correlated with reductions in soil-borne diseases (Graham, 1988; Graham and Menge, 1982). So it is expected that mycorrhizae alter exudation which in turn indirectly alter the activities of microbes. At this point, the rhizosphere becomes the 'mycorrhizosphere' (Linderman, 1988;

Srivastava et al., 1996). No longer is the sphere of influence restricted to zones of soil around roots, it now occurs around hyphae extending from the root surface as well. The 'mycosphere' exerts its own selective influence on microbial activities in the surrounding soil. Therefore mycorrhizae may be a primary determinant in microbial management and biological control of soil-borne plant pathogens. Mycorrhizae-mediated effects on host nutrition indirectly influence these interactions in most cases.

VA mycorrhizae are known to benefit host plants in several ways. They have been shown to help plants acquire mineral nutrients from the soil, especially immobile elements such as P, Zn and Cu and mobile ions such as S, Ca, K, Fe, Mg, Mn, Cl, Br and N (Mukerji et al., 1991, 1996; Tinker, 1984). Mycorrhizae have also been shown to increase water uptake and/or otherwise alter the plant's physiology to reduce the stress response to soil drought (Safir and Nelsen, 1985). Mycorrhizal fungi can also reduce the plant's response to other soil stresses such as high salt levels, toxicities associated with mine spoils or land fills, heavy metals or due to minor elements, for example manganese toxicity. Some mycorrhizal fungi produce metabolites which can alter plants' ability to produce roots from cuttings or alter root regeneration and morphology resulting in increased absorptive surface area and feeder root longevity (Linderman and Call, 1977). These fungi also improve soil particle aggregation resulting in soil stability (Sutton and Sheppard, 1976).

VAM FUNGAL INTERACTIONS WITH PLANT PATHOGENS

Reported microbial interactions in the mycorrhizosphere may involve a variety of bacteria and fungi with specific functional capabilities that may influence plant growth. This may include microbes such as strict or facultative anaerobes, extracellular chitinase producers, phosphate solubilizers, siderophore producers, antibiotic producers, hormone producers, pathogen suppressors, plant growth promoters, exopathogens, mycorrhiza suppressors etc. The present paper deals with the interactions of VA mycorrhizal fungi with soil-borne root pathogens.

Interactions of VAM with Soil-borne Fungal Pathogens

Research on the potential of VAM fungi to limit harm due to pathogens has been largely restricted to greenhouse studies conducted in sterilized, phosphorus-deficient soils to maximize plant growth stimulation by VAM fungi. Because the interactions varied with the specific host-symbiont- pathogen combination, generalizations on the effect of VAM fungi on disease were difficult to interpret (Bali, 1991; Bali and Mukerji, 1988, 1991; Mukerji et al., 1996).

Reduction in disease severity in cotton-*Verticillium* wilt and citrus-*Phytophthora* root rot on VAM formation has been due to the enhanced uptake of phosphorus (Davis and Menge, 1980; Davis *et al.*, 1979). It was found that increased phosphorus nutrition inhibits the release of zoospores from sporangia. Increases in the level of host resistance in mycorrhizal wheat to 'take-all disease' were also attributed to improved phosphorus nutrition (Graham and Menge, 1982). High levels of VAM root colonization in plants grown in phosphorus-deficient soils or the addition of phosphorus to soils equally suppressed disease severity. Root necrosis and *Fusarium* propagule density on mycorrhizal root systems were reduced at all P levels even if VAM colonization varied from 8% to 65% (Caron et al., 1986).

Initial inoculum density of pathogens and the symbiont also affects the results of the experiments (Wallace, 1983). Very high inoculum density of the pathogens may severely stunt or kill the plants as mycorrhizal fungi are not given an opportunity to colonize roots and stimulate growth (MacGuidwin et al., 1985). The sequence in which plants are inoculated with a pathogen relative to the time of VAM fungal inoculations may also affect the nature of the interactions (Caron et al., 1986; Hussey and Roncadori, 1982). Although pre-inoculation with VAM fungi gives the desired effect it is an artificial system. In nature both VAM fungi and pathogens are present in the soil at the time of seed germination, hence both have equal opportunity to colonize the root system.

The use of varying inoculum densities of VAM fungi has not been considered an important factor. Very low VAM fungal inoculum densities, i.e. 0.5 to 5.0 spores per gram of soil, have been shown to produce optimal growth responses and maximum root colonization levels. Moreover, high levels of VAM fungal root colonization have not reduced the degree of root infection by fungal pathogens (Davis and Menge, 1980; Davis et al., 1979; Ross, 1971).

As both VAM fungi and plant pathogens occupy the same root tissue, direct competition for space has been postulated as a mechanism of pathogen inhibition by VAM fungi (Davis and Menge, 1980, Linderman, 1985, 1988). The competition between VAM fungi and *Phytophthora parasitica* has been studied in citrus using split root systems. On the split root system the amount of mycorrhiza-colonized root tissue was reduced only when VAM fungus and *P. parasitica* were in direct association. However this hypothesis did not receive much attention because root infection precedes mycorrhizal colonization and many pathogens infect the root tip where VAM structures do not occur (Garrett, 1970; Harley and Smith, 1983). Several studies have been conducted using VAM fungus - host plant - fungal pathogens and a general overview of these is provided in Table 1.

Andrea Torres-Barragan et al. (1996) studied the use of arbuscular mycorrhizae to control onion white rot under field conditions. They reported that mycorrhizal inoculations delayed the white rot epidemic by two weeks. Sufficient protection against disease was provided for 11 weeks after transplantation in comparison to non-mycorrhizal plants. Plants inoculated with mycorrhizae showed an increase in yield of 22%. Calvet et al. (1993) showed the stimulatory effect of *Trichoderma* sp. on VAM fungi and subsequent control of *Pythium ultimum* in marigold. The interaction between *Trichoderma koningii, Fusarium solani* and *Glomus mosseae* was studied on maize and lettuce, with or without mycorrhiza *Glomus mosseae* (McAllister et al., 1994a,b). He reported a decrease in the populations of both *Trichoderma* and *Fusarium* sp. However *F. solani* has no effect on mycorrhizal development, whereas *G. mosseae* was inhibited in its extramatrical stage by *T. koningii*. Newsham et al. (1993) conducted studies which showed that root pathogens *Embellisia* spp., *Fusarium sascysporum* and *Phoma* spp. reduced seed production in winter grass *Vulpia ciliata*. Using analysis of covariance, the authors showed that the frequency of VAM colonization was a significant covariate, counteracting the negative effect of pathogens.

VAM in Biocontrol of Shoot and Leaf Diseases

Somewhat different types of results were reported when VAM fungi were studied in relation to diseases of aerial parts of the plants. Very little work has been done in this aspect of disease management. Schoenbeck and Dehne (1981) observed that mycorrhizal plants were more susceptible to diseases in comparison to non-mycorrhizal plants. When young leaves of cucumber were infected with powdery mildew fungus, the increase in susceptibility of the shoot was due to enhanced development of pathogens rather than to an increased frequency of infection. This influence seems to be correlated to nutritional aspects of mycorrhiza, disease interactions and higher physiological activities in the host plants. Genetically-resistant plants remained resistant but their susceptibility was modified by symbiosis. A greater number of chlamydospores of VAM fungi were observed in the rhizosphere of healthy plants than in the diseased ones (Zaidi and Mukerji, 1983). Larger numbers of arbuscules were formed in the roots of diseased plants than in the roots of healthy ones.

Table 1. Interaction between VAM fungi, and soil and root borne plant pathogens (fungus).

Pathogen	Host	Effect of mycorrhizal plants		References
		(-) Dec. Damage	(+) Inc. Interaction	
Aphanomyces euteiches	Peas	-	-	Rosendahl, 1985
Bipolaris sorokiniana	Barley	-	-	Boyetchko and Tewari, 1990
Cylindrocarpon destructans	Strawberry	-	-	Paget, 1975; Traquair, 1995
Cylindrocladium scoparium	Yellow poplar	-	-	Barnard, 1977; Chakravarty and Unestam, 1987
Fusarium avenaceum	Clover	-	-	Dehne, 1982
F. oxysporum	Cucumber	-	-	Schoenbeck, 1979
F. oxysporum	Easter lily	+	-	Ames and Linderman, 1978
F. oxysporum	Tomato	-	-	Dehne and Schoenbeck, 1975
F. oxysporum	Onion	-	-	Dehne, 1982
F. oxysporum	Tomato	-	-	Al-Momany and Al-Radded, 1988
F. oxysporum	Tomato	-	-	Ramraj et al., 1988
F. oxysporum	Capsicum	-	-	Ramraj et al., 1988
F. oxysporum	Asparagus	-	-	Wacker et al., 1990
F. oxysporum	Tomato	-	-	Raman and Gnanaguru, 1996
F. oxysporum lycopersici	Cucumber	-	-	Dehne and Schoehbeck, 1979
F. oxysporum lycopersici	Tomato	+	-	McGraw and Schenck, 1981
F. solani	Soybean	-	-	Zambolim and Schenck, 1983
Fusarium spp.	Tomato	-	-	Caron et al., 1986
Gaeumannomyces graminis	Wheat	-	-	Graham and Menge, 1982
Macrophomina phaseolina	Soybean	No effect		Stewart and Pfleger, 1977
Olpidium brassicae	Tobacco	-	-	Schoenbeck and Dehne, 1977
O. brassicae	Lettuce	-	-	Schoenbeck and Schinzer, 1972
Phoma terrestris	Onion	-	-	Becker, 1976
Phytophthora cinnamomi	Lawson pine	-	-	Bartschi et al., 1981
P. cinnamomi	Avocado	+	+	Davis et al., 1978
P. cinnamomi	Citrus	-	-	Davis et al., 1978
P. cinnamomi	Alfalfa	-	-	Davis et al., 1978
P. drechsleri f.sp. *cajani*	Pigeon pea	-	-	Bisht et al., 1985
P. megasperma var. *sojae*	Soybean	-	-	Ross, 1971
P. palmivora	Papaya	-	no	Ramirez, 1974
P. parasitica	Citrus	-	-	Schenck and Nicolson, 1977
P. parasitica	Citrus	+	+	Davis et al., 1978

P. parasitica	Citrus	-	-	Davis and Menge, 1981
Pyrenochaeta lycopersici	Tomato	-	-	Bochow and Aboushaar, 1990
P. terrestris	Onion	-	-	Becker, 1976
P. terrestris	Onion	-	-	Schoenbeck and Dehne, 1981
Pythium ultimum	Soybean	-	-	Christie et al., 1978
P. ultimum	Cucumber	-	-	Rosendahl and Rosendahl, 1990
P. ultimum	Poinsettia	-	-	Harley and Wilson, 1959
P. ultimum	Soybean	No effect		Kaye et al., 1984; Chou and Schmitthenner, 1974
Rhizoctonia solani	Poinsettia	-	-	
R. solani	Cotton	-	-	Mathre, 1968
R. solani	Citrus	-	-	Menge et al., 1977
R. solani		-	-	Khadge et al., 1990
Sclerotium rolfsii	Wheat	-	-	Harlapur et al., 1988
S. rolfsii	Peanut	-	-	Krishna and Bagyaraj, 1983
Thielaviopsis basicola	Tobacco	-	-	Tosi et al., 1988
T. basicola	Tobacco	-	-	Giovannetti et al., 1991
Verticillium dehliae	Chrysanthemum	-	-	Pegg and Jouglaekha, 1981
V. dehliae	Cotton	+	+	Davis et al., 1979
V. spp.	Alfa alfa	-	-	Hwang et al., 1992
Verticillium spp.	Tomato	-	-	Baath and Hayman, 1983

Vigorously growing plants are also more prone to viral diseases. In mycorrhizal plants this is due to the better availability of nutrients. Schoenbeck and Schinzer (1972) showed similar effects in the case of tobacco mosaic virus (TMV)-infected tobacco plants. A higher concentration of the virus was observed in mycorrhizal roots at the arbuscular stage of endophyte development. This indicates that high metabolic activity in host cytoplasm of arbuscular cells is favourable for the accumulation and multiplication of viral particles (JabajiHare and Stobbs, 1984).

Nemec and Myhre (1984) showed that the growth of *Citrus macrophylla* inoculated with tristeza virus isolate T-3 and citrus urgose virus (CLRV-2) was not reduced by virus infection in *G. etunicatum* in low phosphorus soil or phosphorus-amended soil with 210 µg P per g soil, 98 days after inoculation. However, in sour orange infected with virus T-3 growth was significantly reduced in both soils compared with non-viral treatments. Jayaraman and Kumar (1995) showed in *Vigna radiata* plants that yellow mosaic virus reduced mycorrhizal formation and spore production by *Gigaspora gilmorei, Acaulospora morrowae, G. fasciculatum* and *G. constrictum*.

CONTROL OF WILT DISEASES OF CROP PLANTS USING VAM FUNGI

In this paper we describe studies on the effect of inoculations with VAM fungus *Glomus macrocarpum* on the wilt diseases of three test plants, namely cotton, jute and flax. Test pathogens used for the experiment were *F. oxysporum* f.sp. *vasinfectum* for cotton, *F. solani* for jute and *F. oxysporum* f.sp. *lini* for flax. The experiment was conducted in fumigated soil. Soil was amended so that half the pots contained VA mycorrhizal inoculum and the other half were non-mycorrhizal. The mycorrhizal inoculum had 85 spores of *G. macrocarpum* per gram of soil. 10 g of soil was used in each pot. Surface sterilized seeds of the test plants, i.e. *Gossypium hirsutum, Corchorus ollitorius* and *Linum usitatissimum,* were sown in all the pots. Seven days after seed germination half of the mycorrhizal and half of the non-mycorrhizal seedlings were inoculated with spore suspension of the wilt fungi. 10 ml of spore suspension were added to the rhizosphere of the plants. Inoculum densities of the wilt pathogens used were as follows :

Fusarium oxysporum f.sp. *vasinfectum*: 5×10^7 spores ml^{-1}
Fusarium solani: 7.5×10^6 spores ml^{-1}
Fusarium oxysporum f.sp. *lini*: 6.9×10^6 spores ml^{-1}

Observations in terms of plant height, dry matter, percent VAM colonization and number of VAM fungal spores in the rhizosphere of the plants were taken at intervals of 15 days. The following four treatments were given to each of the test plants:

(i) G^+F^+ -   Inoculated with both *G. macrocarpum* and *Fusarium* sp.
(ii) G^+F^- -   Inoculated with *G. macrocarpum* only
(iii) G^-F^+ -   Inoculated with *Fusarium* only
(iv) G^-F^- -   Uninoculated controls

Inoculations with *Glomus macrocarpum* resulted in increased plant height and dry matter content in the test plants (Figure 1a-f). The dry matter increase was of the order of two to three times. The plants inoculated with pathogen only (G^-F^+ treated plants) showed minimum plant height and dry matter content. However, better plant growth was observed in plants inoculated with both VAM fungi and pathogens (G^+F^+ treated plants) as compared with G^-F^+ treated plants (i.e. non-mycorrhizal plants inoculated with wilt pathogen). Hence, it is evident that mycorrhizal colonization of the host plants led to a reduction in the deleterious effect of the pathogens. Wilt-infested plants were necrotic and had fewer roots in comparison to mycorrhizal plants. All wilt-infested plants (G^-F^+) showed reduced mycorrhizal colonization. The percentage of VAM fungal colonization was higher in G^+F^- treated plants than in G^+F^+ treated plants. Hence, the number of VAM fungal spores in the rhizosphere of the G^+F^- treated plants was greater than that in the rhizosphere of G^+F^+ treated plants. Maximum yield was observed in G^+F^- treated plants followed by G^+F^+ and G^-F^-. Minimum yield was observed in G^-F^+ treated plants.

Hence, it can be seen from the above studies that colonisation by VAM fungus *Glomus macrocarpum* was associated with a reduction in the deleterious effects of wilt pathogens on plant height, dry matter content and yield in the plant species studied, i.e. cotton, jute and flax.

MECHANISM OF DISEASE CONTROL DUE TO VAM FUNGI

VA mycorrhizal fungi colonize the root system rapidly and compete for nutrients and water, reducing the quantity available to pathogens and thus restricting pathogenic infection. The VA mycorrhizal fungi and plant pathogenic fungi inhabit the same plant rhizosphere and are in direct competition for space and this may be one of the major reasons for disease resistance of mycorrhizal roots. It is well established that the VAM fungi may act as a substitute or supplementary strategy for the host plant under situations that are deleterious to

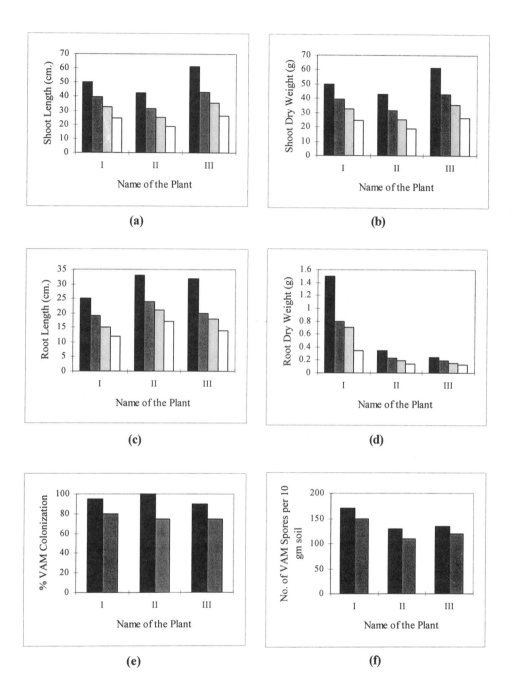

Figure 1 : Effect of soil amendments on the (a) Shoot Length, (b) Shoot Dry Weight, (c) Root Length, (d) Root Dry Weight, (e) Percent VAM Colonization, (f) Number of VAM Spores in the Rhizosphere, of the three test plants ie. Cotton (I), Jute (II) and Flax (III).

G⁺F⁻   G⁺F⁺   G⁻F⁻   G⁻F⁺

root growth. VAM fungi can compensate for disease and/or counteract toxicities if the plants are already inoculated with them.

Mycorrhizal plant roots remain functional for a longer period than non-mycorrhizal roots and are less susceptible to attack by certain type of pathogens. The higher nutritional status of mycorrhizal plants is known to induce resistance against invasion of roots by soil-borne pathogens (Linderman and Paulitz, 1990). Higher arginine levels and chitinase activity of mycorrhizal tissue inhibits development of the pathogen inside the host tissue (Dehne et al., 1978). The increased lignification and high phenol content of VA mycorrhizal roots inhibits the entry of pathogens into the root (Friend, 1981; Grandmaison et al., 1993). High OD phenol content in the roots of peanut helps in imparting resistance against the root pathogen *Sclerotium rolfsii* (Krishna and Bagyaraj, 1983). The presence of cell wall bound phenols in roots of VA mycorrhizal plants have been reported by Codignola et al. (1989).

Altered microbial equilibrium due to enhanced phosphorus uptake is known to favour certain saprophytic fungi and inhibit several soil-borne root pathogens, e.g. species of *Fusarium, Alternaria, Rhizotonia* etc. (Bali, 1991; Bansal and Mukerji, 1994). VAM fungi have an extremely wide host range, but plants select certain indigenous endophytes in preference to others. Different species of mycorrhizal fungi may differ in their ability to control plant diseases. Therefore, in general it is suggested that initial inoculum with a mixture of VAM fungi containing species best suited for different stages of plant growth will be more beneficial than inoculation with a single VAM fungal species (Kumar, 1990). For long term benefits a suitable host-symbiont combination has to be found out so as to produce desirable effect of improving plant growth and inducing resistance to pathogens.

CONCLUSIONS

Different species of VA mycorrhizal fungi may differ in their ability to control plant diseases. For useful results, a suitable host-symbiont combination has to be found so as to produce the desirable effect of improving plant growth and inducing resistance to pathogens under given environmental conditions. Since pre-inoculation of plants with VAM fungi has been found to be effective in controlling the severity of disease (Zambolin and Schenck, 1983), efforts should be directed to finding out suitable techniques for large-scale inoculum production which could be commercially exploited. In the biocontrol of diseases with the help of VAM fungi, the results being host mediated are difficult to predict. But the future looks promising for commercial exploitation of VAM fungi as potential biofertilizer and biocontrol agent.

ACKNOWLEDGEMENTS

One of the authors (MS) acknowledges financial assistance from CSIR.

REFERENCES

Al-Momany, A. and Al-Radded, A., 1988, Effect of vesicular-arbuscular mycorrhizae on *Fusarium* wilt of tomato and pepper, *Alexandria J. Agric. Res.* 33: 249-261.

Ames, R.N. and Linderman, R.G., 1978, The growth of Easter Lily (*Lillium longiflorum*) as influenced by vesicular-arbuscular mycorrhizal fungi, *Fusarium oxysporum* and fertility level *Can. J. Bot.* 56 : 2778-2782.

Andrea Torres-Barragan, Emma Zavaleta-Mejia, Carmen Gonzalez-Chavez, and Ronald Ferrera-Cerrato, 1996, The use of arbuscular mycorrhizae to control onion white rot (*Sclerotium cepivorum* Berk.) under field conditions, *Mycorrhiza* 6: 253-257.

Baath, P. and Hayman, D.S. 1983, Plant growth responses to vesicular arbuscular mycorrhiza XIV. Interaction with *Verticillium* wilt on tomato plants, *New Phytol.* 95: 419-425.

Bali, M., 1991, *Responses of certain economically important plants to VAM fungi, Ph.D. Thesis*, Univ. of Delhi, India.

Bali, M. and Mukerji, K.G., 1988, Effect of VAM fungi on Fusarial wilt of cotton and jute. Mycorrhizae for green Asia. Proc. 1st Asian Conf. on Mycorrhizae, pp.233-234.

Bali, M. and Mukerji, K.G., 1991, Interactions between VA mycorrhizal fungi and root microflora of jute, in: *Plant Roots and their Environment* (McMichael and H. Persson, eds.), pp.396-401, Elsevier, Amsterdam.

Bansal, M. and Mukerji, K.G., 1994, Positive correlation between VAM induced changes in root exudation and mycorrhizosphere mycoflora, *Mycorrhiza* 5: 39-44.

Barber, D.A. and Martin, J.K., 1996, The release of organic substances by cereal roots into soil, *New Phytol.* 76: 69-80.

Barnard, E-L. 1977, The mycorrhizal biology of *Liriodendron tulipifera* L. and its relationship to *Cylindrocladium* root rot, Ph.D. Diss. Duke Univ., Durham. NC. p.147.

Bartschi, H., Gianinazzi-Pearson, V. and Veigh, I., 1981, Vesicular arbuscular mycorrhiza formation and root rot disease (*Phytophthora cinnamoni*) development in *Chamaecyparis lawsoniana*, *Phytopathol. Z.* 102: 213-219.

Becker, W.N. 1976, Quantification of onion vesicular-arbuscular mycorrhizae and their resistance to Pyrenochaeta terrestris,. *Ph.D. Thesis*, Univ. of Illinois. Urbana. pp.72.

Bisht, V.S., Krishna, K.R. and Nene, Y.L., 1985, Interactions between vesicular-arbuscular mycorrhiza and *Phytophthora drechsleri* f.sp. *cajani*. *International Pigeonpea Newsletter*, 4 : 63-64.

Bochow, H., and Abou-Shaar, M., 1990, On the phytosanitary effect of VA mycorrhiza in tomatoes to the corky-root disease. *Zentralblatt fur mikrobiologie*, 145: 171-176.

Boyetchko, S.M., and Tewari, J.P., 1990, Effect of phosphorus and VA mycorrhizal fungi on common root rot of barley, *Proc. 8th NACOM*. Innovation and Hierarchial Integration. Jackson, Wyoming. 33 p.

Calvet, C., Barea, J.M. and Pera, J., 1993, Growth response of marigold (*Tagetes erecta* L.) to inoculation with *Glomus mosseae*, *Trichoderma aureoviride* and *Pythium ultimum* in a peat-perlite mixture, *Plant and Soil* 148: 1-6.

Caron, M., Fortin, A. and Richard, C. 1986, Effect of phosphorus concentration and *Glomus intraradices* on *Fusarium* crown and root rot of tomatoes, *Phytopathol.* 76: 942-946.

Chakravarty, P. and Unestam, Torgny, 1987, Mycorrhizal fungi prevent disease in stressed pine seedlings, *Phytopathol.* 118: 335-340.

Chou, L.G. and Schmitthenner, A.F., 1974, Effect of *Rhizobium japonicum* and *Endogone mosseae* on the soybean root rot caused by *Pythium ultimum* and *Phytophthora megasperma* var. *sojae*, *Plant Dis. Reporter*, 58: 221-227.

Christie, P., Newman, E.I., and Campbell, R., 1978, The influence of neighbouring grassland plants on each others endomycorrhizas and root surface microorganisms, *Soil Biol. Biochem.* 10: 52-57.

Codignola, A., Verotta, L., Spanu, P., Maffei, M., Scannerini, S. and Bonfante-Fasolo, P., 1989, Cell wall bound phenols in roots of vesicular-arbuscular mycorrhizal plants. *New Phytol.* 112 : 221-228.

Davis, R.M. and Menge, J.A., 1980, Influence of *Glomus fasciculatum* and soil phosphorus on *Phytophthora* root rot of citrus, *Phytopathol.* 70: 447-452.

Davis, R.M. and Menge, J.A., 1981, *Phytophthora parasitica* inoculations and intensity of vesicular - arbuscular mycorrhizae in citrus, *New Phytol.* 87: 705-711.

Davis, R.M., Menge, J.A. and Erwin, D.C., 1979, Influence of *Glomus fasciculatus* and soil phosphorus on *Verticillium* wilt of cotton, *Phytopathol.* 69: 453-456.

Davis, R.M., Menge, J.A. and Zentmeyer, G.A., 1978, Influence of vesicular-arbuscular mycorrhizae on *Phytophthora* root rot of three crop plants, *Phytopathol.* 68: 1614-1619.

Dehne, H.W., 1982, Interaction between vesicular arbuscular mycorrhizal fungi and plant pathogens, *Phytopathol.* 72: 115-121.

Dehne, H.W. and Schoenbeck, F., 1975, The influence of endotrophic mycorrhiza on the *Fusarium* wilt of tomato. *Z. Pflanzenkr. Pflanzenschutz.* 82: 630-636.

Dehne, H.W. and Schoenbeck, F., 1979, The influence of endotrophic mycorrhizae on plant disease. 1. Colonization of tomato plants by *Fusarium oxysporum* f.sp. *lycopersici*, *Phytopathol. Z.* 95: 105-112.

Dehne, H.W., Schoenbeck, F. and Baltruschat, H., 1978, The influence of endotrophic mycorrhiza on plant diseases. 3. Chitinase activity and ornithine cycle. *Z. Pflanzenkrankh. Pflanzenschutz.* 85: 666-671.

Friend, J., 1981, Plant phenolics, lignification and plant disease, in: *Progress in Phytochemistry vol.7.* (J.B. Harbone and T. Swain, eds.), pp. 197-261, Pergamon Press, New York.

Garrett, S.D., 1970, *Pathogenic - Root Infecting Fungi*, Cambridge University Press, Cambridge. pp.294.

Giovannetti, M., Tosi, L., Torre Gdella and Zazzerini, A., 1991, Histological, physiological and biochemical interactions between vesicular-arbuscular mycorrhizae and *Thielaviopsis basicola* in tobacco plants, *J. Phytopathol.* 131: 265-274.

Graham, J.H., 1988, Interactions of mycorrhizal fungi with soil-borne plant pathogens and other organisms: an introduction, *Phytopathol.* 78: 365-366.

Graham, J.H. and Menge, J.A., 1982, Influence of arbuscular mycorrhizae and soil phosphorus on take - all disease of wheat, *Phytopathol.* 72: 95-98.

Graham, J.H., Leonard, R.T. and Menge, J.A., 1981, Membrane mediated decrease in root exudation responsible for phosphorus inhibition of vesicular-arbuscular mycorrhiza formation, *Plant Physiol.* 68: 548-552.

Grandmaison, J., Olah, G.M., Vancalsteren, M.R. and Furlan, V., 1993, Characterization and localization of plant phenolics likely to be involved in the pathogen resistance expressed by endo-mycorrhizal roots, *Mycorrhiza* 3: 155-164.

Harlapur, S.I., Kulkarni, S. and Hegde, R.K., 1988, Studies on some aspects of root rot of wheat caused by *Sclerotium rolfsii, Plant Pathology Newsletter* 6: 49-54.

Harley, J.L. and Smith, S.E., 1983, *Mycorrhizal Symbiosis*, Academic Press, London, pp. 483.

Harley, J.L. and Wilson, J.M. 1959, The absorption of potassium by beech mycorrhizas, *New Phytol.* 58: 281.

Hussey, R.S. and Roncadori, R.W. 1982, Vesicular arbuscular mycorrhizae may limit nematode activity and improve plant growth. *Plant Dis.* 66: 9-14.

Hwang, S.F., Chang, K.F. and Chakravarty, P., 1992, Effect of vesicular-arbuscular mycorrhizal fungi on the development of *Verticillium* and *Fusarium* wilts of *Alfa alfa, Plant Dis.* 76: 239-243.

Jabaji Hare, S.H. and Stobbs, L.W., 1984, Electron microscopic examination of Tomato roots coinfected with *Glomus* spp. and tobacco mosaic virus, *Phytopathol.* 74: 277-279.

Jayaraman, J. and Kumar, D., 1995, Influence of mung bean yellow mosaic virus on mycorrhizal fungi associated with *Vigna radiata* var P516, *Indian Phytopathol.* 48: 108-110.

Kaye, J.W., Pfleger, F.L. and Stewart, E.L. 1984, Interaction of *Glomus fasciculatum* and *Pythium ultimum* on green house grown poinsettia, *Can. J. Bot.* 62: 1575-1581.

Khadge, B.R., Ilag, L.L. and Mew, T.W. 1990, Interaction study of *Glomus mosseae* and *Rhizoctonia solani, Proc. Nat. Conf. on Mycorrhiza: Current Trends in Mycorrhizal Research*, pp. 94-95.

Krishna, K.R. and Bagyaraj, D.J., 1983, Interactions between *Glomus fasciculatum* and *Sclerotium rolfsi* in peanuts, *Can. J. Bot.* 61: 2349-2354.

Kumar, D. 1990, Emerging concepts of mycobiont succession in mycorrhizal associations, *Mycorrhiza News.* 2: 5-6.

Linderman, R.G., 1985, Microbial interactions in the mycorrhizosphere, in: *Proc. 6th North Am. Conf. Mycorrhiza* (R. Molina, ed.), pp. 117-120. Forest Research Lab. Corvallis, OR.

Linderman, R.G., 1988, Mycorrhizal interactions with the rhizosphere microflora: The mycorrhizosphere effect. *Phytopathol.* 78: 366-371.

Linderman, R.G. and Call, C.A. 1977, Enhanced rooting of woody plant cuttings by mycorrhizal fungi, *J. Am. Soc. Hort. Sci.*, 102: 629-632.

Linderman, R.G. and Paulitz, T.C., 1990, Mycorrhiza-rhizobacteria interactions, in: *Biological Control of Soil Borne Plant Pathogens* (D. Hornby, ed.), pp.261-283, CAB International, Wallingford, U.K.

MacGuidwin, A.E., Bird, G.W., and Safir, G.R., 1985, Influence of *Glomus fasciculatum* on *Meloidogyne hapla* infecting *Allium cepa, J. Nematol.* 17: 389-395.

Mamta, Sharma and Mukerji, K.G., 1982, Mycorrhiza - tool for biocontrol in: *Recent Developments in Biocontrol of Plant Diseases* (K.G. Mukerji, J.P. Tewari, D.K. Arora and G. Saxena,eds.), pp.52-80. Aditya Books Pvt. Ltd., Delhi, India.

Mathre, D.E., 1968, Photosynthetic activities of cotton plants infected with *Verticillium albo-atrum, Phytopathol.* 58: 137-144.

McAllister, C.B., Garcia-Romera, I., Godeas, A. and Ocampo, J.A., 1994a, Interactions between *Trichoderma koningii, Fusarium solani* and *Glomus mosseae*: effects on plant growth, arbuscular mycorrhizas and the saprophyte inoculants, *Soil Biol. Biochem.* 26: 1363-1367.

McAllister, C.B., Garcia-Romera, I., Godeas, A. and Ocampo, J.A., 1994b, *In vitro* interactions between *Trichoderma koningii, Fusarium solani* and *Glomus mosseae, Soil Biol. Biochem.* 26: 1369-1374.

McGraw, A.C. and Schenck, N.C. 1981, Effect of two species of vesicular arbuscular mycorrhizal fungi on the development of *Fusarium* wilt of tomato, *Phytopathol.* 71: 894-899.

Menge, J.A., Nemec, S., Davis, R.M. and Minassian, V., 1977, Mycorrhizal fungi associated with citrus and their possible interactions with pathogens, *Proc. Intl. Soc. of Citriculture,* 3: 872.

Mukerji, K.G., Jagpal, R., Bali, M. and Rani, R., 1991, The importance of mycorrhiza for roots, in: *Plant Roots and their Environment* (McMichael and Persson, eds.), pp. 290-308, Elsevier, Amsterdam.

Mukerji, K.G., Chamola, B.P. and Sharma, Mamta, 1996, Mycorrhiza in control of plant pathogens, in: *Management of Threatening Plant Diseases of National Importance* (V.P. Agnihotri and A.K. Sarabhoy, eds.), pp. 1-18, M.Phil, Delhi.

Mukerji, K.G., Chamola, B.P., Kaushik, A., Sarwar, S. and Dixoni, R.K. 1996, Vesicular arbuscular mycorrhiza: a potential biofertilizer for nursery raised multipurpose tree species in tropical soils, *Ann. For* 4(1): 12-20.

Nemec, S. and Donald, Myhre, 1984, Virus *Glomus etunicatum* interactions in citrus root stocks. *Plant Dis.* 68: 311-314.

Newsham, K.K., Fitter, A.H. and Watkinson, A.A., 1993, The role of mycorrhizal and pathogenic fungi in the fitness of a winter annual grass, *9th NACOM*, Aug.8-12. pp.38.

Paget, D.K., 1975, The effect of *Cylindrocarpon* on plant growth responses to vesicular arbuscular mycorrhiza, in: *Endomycorrhizas* (Sanders, F.E., Mosse, B., Tinker, P.B., eds.), pp.593-606, Academic Press, London.

Pegg, G.F. and Jouglaekha, N. 1981, Assessment of colonization in chrysanthemum grown under different photoperiods and infected with *Verticillium dahliae*, *Trans. Br. Mycol. Soc.*, 76: 353-355.

Raman, N. and Gnanaguru, M., 1996, Biological control of *Fusarium* wilt of tomato by *Glomus fasciculatum*, in: *Current Trends in Life Sciences Vol.XXI : Recent Developments in Biocontrol of Plant Pathogens.* (K. Manibhushan Rao and A. Mahadevan, eds.), pp.21-25, Today and Tomorrow's Printers and Publishers, New Delhi, India.

Ramirez, B.N., 1974, Influence of endomycorrhizae on the relationship of inoculum density of *Phytophthora palmivora* in soil to infection of papaya roots, *M.Sc. Thesis*, Univ. of Florida. Gainsville, 45 pp.

Ramraj, B., Shanmugam, N. and Dwarkanath Reddy, A. 1988, Biocontrol of *Macrophomina* root rot of cowpea and *Fusarium* wilt of tomato by using VAM fungi, in: *Mycorrhizae for Green Asia*, pp.250-251. Proc. 1st Asian Conf. on Mycorrhizae.

Ratnayake, M., Leonard, R.T. and Menge, J.A., 1978, Root exudation in relation to supply of phosphorus and its possible relevance to mycorrhizal formation, *New Phytol.* 81: 543-552.

Rosendahl, S., 1985, Interactions between the vesicular-arbuscular mycorrhizal fungus *Glomus fasciculatum* and *Aphanomyces euteiches* root rot of peas (*Pisum sativum*), *Phytopathol.* 114: 31-40.

Rosendahl, C.N. and Rosendahl, S. 1990, The role of vesicular-arbuscular mycorrhiza in controlling damping-off and growth reduction in cucumber caused by Pythium ultimum, *Symbiosis* (Rehovot) 9: 363-366.

Ross, J.P. 1971, Effect of phosphate fertilization on the yield of mycorrhizal and non-mycorrhizal soybeans, *Phytopathol.* 61: 1400-1404.

Safir, G.R. and Nelsen, C.E. 1985, VA mycorrhizas plant and fungal water relations, in: *Proc. 6th North. Am. Conf. on Mycorrhiza* (R. Molina, ed.), pp. 471, Corvallis, OR.

Schenck, N.C. and Nicolson, T.H., 1977, A zoosporic fungus occurring on species of *Gigaspora margarita* and other vesicular arbuscular mycorrhizal fungi, *Mycologia* 69: 1049-1052.

Schoenbeck, F., 1979, Endomycorrhiza in relation to plant diseases, in: *Soil Borne Plant Pathogens* (Schippers, B. and Gams, S., eds.), pp.271. Academic Press, London.

Schoenbeck, F. and Dehne, H.W. 1977, Damage to mycorrhizal and non-mycorrhizal cotton seedlings by *Thielaviopsis basicola*, *Plant Dis. Rep.*,61: 266.

Schoenbeck, F. and Dehne, H.W., 1981, Mycorrhiza and plant health, *Gesunde. Pflanzen.* 33: 186.

Schoenbeck, F. and Schinzer, U., 1972, Investigations on the influence of endotrophic mycorrhiza on TMU lesion formation in *Nicotiana tabaccum* L. var. *xanthi*, *Phytopathol. Z.*, 73: 78-84.

Srivastava, Deepti, Kapoor, Roopam, Srivastava, S.K. and Mukerji, K.G., 1996, Vesicular arbuscular mycorrhiza-an overview, in: *Concepts in Mycorrhizal Research* (K.G. Mukerji, ed.), pp. 1-39, Kluwer Academic Publishers, Netherlands.

Stewart, E.L. and Pfleger, F.L., 1977, Development of poinsettia as influenced by endomycorrhizas, fertilizer and root rot pathogens *Pythium ultimum* and *Rhizoctonia solani*, *Florists Rev.* 159: 37-43.

Sutton, J.C. and Sheppard, B.R., 1976, Aggregation of sand dune soil by endomycorrhizal fungi, *Can. J. Bot.*, 54: 326-333.

Tinker, P.B., 1984, The role of microorganisms in mediating and facilitating the uptake of plant nutrients from soil, *Plant and Soil*, 76: 77-91.

Tosi, L., Giovannetti, M., Zazzerini, A. and Torre G. della, 1988, Influence of mycorrhizal tobacco roots incorporated into the soil on the development of *Thielaviopsis basicola*, *J. Phytopathol.* 122: 186-189.

Traquiar, J.A., 1995, Fungal biocontrol of root diseases : Endomycorrhizal suppression of *Cylindrocarpon* root rot, *Can. J. Bot.* 73: 89-95.

Wacker Tracy, L., Safir, G.R. and Stephjens, G.T., 1990, Effect of *Glomus fasciculatum* on the growth of Asparagus and the incidence of *Fusarium* Root Rot, *J. Amm. Soc. Hort. Sci.* 115: 550-554.

Wallace, H.R., 1983, Interactions between nematodes and other factors on plants, *J. Nematol.*, 15: 221-227.

Zaidi, R. and Mukerji, K.G., 1983, Incidence of vesicular-arbuscular mycorrhiza (VAM) in diseased and healthy plants, *Indian J. Plant. Pathol.* 1: 24-29.

Zambolim, L. and Schenck, N.C., 1983, Reduction of the effect of pathogenic root-infecting fungi on soybean by the mycorrhizal fungus *Glomus mosseae*, *Phytopathol.* 73: 1402-1409.

VESICULAR ARBUSCULAR MYCORRHIZAE IN INCREASING THE YIELD OF AROMATIC PLANTS

Rupam Kapoor and K.G. Mukerji

Applied Mycology Laboratory
Botany Department
Delhi University
Delhi-110 007
India

INTRODUCTION

The mycorrhizae represent one of nature's best gifts to mankind in augmenting yield increase in plants. The increase in effective nutrient absorbing surface provided by vesicular arbuscular mycorrhizal (VAM) fungi is primarily responsible for the increase in uptake of soil nutrients by mycorrhizal plants. Because mycorrhizal fungi increase the efficiency of fertilizer use, they are referred to as 'biofertilizers' and can be substituted for substantial amounts of some fertilizers (Bansal and Mukerji, 1994).

Recent researchers have amply demonstrated the beneficial role of VAM fungi in plant growth and health and it appears that they are essential for the survival of plant species in many ecosystems (Allen, 1991; Read et al., 1992). Several reports are available on the extent of mycorrhiza formation and its effect on plant growth and biomass production in different genotypes (Krishna et al., 1985), varieties (Hall, 1978; O'Bannon et al., 1980) and cultivars (Tilak and Murthy, 1987; Vierheilig and Ocampo, 1991) of economically important crops and ornamental plants (Aboul-Nasr, 1996). Many VAM inoculation experiments attempted for studying the effect on the growth and yield of annual crops showed promising results (Black and Tinker, 1977; Jakobsen, 1986; Owusu-Bennoah and Mosse, 1979). For certain crops (citrus, bell pepper) the use of mycorrhizal fungi is already known to be commercially and economically feasible (Johnson and Menge, 1982; Powell, 1982). However, the role of VAM fungi in the growth and yield of aromatic plants remains poorly understood which limits the advances towards their practical applications.

MEDICINAL AND AROMATIC PLANTS

Medicinal and aromatic plants occupy an important position in the world today because of renewed interest in drugs from natural sources. All distinctly aromatic plants contain essential oils. They occur in some 60 families and are particularly characteristic of Lauraceae, Myrtaceae, Umbelliferae, Labiatae and Compositae.

Essential oils are distinguished from fatty oils by the fact that they evaporate or volatilize in contact with the air and possess a pleasant taste and strong aromatic odour. Essential oils are very complex in their chemical nature. The two principal groups are the terpenes, which are hydrocarbons (based on an integral number of C_5 units, 'isoprenoids') and the oxygenated and sulphurated oils. The amount of oil varies from an infinitesimal quantity to as much as 1 to 2 percent. The oils are secreted in internal glands or in hair like structures: almost any organ of a plant may be a source of the oil: flowers (rose), fruits (orange), leaves (mint), bark (cinnamon), root (ginger), wood (cedar), or seeds (umbellifers), and many resinous exudations as well.

PHYSIOLOGICAL AND ECONOMIC SIGNIFICANCE OF ESSENTIAL OILS

The physiological significance of these oils as far as the plant is concerned is not obvious. They probably represent by-products of metabolisms. The characteristic flavour and aroma that they impart in air are probably of advantage in attracting insects and other animals which play a role in pollination or the dispersal of the fruits and seeds. When present in high concentrations, these odours may serve to repel predators (Swain, 1977). The oils may also have some antiseptic and bactericidal value. There is some evidence that they may play an even more vital role as hydrogen donors in oxido-reduction reactions, as potential sources of energy, or in affecting transpiration and other physiological processes (Loomis and Corteau, 1973).

Essential oils have very varied industrial applications. Because of their odour and high volatility they are extensively used in the manufacture of perfumes, soaps and other toiletries. Many are used as flavouring materials or essences for candy and ice-cream, in cooking and for cordials, liquors and non-alcoholic beverages. Still others have therapeutic, antiseptic, or antibacterial properties and so are valuable in medicine and dentistry.

In fact, nearly all the essential oils are used for different medicinal purposes. Some of the oils are used as clearing agents in histological works, as solvents in paint and varnish industries, as insecticides and deodorants, in the manufacture of various synthetic odours and flavours, and in such widely diversified products as chewing gum, tobacco, shoe polish, library paste, printer's ink, toothpaste and fish glue.

ESSENTIAL OIL PRODUCTION

Essential oils are extracted from the plant tissues in various ways depending on the quantity and stability of the compound. There are three principal methods: distillation, expression and extraction by solvents. However, there are certain intrinsic factors (genotype and ontogeny) and extrinsic factors (water and nutrients) that strongly influence oil production and optimisation.

Intrinsic Factors

The lack of similarity in oil composition between phenotypes grown in the same environment is a manifestation of genotypic differences. Thus in plants at the same stage of development (ontogeny) and of the same genotype, extrinsic factors can assume a quantitative modifying effect which can cause both quantitative and qualitative variation in oil content. It has been reported that the formation of active principle occurs predominantly during periods of vigorous growth and during times of intensive metabolic process such as when a plant is flowering or fruiting (Dey and Choudhari, 1981; Pareek et al., 1982).

An example of change in chemical composition of the essential oil in *Coriandrum sativum* during its growth period demonstrates the effect of plant age and metabolism on the composition of essential oil. Similar compositional changes have been observed with *Anethum graveolens* (Hornok, 1983).

Extrinsic Factors

The yield of oil obtained from a specific clone can often be influenced by changing environmental conditions. The most important extrinsic factors affecting essential oil production are nutrients and water.

Duhan et al. (1974) observed that date of sowing and nitrogen fertilisation influence the quality of the oil in *Anethum graveolens*. Evidence for the positive effect of phosphorus on oil yield was also obtained by several workers on basil and mints (Gulati, 1976; Rao et al., 1983; Singh and Duhan 1984; Singh et al., 1989).

The magnitude and overall effect of a macronutrient on oil yield is also dependent upon available water and stage of development. Studies on water requirements and effect of water on yield of essential oil have indicated that the optimum soil water content for good oil yield should be 80-90 per cent (Clark and Minary, 1980).

Robert et al. (1986) reported that mesophytic plants such as *Carum carvi, Anethum graveolens* and *Ocimum basilicum* produce a decreased oil yield under moisture stress. They require a regulated water supply throughout their growth cycle to maximise oil yield.

VAM ASSOCIATION IN AROMATIC PLANTS

In earlier reports, VAM association has been reported to be absent in medicinal plants and the possible reasons for its absence were attributed to the presence of various secondary substances in those plants (Mohan Kumar and Mahadevan, 1984). However, in the recent past, many medicinal and essential oil-bearing plants have been reported to harbour VA mycorrhizae in their root systems (Rao et al., 1989; Gupta et al., 1990; Srivastava and Basu, 1995).

An investigation was carried out to study the effect of VAM fungal (*Glomus macrocarpum*) inoculation on *Coriandrum sativum*. The fungus established itself completely within thirty days of inoculation and all the structures characteristic of VAM fungi were observed (Figure 1). VAM inoculation caused an appreciable increase in growth and biomass, seed yield and percent essential oil content of seeds (Figure 2). Similar results were obtained by other workers working on VAM associations in various aromatic plants. VAM inoculated plants exhibited a two to six times increase in growth and biomass production of aromatic grasses, namely palmarosa (*Cymbopogon martinii* var. *motia*), lemongrass (*Cymboprogen wintereanus*) (Janardhanan and Khaliq, 1995). Mago (1994) studied the influence of *Glomus fasciculatum* on seven cultivated *Mentha* species. Results indicated that VAM association enhanced the shoot biomass of all inoculated plants significantly. There was a significant increase in the essential oil content of the plants.

It was also observed in the case of the mints (Mago 1994) and umbellifers (Kapoor, 1997) that VAM fungal colonization enhances the quality of the essential oil by increasing the concentration of the desired constituents.

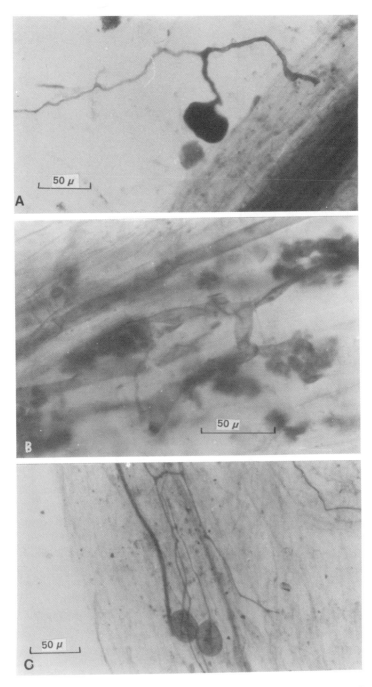

Figure ∴. VAM colonisation in *Coriandrum sativum*. A. Germinating chlamydospore. B. Intercellular arbuscules. C. hyphae with vesicles

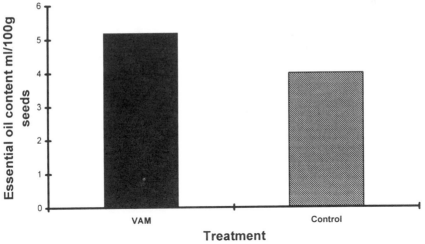

Figure 2. Top: Effect of VAM inoculation on growth of plants. Bottom: Essential oil content of coriander seeds.

Much work has been done by several workers on the fertilizer requirements of essential oil bearing plants and they concluded that the level of phosphorus in soil affects the quantity and quality of essential oil. Singh and Duhan (1984) working on *Mentha peperita* found that phosphorus application increased the oil yield significantly. Singh and Randhawa (1990) reported that phosphorous application enhances the quantity and quality of *Anethum graveolens* oil. Results on the positive effects of phosphorus were also obtained by several

201

other workers (Gulati, 1976; Singh et al., 1979; Rao et al., 1983; Singh et al., 1989; Virmani and Dutta, 1970).

Since essential oils consist of isoprenoid units, its biosynthesis requires acetyl-CoA, ATP and NADPH. Hence, the biosynthesis of essential oil is dependent on the inorganic phosphorus content in the plant. The most well known benefit of VAM to the host plant is the increased absorption of phosphorus (Abbott and Robson, 1984; Gianinazzi and Gianinazzi, 1981). The rate of phosphorus uptake per unit length of root is higher in mycorrhizal plants (Hale and Sanders, 1982; Sanders, 1975; Smith, 1982; Smith et al., 1986). On soils with low phosphorus availability only a small fraction of phosphorus occurs in available (solution) form. Generally, the phosphorus concentration in the soil solution is so low and diffusion so slow that the kinetics lead to increased plant growth rates and higher amounts of total phosphorus in both root and shoot tissues. Mycorrhizal roots exploit the soil profile with hyphae extension beyond the depletion zone surrounding the absorbing root and its root hairs (Clarkson, 1985; Owusu-Bennoah and Wild, 1979). It was also observed that on external application of phosphorus the increase in essential oil content was comparable to that due to VAM fungal colonisation (Kapoor, 1997).

CONCLUSIONS

The use of VAM fungal association with aromatic plants has significant economic importance due to its ability to support better growth, biomass production and essential oil yield as a result of enhanced phosphorus uptake, enabling a more rational use of fertilizers.

ACKNOWLEDGEMENTS

Rupam Kapoor acknowledges CSIR, New Delhi, for financial support.

REFERENCES

Abbott, L.K. and Robson, A.D., 1984, The effect of mycorrhiza on plant growth, in: *VA Mycorrhiza.* (L. Powell and D.J. Bagyaraj, eds.), pp. 113-130, CRC Press, Boca Raton, Florida.

Aboul-Nasr, D., 1996, Effect of vesicular arbuscular mycorrhizas on *Tagetus erecta* and *Zinnia elegans, Mycorrhiza* 6: 61-64.

Allen, M.F., 1991, *The Ecology of Mycorrhizae,* Cambridge University Press, Cambridge.

Bansal, M. and Mukerji, K.G., 1994, Positive correlation between VAM induced changes in root exudation and mycorrhizosphere mycoflora, *Mycorrhiza* 5: 39-49.

Black, R.L.B. and Tinker, P.B., 1977, Interactions between effects of vesicular-arbuscular mycorrhiza and fertiliser phosphorus on yields of potatoes in the field, *Nature* 267: 510-511.

Clark, R.J. and Minary, R.C., 1980, The effect of irrigation and nitrogen on the yield and composition of peppermint oil *(Mentha peperita* L.), *Aust. J. Agric. Res.* 31: 489-498.

Clarkson, D.T., 1985, Factors affecting mineral nutrition acquisition by plants, *Ann. Rev. Pl. Physiol* 36: 77-115.

Dey, B.B. and Choudhari, M.A., 1981, Changes in the composition of essential oil of *Ocimum sanctum* L. during reproductive development with special reference to the effect of growth regulators, *Pafai,* pp. 12-15.

Duhan, S.P.S., Bhattacharya, A.K. and Gulati, B.C., 1974, Effect of date of sowing and nitrogen on the yield of seed and quality of oil of *Anethum graveolens, Indian J. Pharm.* 36: 5-7.

Gianinazzi-Pearson, V. and Gianinazzi, S., 1981, Role of endomycorrhizal fungi in phosphorus cycling in the ecosystem, in: *The Fungal Community: Its Organisation and Role in the Ecosystem* (D.T. Nicklow and G.C. Carrol, eds.), Marcel Dekker Inc. New York.

Gulati, S., 1976, Effect of fertilisers on *Mentha arvensis, Indian Perfumer* 2: 12-18.

Gupta, M.L., Janardhanan, K.K., Chattopadhyay, A and Hussain, A., 1990, Association of *Glomus* with *Plamrosa* and its influence on growth and biomass production, *Mycol. Res.* 94: 561-563.

Hale, K.A. and Sanders, F.E., 1982, Effect of benomyl on vesicular-arbuscular mycorrhizal infection of red clover and consequences for phosphorus inflow, *J. Pl. Nutr.* 5: 1355-1367.

Hall, R.I., 1978, Vesicular arbuscular mycorrhizas on two varieties of maize and one of sweet corn, *NZ. J. Agric. Res.* 21: 517-519.

Harley, J.L. and Smith, S.E., 1983, *Mycorrhizal Symbiosis*, Academic Press, New York.

Hornok, L., 1983, Essential oil composition of various stages of plant development in *Anethum graveolens, Acta Hort.* 132: 237-239.

Jakobsen, I., 1986, Vesicular-arbuscular mycorrhiza in field grown crops. Mycorrhizal infection and rates of phosphorus inflow in pea plants, *New Phytol.* 104: 573-581.

Janardhanan, K.K. and Khaliq, A., 1995, Influence of vesicular-arbuscular mycorrhizal fungi on growth and productivity of German Chamomile in alkaline Usar soil, in: *Proceedings of the Third National Conference on Mycorrhizae*, pp. 410-412.

Johnson, G.R. and Menge, J.A., 1982, Mycorrhizae may save fertiliser dollars. *Am. Nurseryman* 155: 79-87.

Kapoor, R., 1997, VAM in relation to growth and essential oil yield in umbelliferous plants, *Ph.D. Thesis*, Delhi University, India.

Krishna, K.R., Shetty, K.G., Dart, P.J. and Andrews, D.J., 1985, Genotype dependent variation in mycorrhizal colonisation and response to inoculation of pearl millet, *Pl. Soil* 86: 113-125.

Loomis, W.D. and Corteau, R., 1973, Essential oil biosynthesis, *Rec. Adv. Phytochemist* 6: 197-203.

Mago, P., 1994, VAM in relation to productivity of certain members of Lamiaceae, *Ph.D. Thesis*, Delhi University, India.

Mohan Kumar, V. and Mahadevan, A., 1984, Do secondary substances inhibit mycorrhizal association? *Curr. Sci.* 53: 377-378.

O'Bannon, J.H., Evans, D.W. and Peaden, R.N., 1980, Alfalfa varietal response to seven isolates of vesicular arbuscular mycorrhizal fungi, *Can. J. Pl. Sci.* 60: 859-863.

Owusu-Bennoah, E. and Mosse, B., 1979, Plant growth responses to vesicular-arbuscular mycorrhiza XI. Field inoculation responses in barley, lucerne and onion, *New Phytol.* 83: 671-679.

Owusu-Bennoah, E. and Wild, A., 1979, Autoradiography of the depletion zone of phosphate around onion roots in the presence of vesicular-arbuscular mycorrhiza, *New Phytol.* 82: 133-140.

Pareek, S.K., Maheswari, M.L. and Gupta, R., 1982. Oil content and its composition at different growth stages in *Ocimum sativum L., Indian Perfumer* 26: 2-4.

Powell Cl. L., 1982, Mycorrhizal fungi and blueberries: how to introduce them, *NZ. J. Agric. Res.* 143: 33-35.

Rao, B.R., Rao, R., Prakasa, E.V.S. and Singh, S.P., 1983, Influence of NPK fertilisation on the herbage yield, essential oil content and essential oil yield of bergamot mint (*Mentha citrata*), *Indian Perfurmer* 27: 77-79.

Rao, G.V.S., Suresh, C.K., Suresh, N.S., Mallikarjunaiah, R.B. and Bagyaraj, D.J., 1989, Vesicular-arbuscular mycorrhizae in medicinal plants, *Indian Phytopath.* 42: 476-477.

Read, D.J., Lewis, D.H., Fitter, A.H. and Alexander, I.J., 1992, *Mycorrhiza in Ecosystems,* CAB International, Oxon, U.K.

Robert, J.W., Habib, G.C. and Steward, L.A., 1986, Production of secondary metabolites in plant cell cultures, in: *Progeneration of Aromas* (Katza, ed.), pp. 347-359.

Singh, A. and Randhawa, G.S., 1990, Studies on some agronomic inputs affecting oil content, oil and herb yield of dill (*Anethum graveolens* L.), *Indian Perfumer* 34: 108-114.

Singh, A., Singh, D.V. and Thakur, R.S., 1989, Spearmint cultivation in India - past and present scenario, *Pafai* 1989: 39-41.

Singh, V.P. and Dunham, S.P.S., 1984, Nitrogen and phosphorus fertilisation of *Mentha piperita* L. in Tarai region of Nainital, *Indian Perfumer* 25: 82-83.

Singh, V.P., Bisht, H.S. and Dunham, S.P.S., 1979, Studies on the split application of nitrogen through soil and foliage on the herb and oil yield of *Mentha citrata* oil, *Indian Perfumer* 27: 24-27.

Smith, S.E., 1982, Inflow of phosphate into mycorrhizal and non-mycorrhizal *Trifolium subterraneum* at different levels of soil phosphate, *New. Phytol.* 90: 293-303.

Smith, S.E., St John, B.J., Smith, F.A. and Bormley, J.L., 1986, Effect of mycorrhizal infection on plant growth nitrogen and phosphorus nutrition of glasshouse-grown *Allium cepa* L., *New Phytol.* 103: 359-373.

Srivastava, N.K. and Basu, M., 1995, Occurrence of vesicular-arbuscular mycorrhizal fungi in some medicinal plants, in: *Proceeding of Third National Conference on Mycorhiza*. Mycorrhizae: Biofertilisers for the future, pp. 59-61.

Swain, T., 1977, Importance of essential oil for plants producing them, *Ann. Rev. Pl. Physiol.* 28: 479.

Tilak, K.V.B.R. and Murthy, B.N., 1987, Association of vesicular-arbuscular mycorrhizal fungi with the roots of different cultivars of barley *(Hordeum vulgare) Curr. Sci.* 56: 1114-1115.

Vierheilig, H. and Ocampo, J.A., 1991, Receptivity of various wheat cultivars to infection by VA-mycorrhizal fungus influenced by inoculum potential and the relation of VAM effectiveness to succinic-dehydrogenase activity of the myceliums in the root, *Pl. Soil* 133: 291-296.

Virmani, O.P. and Dutta, S.C., 1970, Essential oil of Japanese mint, *Indian Perfumer* 14: 21-25.

MYCORRHIZAL ROOT LITTER AS A BIOFERTILIZER

K.G. Mukerji and Manju Bansal

Applied Mycology Laboratory
Department of Botany
University of Delhi
Delhi-110 007
India

INTRODUCTION

Litter in general is often considered to be a mere aggregation of dead and fallen plant parts. In the biosphere it is a link between organic and inorganic forms of mineral nutrients. Litter or the organic matter residue of plants is a natural reservoir of fertilizers which, upon decomposition, adds substantially to soil fertility. The litter also improves the physical characteristics of soils, supplies micronutrients in addition to major nutrients and provides nutrients for soil microbes.

The main plant parts which undergo natural geochemical cycles are leaves, branches and roots. Root litter, particularly the fine root litter, is often more important than leaf and branch litter in returning organic material to the plant. Up to 4 to 5 times more material is returned from fine roots than from leaf or branch litter (Fogel, 1985; Ford and Deans, 1977; Harris et al., 1977; Montagnini et al., 1991; Persson 1980; Vogt et al. 1986). Most of the studies on decomposition in forest systems have been carried out on leaf and needle litter (Berg, 1984; Montagnini et al., 1991; Swift et al., 1979) whereas little research has been done on root litter decomposition.

Studies on root litter are of particular interest in natural ecosystems for various reasons (Persson, 1981). Firstly, root biomass in some systems appears to be much higher in magnitude and is comparable to the litter formation in above ground systems. Secondly, the release of nutrients from root litter may be very similar to nutrient release from leaf and branch litter. Another reason is that the patterns for release or net uptake of nutrients may be very different when the litter is decomposing inside the soil compared on the soil surface as with leaf litter.

An understanding of root decomposition is important to agronomists and foresters. The extension of growth and secondary thickening of roots creates and widens channels within the soil and these channels can persist even after the disappearance of the roots. Such channels modify the physical environment of the soil and improve soil aeration, drainage and water

From Ethnomycology to Fungal Biotechnology
Edited by Singh and Aneja, Plenum Press, New York, 1999

holding capacity (Vogt et al., 1991). Decomposing roots also provide energy and nutrients to support the growth of microorganisms, such as bacteria and fungi, that can themselves improve soil structure. Dead roots can act as reservoirs of infection for soil-borne plant pathogens and knowledge of conditions which promote the disappearance of residues harbouring these pathogens is of obvious advantage to plant pathologists. Another aspect which is of practical interest is knowledge of the withdrawal (immobilization) and release (mineralization) of plant nutrients during root growth and decay.

The mycorrhiza is a physiological and physical manifestation of a symbiotic association between fungi and plant roots. Mycorrhizae are the normal nutrient- and water-absorbing organs of the vast majority of vascular plants (about 90%). The fungi in vesicular arbuscular mycorrhizae (VAM) form two types of branches: (i) outside the root in soil, i.e. extramatrical hyphae absorbing nutrients from soil, and (ii) bush-like, highly branched structures, i.e. arbuscules in cells which absorb nutrition and store it in vesicles, especially phosphorus as polyphosphate granules.

Mycorrhizae are an integral part of the root system. In forest ecosystems mycorrhizae have scarcely been studied in relation to fine root productivity and turnover. Studies on mycorrhizae and fine root production are so scarce that Fogel (1983) proposed that "it is highly probable that mycorrhiza has not been included in the fine root biomass estimates available, despite their importance in nutrient absorption". Most studies on the mycorrhizal contribution to fine root production concern ectomycorrhiza with very little work on VAM (Trappe and Fogel, 1977).

BIOCHEMISTRY OF ROOT LITTER DEGRADATION

The ability of fine roots to add to soil fertility mainly depends on their chemical composition, which varies with root diameter both within and between species. Similarly, nutrient concentration within fine roots varies inversely with root diameter (Goldfarb et al., 1990; Vogt et al., 1991). Increase in concentrations of phosphorus, nitrogen, carbon, magnesium, sulphur and sodium with decreasing root diameter is known for *Pinus sylvestris* (Berg, 1984) and *Acer saccharum* (Goldfarb et al., 1990). This explains the better capabilities of fine roots as litter than coarse roots.

Biochemical changes during root litter decomposition follow the same pattern as those during leaf litter decomposition. For example, in Scots Pine the quantitative organic chemical composition pattern of roots was similar to that of needles and followed the same decomposition pattern (Berg, 1984; Berg et al., 1980). The concentrations of sulphuric acid and lignin were fairly similar in the two plant parts (i.e. roots and needles), the only variation reported was in the glucan fraction of carbohydrates.

A generalized graph of the degradation of organic chemical components of Scots pine (*P. sylvestris*) root litter of 2-3 mm diameter is shown in Figure 1 (Berg, 1984). First, a relatively fast decomposition of soluble compounds took place combined with low leaching. Some of the cellulose and part of the arabinose were degraded early due to their easily hydrolysed structures, whereas some other hemicelluloses such as xylans, galactan and mannans did not decrease at all or showed only a slight decrease during the three year period of study. The absolute amount of the analytical lignin fraction increased initially due to humification. Thus concentration of lignin and humification products increased during the whole period and was related to mass loss.

Changes in organic chemical composition of decomposing roots (2-3 mm diameter) revealed a pattern similar to that for needles (Berg, 1984; Berg et al., 1982) though there was very little leaching associated with decomposition of roots (< 1%) compared to that of needles. The total mass loss during fine root litter decomposition could be fairly well described by a first-order kinetic reaction for all diameter classes, with disintegration constant (k) varying between 0.195 and 0. 149 yr^{-1} (Berg, 1984). Sandhu et al. (1990) compared ash free mass in the case of *Leucaena leucocephala* and reported that loss of ash free mass was maximum for fruits followed by roots, with a decreasing order as fruit>root>twigs>leaves.

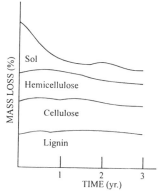

Figure 1. A generalised graph of degradation of organic chemical components of root litter of 2-3 mm diameter (from Berg et al., 1984).

THE CONCEPT OF SOIL FERTILITY IN RELATION TO ROOT LITTER

Although there is no generalised concept of the term 'fertility', it is believed that a fertile soil must supply in reasonable amounts and in suitable balance all the nutrients which plants take from the soil. The exact nutrient requirement of an individual plant is not known, but the available forms of nitrogen, phosphorus and potassium are needed beyond doubt. The fertility status of soil is reflected in the growth and yield of the plant it is supporting. Both living roots and decomposing roots enhance the capacity of the soil to produce nutrients through increasing microbial activity.

The decomposition of litter, besides returning nutrients to the soil, improves soil properties such as texture, moisture and pH, all of which favour microbial activity directly. The influence of root litter decomposition on soil properties varies with the type of litter. The chemical changes occurring in soil during the initial stages of litter breakdown consist of decomposition of soluble carbohydrates, starches, pectins and soluble nitrogenous compounds. This adds to the nitrogen, phosphorus, potassium, organic carbon and organic matter status of the soil. The role of various fungi in the decomposition of cellulose and pectin has been reviewed recently by some workers (Buswell, 1991; Markham and Bazin, 1991).

In comparison to structural roots, the fine root turnover is more rapid and contributes more carbon to organic matter accumulation. Vogt et al. (1986) reported that evergreen species add twice as much root material to detritus in comparison to deciduous species growing at the same latitude. Their results strongly imply that roots contribute more than foliage to soil organic matter accumulation in evergreen sites.

Very few studies have examined root decomposition in the field. Numerous observations in the literature would suggest that the litter bag technique underestimates fine root decomposition (Lyr and Hoffmann, 1967; Santantonio and Hermann, 1985). Estimates of the proportion of total carbon added annually to the soil from root mortality and decay have varied from 25-36% (McClaugherty et al., 1982; Vogt et al., 1991), 42.2% (Edward and Harris, 1977), 59-67% (Vogt et al., 1986), 54-81% (Gholz et al., 1986) and 78-84% (Fogel, 1983). However, more data are needed from many sites in order to clearly establish the contribution of fine root litters to soil fertility.

DEGRADATION OF MYCORRHIZAL ROOTS

In general, the contribution of mycorrhizal fine roots to primary production for any ecosystem may be between 6 to 78% (Fogel, 1983; Fogel and Hunt, 1979). Based on external morphology of root tips, Alexander and Fairlay (1983) reported 90-97% of the total root tips to be mycorrhizal. Mycorrhizae account for an additional 8% of tree biomass in young Douglas fir stands (Gottsche, 1972). The mycorrhizal biomass is larger in Douglas fir *(Pseudotsuga menziesii)* than in pine and spruce (Fogel, 1983) as shown in Table 1.

Table 1: Contribution of mycorrhizae to fine root productivity in different tree species.

Tree Species	Location	Stand Age	Production (22 mm dia) % Total	Mycorrhiza % Total
Pinus strobus	Michigan	35	94	6
Quercus carya	Michigan	60	88	12
Pseudotsuga	Michigan	30	22	78
menziesii	Oregon	50	49	51

Mycorrhizae are believed to make an important contribution to the turnover of fine roots (Bansal and Mukerji, 1994 a,b; Fogel, 1985; Fogel and Hunt, 1979; Harris et al., 1977). In an ecosystem, given large inputs of fine roots and mycorrhiza for decomposition, it is highly crucial to determine mycorrhizal biomass. Two basic queries about the fate and role of mycorrhizal and nodular root litter (biological nodules in legumes) in below-ground ecosystems are as follows: (i) What is the fate of microbes in symbiotic association when fine roots undergo turnover? (ii) What is the difference in amount of nutrients added to the soil after the turnover of fine roots associated symbiotically with various microbes in comparison to roots without it?

Studies on degradation and turnover of endomycorrhizal roots are particularly neglected. In general, however, endomycorrhizal roots are expected to contribute more to the nutrient pool of soils as these roots have higher amounts of nitrogen (in legumes) and phosphorus than their non-mycorrhizal counterparts (Harley and Smith, 1983; Srivastava et al., 1996).

Degradation of mycorrhizal roots should certainly be different from that of non-mycorrhizal roots as the rhizosphere chemistry of the former (mycorrhizosphere) is considerably altered by VAM fungi (Bansal and Mukerji, 1995; Garbaye, 1991; Paulitz and Lindernan, 1991). Also the qualitative and quantitative composition of root exudates is altered considerably upon associating with mycorrhizal fungi. Both endo- and ectomycorrhizae are important in this context. These changes in root exudates are generally expressed in the microflora of the rhizosphere (Curl and Truelove, 1986).

NUTRIENT STATUS OF LITTER TYPES

The nutrient composition of fine roots varies with plant type (Berg, 1984), root diameter (Goldfarb et al., 1990) and soil fertility status (Alstrom et al., 1988; Satchell, 1974), besides other plant and soil factors. Large differences are reported in annual production of fine roots in herbaceous and tree species (Fogel, 1985). For example, fine root production values for

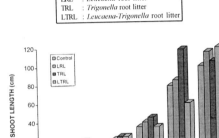

LRL	: *Leucaena* root litter
TRL	: *Trigonella* root litter
LTRL	: *Leucaena-Trigonella* root litter

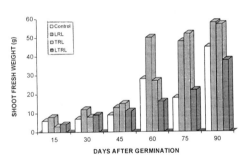

Figure 2. Effect of different treatments of root litters on shoot length of *Zea mays*.

Figure 3. Shoot fresh weight of *Zea mays* treated with different root litters.

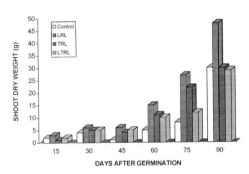

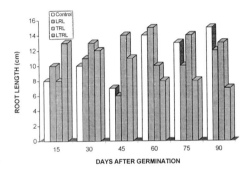

Figure 4. Effect of different treatments of root litters on shoot dry weight of *Zea mays*.

Figure 5. Influence of root litters on root length of Zea mays.

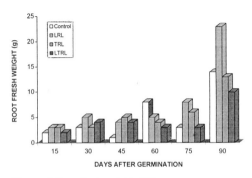

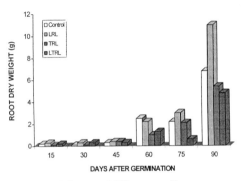

Figure 6. Root fresh weight of *Zea mays* as affected by different treatments.

Figure 7. Influence of different root litters on root dry weight.

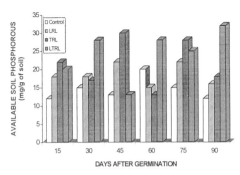
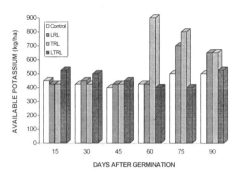

Figure 8. Soil available phosphorus as influenced by application of different root litters.

Figure 9. Soil available potassium as influenced by application of different root litters.

coniferous trees is in the range of 4.1 to 11.0 Mg/ha, whereas for herbs it is l.2 to 4.2 Mg/ha. Also the importance of fine root input in contributing forest floor organic matter increases appreciably with plant age (Vogt et al., 1983).

Bansal and Mukerji (1994 a,b) compared the efficacy of *Leucaena* root litter with that of *Trigonella* root litter and reported that *Leucaena* root litter is better in increasing growth of *Zea mays*. Litters increase the shoot length and biomass, but decrease root length and biomass. Thus root litter tends to promote more energy allocation to above-ground plant parts than to the below-ground ecosystems (Figures 2-7). The growth of *Zea mays* increased 1.5 to 5-fold with the application of root litter of mycorrhizal plants.

The root litter has little effect on soil pH or organic carbon but increases the available phosphorus tremendously (Figures 8 and 9). This increase may be due to rapid release of phosphorus from iron compounds or calcium phosphate released from root litter after microbial decomposition (Bansal and Mukerji, 1994 a,b; Vogt et al., 1991). Mycorrhizal roots have better fertilizing capacity than non-mycorrhizal roots.

CONCLUSIONS

The role of VAM fungi in fine root biomass dynamics, in fine root litter degradation and in uptake of nutrients from root litter was studied in *Leucaena leucocephala* using *Zea mays* as a test plant. A total fine root biomass of 42.97 to 88.13 dw m was produced in a young stand of *Leucaena leucocephala,* of which 65% was endomycorrhizal. The mycorrhizal fine root network was found to be an effective biofertilizer contributing a several-fold increase to the growth of *Zea mays*. Root litter with VAM association may also serve as an inoculum for VAM colonisation of maize roots.

ACKNOWLEDGEMENTS

This work is supported by the grant received from DBT.

REFERENCES

Alstrom, K., Persson, H. and Borjesson, T., 1988, Fertilization in mature Scots pine (*Pinus sylvestris* L.) stand effects and fine roots, *Pl. Soil* 106: 179-190.

Alexander, I.J. and Fairley, R.I., 1983, Effect of N fertilization on population of pine roots and mycorrhizas in spruce humus, *Pl. Soil* 71: 49-53.

Bansal, M. and Mukerji, K.G., 1994a, Dead fine roots - a neglected biofertilizer, in: *Plant Nutrition from Genetic Engineering to Field Practice* (N.J. Barrow, ed.), pp 547-550, Kluwer Academic Publishers, Netherlands.

Bansal, M. and Mukerji, K.G., 1994b, Efficiency of root litter as a biofertiliser, *Biol. Fertil. Soils* 18: 228-230.

Bansal, M. and Mukerji, K.G., 1995, Positive correlation between VAM induced changes in root exudation and mycorrhizosphere mycoflora, *Mycorrhiza* 5: 39-44.

Berg, B., 1984, Decomposition of root litter and some factors regulating the process long term root litter decomposition in Scots pine forest, *Soil Biol Biochem.* 16: 609-617.

Berg, B., Hannus, K., Popoff, T. and Theander, O., 1980, Chemical components of Scots pine needles and needle litter and inhibition of fungal species by extractives, in: *Structure and Function of Northern Coniferous Forests - an Ecosystem Study*, (T. Persson, ed.), Ecological Bulletin, Stockholm 32: 391-400.

Berg, B., Hannus, K., Popoff, T. and Theander, O., 1982, Changes in organic chemical components during decomposition in Scots pine litter, *Oikos* 38: 291-296.

Buswell, J.A., 1991, Fungal degradation of lignin, in: *Handbook of Applied Mycology. Vol. I. Soil and Plants,* (D.K. Arora, B. Rai, K.G. Mukerji and G.R. Knudsen, eds.), pp. 425-480, Marcel Dekker, New York.

Curl, E.A. and Truelove, B., 1986, *The Rhizosphere*, Springer-Verlag, Berlin, Heidelberg, New York, Tokyo.

Edwards, N. I and Harris W.F., 1977, Carbon cycling in a mixed deciduous forest floor, *Ecology* 58: 431-437.

Fogel, R., 1983, Root turnover and productivity of coniferous forests, *Pl. Soil* 71: 75-85.

Fogel, R., 1985, Roots as primary producers in below ground ecosystems, in: *Ecological Interactions in Soil*, (A.H. Fitter, D. Atkinson, D.J. Read and M.B. Usher, eds.), pp. 23-26, Blackwell Scientific Publications, Oxford, London.

Fogel, R. and Hunt, G., 1979, Fungal and arboreal biomass in Western Oregon Douglas fir ecosystem: Distribution patterns and turnover, *Can J. For. Res.* 9: 245-256.

Ford, E.D. and Deans, J.D., 1977, Growth of sitka spruce plantation: Spatial distribution and seasonal fluctuations of length, weights and carbohydrate concentration of fine roots, *Pl. Soil* 47: 463-485.

Garbaye, J., 1991, Biological interactions in the mycorrhizosphere, *Experimentia* 47: 370-375.

Gholz, H.L., Hendry, L.C. and Cropper, W.P.Jr., 1986, Organic matter dynamics of fine roots in plantations of slash pine *(Pinus elliotti)* in North Florida, *Can. J. For. Res* 16: 529-538.

Goldfarb, D., Hendrick, R. and Pregitzer, K., 1990, Seasonal nitrogen and carbon concentrations in white brown and woody fine roots of sugar maple (*Acer saccharum* Marsh), *Pl. Soil* 126: 144-148.

Gottsche, D., 1972, Vertcilung von Feinwurzcln und Mykorrhizen in Bodenprofil cines Buchen Und fichten bestandes in solling Mitt Bund Forsch, *Anst. Forest, U. Holzw.* 88, Hamburg.

Harley, J. L. and Smith, S.E., 1983, *Mycorrhizal Symbiosis*, Academic Press, London, New York.

Harris, W.F., Kinerson. R.S. Jr. and Edwards, N.T., 1977, Comparision of below ground biomass of natural deciduous forest and lablolly pine plantations, *Pedobiologia* 17: 369-381.

Lyr, H. and Hoffmann, G., 1967, Growth rates and growth periodicity of tree roots, *Int. Rev. of For. Res.* 2: 181-236.

Markham, P. and Bazin, M.J., 1991, Decomposition of cellulose by fungi, in: *Handbook of Applied Mycology Vol. 1: Soil and Plants* (D.K. Arora, B. Rai, K.G. Mukerji and G.K. Kundson, eds.), pp. 379-424, Marcel Dekker, New York.

McClaugherty, C.A., Aber, J.D. and Mellilo, J.M., 1982, The role of fine roots in the organic matter and nitrogen budgets of two forested ecosystems, *Ecology* 63: 1481-1490.

Montagnini, F., Sancho, F., Ramstad, K. and Stijfhoom, E., 1991, Multipurpose trees for soil restoration in the humid lowland of Costa Rica, in: *Research on Multipurpose Tree Species in Asia* (D.A. Taylor and K.G. MacDicken, eds.), pp. 47-57, Winrock International F/FRED, Thailand.

Paulitz, T.C. and Linderman, R.G., 1991, Mycorrhizal interactions with soil organisms, in: *Handbook of Applied Mycology Vol. 1* (D.K. Arora, B. Rai, K.G. Mukerji and G.K. Knudson, eds.), pp. 77-129, Marcel Dekker, New York.

Persson, H., 1980, Spatial distribution of fine root growth, mortality and decomposition in a young Scots pine stand in central Sweden, *Oikos* 34: 77-87.

Persson, H., 1981, Death and replacement of fine roots in forest ecosystems - a neglected area of research, *Swedish Coniferous Forest Project, Technical Report* 29: 1-25.

Sandhu, J., Sinha, M. and Ambasht, R.S., 1990, Nitrogen release from decomposing litter of *Leucaena leucocephala* in the dry tropics, *Soil Biol. Biochem,* 22: 859-863.

Santantonio, D. and Hermann, R.K., 1985, Standing crop, production and turnover of fine roots on dry, moderate and wet sites of mature Douglas fir in Western Oregon, *Ann. Sci. For.* 42: 113-142.

Satchell, J.E., 1974, Litter-interface of animate/inanimate matter, in: *Biology of Plant Litter Decomposition* (C.H. Dickinson and G.J.F. Pugh, eds.), pp. 91-103, Academic Press, London, New York.

Srivastava, D., Kapoor, R., Srivastava, S.K. and Mukerji, K.G., 1996, Vesicular arbuscular mycorrhiza: an overview, in: *Concepts in Mycological Research*, (K.G. Mukerji, ed.) pp. 1-39, Kluwer Academic Publishers, Netherlands.

Swift, M.J., Heal, O.W. and Anderson, J.M., 1979, Decomposition in terrestrial ecosystems, Blackwell, London.

Vogt, K.A., Gorier, C.C., Gower, S.T., Sprugal, D.G. and Vogt, D.J., 1986, Overestimation of net root production: a real or imaginary problem? *Ecology*, 67: 577-579.

Vogt, K.A., Gorier, C.C., Meier, C.E. and Keyes, M.R, 1983, Organic matter and nutrient dynamics in forest of young and mature *Abies amabilis* stands in Western Washington, as affected by fine root input, *Ecol. Monogr.* 53: 139-157.

Vogt, K.A., Vogt, D.J. and Bloomfield, J., 1991, Input of organic matter to the soil by tree roots and their environments (B.L. McMichael and H. Persson, ed.), pp. 171-185, London, New York, Tokyo.

THE APPLICATION OF VESICULAR ARBUSCULAR MYCORRHIZAL FUNGI IN AFFORESTATION

M. Kaur and K.G. Mukerji

Applied Mycology Laboratory
Department of Botany
Delhi University
Delhi-110 007
India

INTRODUCTION

Desertification is a natural and dynamic process which claims millions of hectares of land annually. In addition to natural environmental factors such as impoverished soils, extreme temperatures, erratic rainfall and high wind velocity, anthropogenic activities such as deforestation, poor agricultural practices, mining, overgrazing and recreational activities create vast stretches of wastelands. Such areas are characterized by loss of vegetation cover, loss of soil structure, increase in soil erosion, loss of available nutrients and organic matter, loss of microbial propagules and /or diminution of microbial activities (Herrera et al., 1993).

The revegetation of such disturbed ecosystems is a priority and management practices are being developed for this purpose. These involve the introduction of native plant species (i.e. reclamation) or exotic species (i.e. rehabilitation). Plants in disturbed habitats of tropical areas, arid lands, coal mines, oil and mine spoils, oil shale lands etc. face many adverse conditions such as high soil salinity, low fertility, drought, acid imbalance and nutrient deficiency.

Although the use of chemical fertilizers accelerates the establishment of vegetational cover it is not a feasible strategy as it encourages the growth of ruderal weeds and is not very cost effective. Besides, at present there is considerable resistance to the use of chemical fertilizers due to their adverse effects on the environment and on soil, plant, animal and human health.

Therefore, the use of biofertilisers in afforestation programmes is both more beneficial and more economical. Different kinds of tree seedlings fortified with microbes and symbiotic organisms like *Rhizobium* and mycorrhizal fungi can be used in degraded, harsh and poor plantation sites. There is a large occurrence of mycorrhizae in forest trees and the symbiosis can be manipulated to enhance productivity in afforestation programmes. The term 'mycorrhiza' describes the association of plant roots with hyphal fungi and is of common occurrence in angiosperms, gymnosperms, pteridophytes and bryophytes (Smith and Read,

1997). Of the various types of mycorrhizal associations, the vesicular arbuscular mycorrhizae (VAM) are the most abundant, with widest host range and geographic ubiquity (Mukerji and Mandeep, 1997). The roots of most of the world's plants are colonized by VA mycorrhizal fungi forming a mutualistic symbiosis that can be considered an integral part of the plants. This symbiosis seems to have ecological and evolutionary significance in the origin and development of plants on earth.

VAM fungi are obligately biotrophic, belong to the order Glomales of Zygomycotina (Morton and Benny, 1990; Morton and Bentivenga, 1994) and are often the most abundant fungi known in soils from all continents (Harley, 1991). The VAM fungi occupy a unique ecological niche as they extend into both the roots and the surrounding substrate forming bridges connecting the two, thereby improving both nutrient acquisition by the plants and also the soil structure (Bethlenfalvay, 1992). There is a flow of inorganic components from the fungus to the plant and organic components from plant to the fungus (Mukerji and Sharma, 1996; Srivastava et al., 1996).

THE MORPHOLOGY OF VESICULAR ARBUSCULAR MYCORRHIZAE

The vesicular arbuscular mycorrhiza is composed of a two-phase mycelial system, an internal mycelium within the cortex of the mycorrhizal roots and the external mycelium in the soil.

The external mycelium surrounding the roots is dimorphic (Mosse, 1959) and consists of (i) permanent, coarse, thick-walled, generally aseptate runner hyphae and (ii) numerous fine, thin-walled and highly dichotomously branched absorbing hyphae which form hyphal networks extending into the soil. There are about 100 cm of absorbing hyphae/root penetration (Sieverding, 1991), therefore the mycosymbiont can intensively explore a large soil volume and extract soil resources for transport to the root. The external hyphae may transport nutrients to the host from a distance of up to 7 cm (Rhodes and Gerdemann, 1975).

The extramatrical hyphae provide the structures capable of colonizing new root tissue (Friese and Allen, 1991). The thick-walled runner hyphae penetrate the root forming appresoria at the entry point (Gianinazzi-Pearson et al., 1991; Giovanetti et al., 1991, 1994) which are lens-shaped multinucleate structures, 20-40 μm long (Garriock et al., 1989).

The intraradical structures formed by VAM are (i) the extracellular hyphae, (ii) the intracellular hyphae forming coils often found in the outer layers of cortical parenchyma, (iii) the intracellular hyphae with numerous ramifications, i.e. the arbusules and (iv) the inter- and intracellular hypertrophied hyphae forming thick-walled, lipid-rich vesicles (Bhandari and Mukerji, 1993). The arbuscules are the primary structures involved in the bidirectional transfer of nutrients between the fungal symbiont and host plant (Cox and Tinker, 1976; Scannerini and Bonfante-Fasolo, 1979).

The partners of the mycorrhizal association influence each other during all levels of metabolic activity from gene expression to morphogenesis. They exist in a state of cellular and physiological compatibility during all stages of development, which are: (i) presymbiotic phase of the VAM fungi, (ii) the contact phase, (iii) the colonization process, and (iv) the development of the host-fungus interface.

The persistence of mycorrhiza formation through evolution as new plant species appeared implies that many of the cellular and molecular mechanisms involved are common across the plant kingdom (Gianinazzi-Pearson et al., 1995). The highly sophisticated morpho-functional integration between the two symbionts can only be explained by synchronized modifications in the gene expression of both the partners (Gianinazzi-Pearson and Gianinazzi, 1989).

BENEFICIAL EFFECTS OF VA MYCORRHIZAL SYMBIOSIS

Vesicular arbuscular mycorrhizae enhance nutrient uptake in both cultivated and native species (Smith and Read, 1997). The VA mycorrhizal symbiosis increases the growth rate of the plants and influences the partitioning of phytomass between the root and shoot. VAM-inoculated plants are not only large but also usually have an increased concentration of minerals, especially phosphorus. The extramatrical hyphae of VAM fungi increase the total absorptive surface area of the root system and acquire nutrients such as P, NH^+, K, Ca, SO_4^{2-}, Cu and Zn from beyond the depletion zone around the root and consequently enhance the uptake of nutrients from the soil. The VAM-induced changes in supply of mineral nutrients from soil also modify the soil fertility, mycorrhizosphere and the aggregation of soil particles (Varma, 1995). The non-nutritional effects of mycorrhizae such as reducing the severity of certain plant diseases, modifying water relations and the soil structure are also potentially important. The benefits of VAM may also extend to alleviation of stresses caused by mineral excesses.

VAM AND NUTRIENT UPTAKE

Phosphorus Uptake

Phosphorus is a major plant nutrient which is required in relatively large amounts and which plays a vital role in all biological functions. The various forms of phosphorus present in soil have very low solubility and the concentration of phosphorus in soluble form is extremely low in forest soils. The organic and insoluble mineral phosphorus which constitute the greater part of phosphorus in soil is not available to plants. The uptake of phosphate by roots is much faster than diffusion of ions to the absorptive surface of the roots and this causes a phosphate depletion zone to develop around the root.

The inoculation of plants with VAM fungi increases the growth of plants by several-fold and generally most of the growth enhancement effects observed on root colonization with VAM fungi are caused by increased P absorption (Bagyaraj and Varma, 1995). The increase in absorption of phosphorus by VA mycorrhizal plants may be because of (i) increased physical exploration of the soil; (ii) increased movement of P into fungal hyphae; (iii) modifications of the root environment; (iv) efficient transfer of P to plant roots; (v) increased storage of absorbed P; and (vi) efficient utilization of P within plants.

The process of phosphate uptake by mycorrhizal fungi consists of three sub-processes, i.e. (i) absorption of phosphate from soil by VAM fungal hyphae; (ii) the translocation of phosphate along the hyphae from external to internal (root cortex) mycelia and (iii) the transfer of phosphate to cortical cells (Barea, 1991).

The plant is able to exploit microhabitats beyond the nutrient-depleted areas (O'Keefe and Sylvia, 1992). The VAM hyphae also take advantage of their geometry and better distribution than the roots to acquire phosphate from transitory, localized and diluted sources of phosphorus. The greater exploration of soil P by VAM fungal hyphae is because (i) they extend away from roots and can translocate P from as far away as 8 cm; (ii) they exploit smaller soil pores due to their smaller diameter than roots; (iii) they have a greater unit surface area of absorption (Jungk and Claasen, 1989).

It has been suggested that mycorrhizal hyphae may have an affinity for absorption of P (Cress et al., 1979) and that mycorrhizal roots have a lower threshold of P concentration for absorption than non-mycorrhizal roots (Karunaratne et al., 1986; Lei et al., 1991; Mosse, 1973).

There are quantitative and qualitative changes in root exudation patterns and differences

BENEFICIAL EFFECTS OF VA MYCORRHIZAL SYMBIOSIS

Vesicular arbuscular mycorrhizae enhance nutrient uptake in both cultivated and native species (Smith and Read, 1997). The VA mycorrhizal symbiosis increases the growth rate of the plants and influences the partitioning of phytomass between the root and shoot. VAM-inoculated plants are not only large but also usually have an increased concentration of minerals, especially phosphorus. The extramatrical hyphae of VAM fungi increase the total absorptive surface area of the root system and acquire nutrients such as P, NH^+, K, Ca, SO_4^{2-}, Cu and Zn from beyond the depletion zone around the root and consequently enhance the uptake of nutrients from the soil. The VAM-induced changes in supply of mineral nutrients from soil also modify the soil fertility, mycorrhizosphere and the aggregation of soil particles (Varma, 1995). The non-nutritional effects of mycorrhizae such as reducing the severity of certain plant diseases, modifying water relations and the soil structure are also potentially important. The benefits of VAM may also extend to alleviation of stresses caused by mineral excesses.

VAM AND NUTRIENT UPTAKE

Phosphorus Uptake

Phosphorus is a major plant nutrient which is required in relatively large amounts and which plays a vital role in all biological functions. The various forms of phosphorus present in soil have very low solubility and the concentration of phosphorus in soluble form is extremely low in forest soils. The organic and insoluble mineral phosphorus which constitute the greater part of phosphorus in soil is not available to plants. The uptake of phosphate by roots is much faster than diffusion of ions to the absorptive surface of the roots and this causes a phosphate depletion zone to develop around the root.

The inoculation of plants with VAM fungi increases the growth of plants by several-fold and generally most of the growth enhancement effects observed on root colonization with VAM fungi are caused by increased P absorption (Bagyaraj and Varma, 1995). The increase in absorption of phosphorus by VA mycorrhizal plants may be because of (i) increased physical exploration of the soil; (ii) increased movement of P into fungal hyphae; (iii) modifications of the root environment; (iv) efficient transfer of P to plant roots; (v) increased storage of absorbed P; and (vi) efficient utilization of P within plants.

The process of phosphate uptake by mycorrhizal fungi consists of three sub-processes, i.e. (i) absorption of phosphate from soil by VAM fungal hyphae; (ii) the translocation of phosphate along the hyphae from external to internal (root cortex) mycelia and (iii) the transfer of phosphate to cortical cells (Barea, 1991).

The plant is able to exploit microhabitats beyond the nutrient-depleted areas (O'Keefe and Sylvia, 1992). The VAM hyphae also take advantage of their geometry and better distribution than the roots to acquire phosphate from transitory, localized and diluted sources of phosphorus. The greater exploration of soil P by VAM fungal hyphae is because (i) they extend away from roots and can translocate P from as far away as 8 cm; (ii) they exploit smaller soil pores due to their smaller diameter than roots; (iii) they have a greater unit surface area of absorption (Jungk and Claasen, 1989).

It has been suggested that mycorrhizal hyphae may have an affinity for absorption of P (Cress et al., 1979) and that mycorrhizal roots have a lower threshold of P concentration for absorption than non-mycorrhizal roots (Karunaratne et al., 1986; Lei et al., 1991; Mosse, 1973).

There are quantitative and qualitative changes in root exudation patterns and differences between the inoculated plants in absorption of anions and cations which cause changes in pH of the rhizosphere. These may be the indirect mechanisms which explain the effect of VAM in increasing the phosphate availability to the plant. The changes induced in the mycorrhizosphere alter the microflora which also alters the availability of both organic and inorganic P to the plants. Mycorrhizal fungi produce phosphatases that allow utilization of organic phosphorus under the humid tropical conditions (Tarafdar and Marschner, 1994).

Nitrogen Uptake

VAM fungi increase nitrogen uptake in plants. A number of mechanisms are suggested for this effect, namely (i) improvement of symbiotic N_2 fixation; (ii) direct uptake of combined nitrogen by VAM fungi; (iii) facilitated 'N transfer', a process by which a part of biologically-fixed nitrogen benefits the non-fixing plants growing nearby; and (iv) enzymatic activities involved in N metabolism (Barea, 1991).

The mycorrhizal hyphae have the capacity to extract nitrogen and transport it from the soil to the plant. They contain enzymes that break down organic nitrogen and contain N-reductase which alters the forms of nitrogen in soil.

The uptake of nitrogen by plants is usually in the form of nitrate which is readily absorbed. In soils where nitrate is the dominant nitrogen source, VAM fungi have only a minor influence in acquisition of nitrogen by plants (Johanson et al., 1992, 1993). In many natural ecosystems a large fraction of available nitrogen in soil is ammonium not nitrate (Van Kessel et al., 1985). Mycorrhizal fungi readily transport NH_4^+ from soil to plant (Smith and Smith, 1990) and this may be important where nitrogen is distributed in discrete patches.

Mycorrhizal fungi enhance nitrogen gain in plant communities by increasing the efficiency of N_2-fixing organisms. Increases in nodulation status and rate of nitrogen fixation with mycorrhiza have been reported in legume-*Rhizobium* symbiosis, in actinorrhizal associations *(Frankia)* and also with free-living nitrogen fixing bacteria such as *Azospirilum* and *Azotobacter* (Barea, 1991; Barea et al., 1992; Michelsen and Sprent, 1994). The increased efficiency of legume-*Rhizobium* symbiosis in the presence of VAM is probably because of the increased supply of phosphate to the roots and nodules (Barea et al., 1988; Reinhard et al., 1994).

VAM hyphae improve nitrogen transfer in communities, since the network of VAM mycelia links different plant species growing nearby and helps overlap the pool of available nutrients for these plants (Barea et al., 1988; Newman et al., 1992).

Other Essential Nutrients

The extramatrical hyphae of VAM take up and transport potassium (K) and VAM colonization affects the concentration and amounts of K in shoots (Marschner and Dell, 1994). The external hyphae of VAM also have a capacity to take up and transport Ca^{2+} and SO_4^{2-}-S (Tester et al., 1992).

VA mycorrhizal plants accumulate great quantities of some micronutrients (Zn, Cu, Co) under conditions of low soil nutrient availability (Faber et al., 1990; Smith 1980). This higher absorption is attributed to the uptake and transport by external hyphae due to wider exploration of soil volume by extended extramatrical hyphae. Tissue concentrations of Cu and Zn are higher in the presence of VAM fungi (Fairchild and Miller, 1988; Li et al., 1991). Mycorrhizal inoculation may increase Zn and Cu concentration in shoots, roots or both (Gildon and Tinker, 1983; Lambert and Weidensaul, 1991).

Uptake and concentration of manganese (Mn) in plants may not be affected by VAM and more often it may be lower in VAM plants (Lambert and Weidensaul, 1991) thus contributing to higher Mn tolerance in plants. There is enhanced iron (Fe) uptake in the presence of mycorrhiza and this may be due to specific Fe chelators. Siderophore activity has been found to be associated with VAM colonization (Szanizzlo et al., 1981).

Water Uptake

VAM fungi play a vital role in the water economy of plants. The VAM fungal colonization may provide a low resistance pathway for radial flow of water across the root cortex, thereby increasing the hydraulic conductivity of the roots and contributing towards better uptake of water by plants (Kothari et al., 1990). The permeability of the cell membrane to water may also be altered by mycorrhizal colonization through improved phosphorus nutrition.

Colonization by VAM can improve the drought resistance of plants (Bethlenfalvay et al., 1988; Osonubi et al., 1991; Sylvia and Williams, 1992). VAM-inoculated seedlings are less susceptible to wilting and transplant shock. Under conditions of drought stress VAM fungi exert their influence by increasing the transpiration rate and lowering stomatal resistance or by altering the balance of plant hormones (Augé et al., 1987a,b; Huang et al., 1985). The changes in leaf elasticity (Augé et al., 1987a), improved leaf water and turgor potential, increased root length and depth (Ellis et al., 1985, Kothari et al., 1990) may also influence water relations and therefore the drought resistance of the plants.

It has been suggested that these changes could be secondary responses to better phosphorus nutrition (Michelsen and Rosendahl, 1990) or mediated via direct mycorrhizal effects (Henderson and Davis, 1990). Allen (1996) has suggested that the likely mechanism of water transport to the host is along the VAM fungal hyphae and since the hyphae structurally connect the soil to the plant, the soil-plant-atmosphere continuum which may be broken due to transpiration demands is maintained by the mycorrhizal hyphae.

Soil Structure Stabilization

The extramatrical hyphae of VAM fungi improve soil structure by binding the soil particles into more stable aggregates (Barea and Jeffries, 1995). This makes VAM fungi particularly useful for reclamation of sand dunes, eroding soils, desert habitats etc.

The VAM fungi can be manipulated to improve soil aggregation. For this purpose the following criteria should be taken into consideration (i) VAM fungi should be selected for the amount of mycelium they produce; (ii) plant traits that favour the development of fungal soil mycelium should be identified; (iii) appropriate host cultivars should be used to manipulate the fungal populations (Miller and Jastrow, 1992).

Interactions with Other Soil Microbes

The rhizosphere of the mycorrhizal plants is called the 'mycorrhizosphere'. The altered root exudation due to mycorrhizal association induces changes in the composition of microbial communities in the mycorrhizosphere (Bansal and Mukerji, 1994a,b, 1996; Mukerji et al., 1998). The interactions that occur between these microbial communities and mycorrhizal fungi may be competitive or mutualistic. VAM fungi modify the interaction of plants with pathogens as well as other symbiotic organisms (Fitter and Garbaye, 1994; Paulitz and Linderman, 1991).

The mycorrhizal fungi enhance nitrogen-fixation by nodule-producing bacteria *(Rhizobium)* and interact positively with free-living or associative nitrogen-fixers such as *Azotobacter, Azospirillum* (Barea et al., 1992; Sharma and Mukerji, 1995). There is a synergistic effect on plant growth on co-inoculation with VAM fungi and PGPR, phosphate-solubilizing bacteria, etc.

The mycorrhizae exert selective pressure on populations of soil microorganisms and affect the incidence and severity of root diseases caused by major fungal and nematode pathogens (Linderman, 1988). They can serve as effective biocontrol agents (Jalali and Jalali, 1991; Kaushik and Mukerji, 1996; Mukerji et al., 1997).

MYCORRHIZAL INOCULATION IN FORESTRY

Mycorrhizal inoculation is useful in forest nurseries and also while transplanting seedlings to the field. The role of mycorrhiza in improving the quality of planting stock in the nursery and in obtaining good seedling survival and growth is widely recognized (Jeffries and Dodd, 1991; Mukerji and Dixon, 1992; Varma, 1995). Mycorrhizal colonization of seedling root systems in the nursery results in vigorous seedling growth and well-nurtured plants. The pre-inoculation of seedlings is a feasible method for introducing inoculant fungi into the field and is applicable wherever transplanting is a part of the normal production system. A potential advantage is that the inoculant fungus is already established in the root system and consequently it has a competitive advantage over the soil-borne species.

Seedlings of several hardwood species normally have an obligate dependence on VAM for normal growth. Stunting of *Citrus* seedlings in fumigated nursery soil can be corrected by inoculation with VAM fungi (Timmer and Leyden, 1978). Several experiments with forest trees have confirmed that mycorrhization leads to improved growth of seedlings (Osonubi et al., 1991; Sarwar, 1996, Singh, 1997) and plays a role in their successful establishment and subsequent growth (Mukerji et al., 1996) (Figure 1).

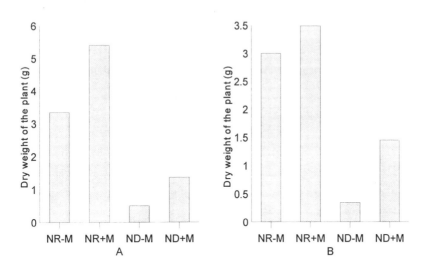

Figure l. Effect of VAM inoculation on dry weights of 90 days old seedlings of (A) *Acacia senegal* and (B) *Prosopis cineraria,* in nutrient deficient (ND) and nutrient rich (NR) soils.

There has been considerable interest in the use of VAM fungi in the revegetation of disturbed habitats. Plant establishment on stressed soils is facilitated by VAM due to their efficient nutrient mobilization mechanism, ability to withstand stressed conditions and developing resistance towards transplant shock. The use of VAM fungi has led to successful afforestation of mining sites (Lumini et al., 1994; Noyd, 1996). In soils stressed with salinity, mycorrhizal plants show better growth response with higher tissue concentration of phosphorus and lower concentration of sodium in shoots than the non-mycorrhizal plants (Rozema et al., 1986; Pfeiffer and Bloss, 1988).

STRATEGIES FOR USE OF VAM FUNGI

Inoculum Production

Indigenous VAM fungi could be used in reforestation programmes on a large scale but the only constraint in this area is lack of technology for mass culture of these fungi. The inability to culture VAM fungi is a major barrier to development of inoculation techniques. VAM fungi are obligate symbionts and are maintained and mass-produced on pot cultures using suitable host plants. Plant-based inoculum formulations include spores and hyphae mixed with a carrier such as expanded clay or pumice, soil pellets and relatively crude soil spore mixtures. The production costs are high and there is no quality control for exclusion of pathogens (Menge et al., 1982). To produce inocula commercially on a large scale, host, substrate and environment have to be manipulated to produce pot cultures yielding mycorrhizal inoculum with high inoculum potential, few contaminants, longer shelf life and easy transportation.

Selection of Efficient Fungal Strains

The mycorrhizal fungi that are selected for inoculation in the field should be able to enhance nutrient uptake by plants and persist in soil. They should be infective, effective, have high colonization potential and should survive in soil. An efficient inoculant strain should possess the following characteristics: (i) ability to form extensive, well distributed mycelium in soil; (ii) ability to form extensive colonization; (iii) efficiency in absorbing nutrients, especially phosphorus, from the soil solution; (iv) should remain effective for a long time in transporting nutrients to the plant (Bagyaraj and Varma, 1995; Gaur, 1997).

Initially, the inoculant fungi can be selected under controlled conditions and later they can be evaluated in the field (Abbott and Robson, 1982; Abbott et al., 1992).

Management of Indigenous VAM Fungal Populations

The activity of indigenous mycorrhizal fungi can be enhanced by adequate cultural practices. In order to promote mycorrhizal ability in soil, factors such as erosion, fungi toxins (e.g. heavy metals, mine wastes, oil) and soil compaction which deplete the fungal spore populations have to be taken into account. Adequate strategies to conserve and multiply efficient indigenous strains of VAM fungi have to be developed.

CONCLUSIONS

Mycorrhizal fungi play a crucial role in plant growth and health. Plants colonized with VAM usually become established on disturbed habitats with relative ease. Thus, VAM fungi offer an environmentally sound, biological alternative to chemical fertilizers for maintaining plant quality and productivity in afforestation programmes.

ACKNOWLEDGEMENTS

The authors thank DBT for financial assistance.

REFERENCES

Abbott, L.K. and Robson, A.D., 1982, The role of vesicular arbuscular mycorrhizal fungi in agriculture and the selection of fungi for inoculation, *Aust. J. Agric. Res.* 33: 389-408.

Abbott, L.K., Robson, A.D. and Gazey, C., 1992, Selection of inoculant vesicular-arbuscular mycorrhizal fungi, in: *Methods in Microbiology: Techniques for Study of Mycorrhiza* (J.R. Norris, D.J. Read and A.K. Varnia, eds.), Academic Press, London.

Allen, M.F., 1996, The ecology of arbuscular mycorrhizas: a look back into the 20th century and a peek into the 21st, *Mycol. Res.* 100: 769-789.

Augé, R.M., Schekel, K.A. and Wample, R.L., 1987a, Rose leaf elasticity changes in response to mycorrhizal colonization and drought acclamation, *Physiol. Plant* 70: 175-182.

Augé, R.M., Schekel, K.A. and Wample, R.L., 1987b, Leaf water and carbohydrate status of VA mycorrhizal rose exposed to drought stress, *Plant Soil* 99: 291-302.

Bagyaraj, D.J. and Varma, A., 1995, Interaction between arbuscular mycorrhizal fungi and plants, in: *Advances in Microbiol Ecology Vol. 14* (J. Gwynfryn Jones, ed.), Plenum Press, New York and London.

Bansal, M. and Mukerji, K.G., 1994a, Positive correlation between VAM induced changes in root exudation and mycorrhizosphere mycoflora, *Mycorrhiza* 5: 39-44.

Bansal, M. and Mukerji, K.G., 1994b, Efficacy of root litter as a biofertilizer, *Biol. Fert. Soil* 18: 228-230.

Bansal, M. and Mukerji, K.G., 1996, Root exudates in rhizosphere biology, in: *Concepts in Applied Microbiology and Biotechnology* (K.G. Mukerji, V.P. Singh and S. Dwivedi, eds.), Aditya Books Pvt. Ltd., New Delhi.

Barea, J.M., 1991, Vesicular-arbuscular mycorrhizae as modifiers of soil fertility, *Adv. Soil. Sci.* 15: 1-40.

Barea, J.M. and Jeffries, P., 1995, Arbuscular mycorrhizas in sustainable soil-plant systems, in: *Mycorrhiza - Structure, Function and Physiology* (A. Varma and B. Hock, eds.), Springer-Verlag, Berlin.

Barea, J.M., Azcòn, R. and Azcòn-Aguillar, C., 1988, The role of rnycorrhiza in improving establishment and function of *Rhizobium*-legume system under field conditions, in: *Nitrogen Fixation by Legumes in Mediterranean Agriculture* (D.P. Beck and L.A. Meterson, eds.), ICARDA, Martinus-Nijhoff, Dordrecht.

Barea, J.M., Azcòn, R., and Azcòn-Aguillar, C., 1992, Vesicular-arbuscular mycorrhizal fungi in nitrogen-fixing systems, in: *Methods in Microbiology, Vol. 24* (J.R. Norris, D.J. Read, A.K. Varma, eds.), Academic Press, London.

Bethlenfalvay, G.J., 1992, Mycorrhizae and crop productivity, in: *Mycorrhiza in Sustainable Agriculture* (G.J. Bethlenfalvay and R.G. Linderman, eds.), ASA Special Publication, Madison, WI.

Bethlenfalvay, G.J., Thomas, R.S., Dakessian, S., Brown, M.S. and Ames, R.N., 1988, Mycorrhizae in stressed environments: effects on plant growth, endophyte development, soil stability and soil water, in: *Arid Lands, Today and Tomorrow* (E.E. Whitehead, ed.), Westview Press, Boulder, CO.

Bhandari, N.N. and Mukerji, K.G., 1993, *The Haustorium*, Research Studies Press Ltd., England.

Cox, G. and Tinker, P.B., 1976, Translocation and transfer of nutrients in vesicular arbuscular mycorrhizas. 1. The arbuscule and phosphorus transfer: a qualitative ultrastructural study, *New Phytol.* 77: 371-378.

Cress, W.A., Throneberry, G.O. and Lindsay, D.L., 1979, Kinetics of phosphorus absorption by mycorrhizal and non mycorrhizal tomato roots, *Plant Physiol.* 64: 484-487.

Ellis, J.R., Larsen, H.J. and Boosalis, M.G., 1985, Drought resistance of wheat plants inoculated with vesicular-arbuscular mycorrhizae, *Plant Soil* 86: 369-378.

Faber, B.A., Zasoki, R.J., Burau, R.G. and Uriu, K., 1990, Zinc uptake by corn as affected by vesicular-arbuscular mycorrhiza, *Plant Soil* 129: 121-130.

Fairchild, G.L. and Miller, M.H., 1988, Vesicular-arbuscular mycorrhizas and soil disturbance-induced reduction of nutrient absorption in maize II. Development of the effect, *New Phytol.* 110: 75-84.

Fitter, A.H. and Garbaye, J., 1994, Interactions between mycorrhizal fungi and other soil organisms, *Plant Soil* 159: 123-132.

Friese, C.F. and Allen, M.F., 1991, The spread of VA mycorrhizal fungal hyphae in soil: inoculum type and external hyphal architecture, *Mycologia* 83: 409-418.

Garriock, M.L., Petersen, R.L., and Ackerley, C.A., 1989, Early stages in colonization of *Allium porrum* (Leek) roots by VAM fungus, *Glomus versiforme, New Phytol.* 122: 85-92.

Gaur, A., 1997, Inoculum production technology development of vesicular-arbuscular mycorrhizae, *Ph.D. Thesis*, University of Delhi, India.

Gianinazzi-Pearson, V. and Gianinazzi, S., 1989, Cellular and genetical aspects of interactions between hosts and fungal symbionts in mycorrhizae, *Genome* 31: 336-341.

Gianinazzi-Pearson, V., Smith, S.E., Gianinazzi, S. and Smith, F.A., 1991, Enzymatic studies on the metabolism of vesicular arbuscular mycorrhiza V. and H^+ ATP hydrolysing enzyme a component of ATP hydrolysing activities in plant-fungus interfaces?, *New Phytol.* 117: 61-74.

Gianinazzi-Pearson, V., Gollotte, A., Lherminier, J., Tisserant, B., Franken, P., Dumas-Gaudat, E., Lemoine, M.C., Tuinen, D. Van and Gianinazzi, S., 1995, Cellular and molecular approaches in the characterization of symbiotic events in functional arbuscular mycorrhizal associations, *Can J. Bot.* 73 (suppl. 1) : S526-S532.

Gildon, A. and Tinker, P.B., 1983, Interactions of vesicular-arbuscular mycorrhizal infection and heavy metals in plants. 1: The effect of heavy metals on development of vesicular-arbuscular mycorrhizal symbiosis, *New Phytol.* 95: 247-261.

Giovannetti, M., Avio, L., Sbrana, C. and Citernesi, A.S., 1993, Factors affecting appresorium development in vesicular arbuscular mycorrhizal fungus *Glomus mosseae* (Nicol. and Gerd.) Gerd. and Trappe, *New Phytol.* 123: 114-122.

Giovannetti, M., Sbrana, C., Citernesi, A.S., Avio, L., Gollotte, A., Gianinazzi-Pearson, V. and Gianinazzi, S., 1994, Recognition and infection process, basis for host specificity of arbuscular mycorrhizal fungi, in: *Impact of Arbuscular Mycorrhizae on Sustainable Agriculture and Natural Ecosystems* (S. Gianinazzi and H. Shcüepp, eds.), Birkhauser, Basel.

Harley, J.L., 1991, Introduction: the state of art, in: *Methods in Microbiology, Vol. 23* (J.R. Norris, D.J. Read and A.K. Varma, eds.), Academic Press, London.

Henderson, J.C. and Davies, F.J., 1990, Drought acclamation and morphology of mycorrhizal *Rosa hybrida* L.cv. 'Ferdy' is independent of leaf element content, *New Phytol.* 115: 503-510.

Herrera, M.A., Salamanca, C.A. and Barea, J.M., 1993, Inoculation of woody legumes with selected arbuscular mycorrhizal fungi and rhizobia to recover desertified Mediterranean ecosystems, *Appl. Environ. Microbiol.* 59: 129-133.

Huang, R.S., Smith, W.K. and Yost, R.S., 1985, Influence of vesicular-arbuscular mycorrhiza on growth , water relations and leaf orientation in *Leucaena leucocephala* (Linn.) De. Wit., *New Phytol.* 99: 229-243.

Jalali, B.L. and Jalali, I., 1991, Mycorrhiza in plant disease control, in: *Handbook of Applied Mycology Vol. I Soil and Plants* (D.K. Arora, B. Rai, K.G. Mukerji and G.R. Knudsen, eds.), Marcel Dekker Inc., New York.

Jeffries, P. and Dodd, J.C., 1991, The use of mycorrhizal inoculants in forestry and agriculture, in: *Handbook of Applied Mycology Vol. 1, Soil and Plants* (D.K. Arora, B. Rai, K.G. Mukerji and G.R. Knudsen, eds.), Marcel Dekker Inc., New York.

Johansen, A., Jakobsen, I. and Jensen, E.S., 1992, Hyphal transport of ^{15}N-labelled nitrogen by a vesicular-arbuscular mycorrhizal fungus and its effect on depletion of inorganic soil N, *New Phytol.* 122: 281-288.

Johansen, A., Jakobsen, I. and Jensen, E.S., 1993, External hyphae of vesicular-arbuscular mycorrhizal fungi associated with *Trifolium subterraneum* L. III Hyphal transport of ^{32}P and ^{15}N, *New Phytol.* 124: 61-68.

Jungk, A. and Claasen, N., 1989, Availability in soil and acquisition by plants as the basis for phosphorus and potassium supply to plants, *J. Pflanzenernachr, Bodenkd.* 152: 151-157.

Karunaratne, R.S., Baker, J.H. and Barker, A.V., 1986, Phosphorus uptake by mycorrhizal and non mycorrhizal roots of soybean, *J. Plant Nutr.* 9: 1303-1313.

Kothari, S.K., Marschner, H. and George, E., 1990, Effect of VA mycorrhizal fungi and rhizosphere microorganisms on root and shoot morphology, growth and water relations in maize, *New Phytol.* 116: 303-311.

Kaushik, A. and Mukerji, K.G., 1996, Mycorrhiza in control of diseases, in: *Disease Scenario in Crop Plants* (V.P. Agnihotri, O. Prakash, R. Kishun and A.K. Mishra, eds.), Int. Books and Periodicals Sup. Ser., Delhi, India.

Lambert, D.H and Weidensaul, T.C., 1991, Element uptake by mycorrhizal soybean from sewage-sludge treated soil, *Soil Sci. Soc. Am. J.* 55: 393-398.

Lei, J., Bécard, G., Catford, J.G. and Piché, Y., 1991, Root factors stimulate ^{32}P uptake and plasmalemma ATPase activity in the vesicular-arbuscular mycorrhizal fungus, *New Phytol.* 118: 289-294.

Linderman, R.G., 1988, Mycorrhizal interactions with the rhizosphere microflora: The mycorrhizosphere effect, *Phytopathology* 78: 366-371.

Li, X-L., Marschner, H. and George, E., 1991, Acquisition of phosphorus and copper by VA-Mycorrhizal hyphae and root to shoot transport in white clover, *Plant Soil,* 136: 49-57.

Lumini, E., Bosio, M., Puppi, G., Isopi, R., Frattegiani, M., Buresti, E. and Favilli, F., 1994, Field performance of *Alnus cordata* Loisel (Italian Alder) inoculated with *Frankia* and VA mycorrhizal strains in mine-spoil afforestation plots, *Soil Biol. Biochem.* 26: 659-661.

Marschner, H. and Dell, B., 1994, Nutrient uptake in mycorrhizal symbiosis, *Plant Soil* 159: 89-102.

Menge, J.A., Jarrell, W.M., Labanauskas, C.K., Ojala, J.C., Huszar, C., Johnson, E.L.V. and Sibert, D., 1982, Predicting mycorrhizal dependency of Troyer citrange on *Glomus fasciculatus* in California citrus soils and nursery mixes, *Soil Sci. Soc. Am. J.* 46: 762-768.

Michelsen, A. and Rosendahl, S., 1990, The effect of VA mycorrhizal fungi, phosphorus and drought stress on growth of *Acacia nilotica* and *Leucaena leucocephala* seedlings, *Plant Soil* 124: 713.

Michelsen, A. and Sprent J.I., 1994, The influence of vesicular-arbuscular mycorrhizal fungi on nitrogen fixation of nursery-grown Ethiopian acacias estimated by ^{15}N natural abundance method, *Plant Soil* 160: 249-257.

Miller, R.M. and Jastrow, J.D., 1992, The role of mycorrhizal fungi in soil conservation, in: *Mycorrhiza in Sustainable Agriculture* (G.J. Bethlenfalvay and R.G. Linderman, eds.), ASA Special Publ, Madison, WI.

Morton, J.B. and Benny, G.L., 1990, Revised classification of arbuscular mycorrhizal fungi (Zygomycetes): a new order, Glomales, two new suborders, Glomineae and Gigasporineae, and two new families, Acaulosporaceae and Gigasporaceae, with an amendation of Glomaceae, *Mycotaxon* 37: 471-491.

Morton, J.B. and Bentivenga, S.P., 1994, Levels of diversity in endomycorrhizal fungi (Glomales, Zygomycetes) and their role in defining taxonomic and non-taxonomic groups, *Plant Soil* 159: 47-60.

Mosse, B., 1959, Observations on the extra-matrical mycelium of a vesicular-arbuscular endophyte, *Trans. Br. Mycol. Soc.* 42: 439-448.

Mosse, B., 1973, Advances in the study of vesicular arbuscular mycorrhiza, *Ann. Rev. Phytopathol.* 11: 171-196.

Mukerji, K.G. and Dixon, R., 1992, Mycorrhizae in reforestation in: *Proceedings International Symposium on Rehabilitation of Tropical Rainforest Ecosystems*, Malaysia.

Mukerji, K.G. and Mandeep, 1997, Mycorrhizal relationships of wetlands and rivers associated plants, in: *Ecology of Wetlands* (S.K. Majumdar, ed.), The Pennsylvania Academy of Sciences, USA.

Mukerji, K.G. and Sharma, M., 1996, Mycorrhizal relationships in forest ecosystems, in: *Forests: A Global Perspective* (S.K. Majumdar, E.W. Miller and F.J. Brenner, eds.), The Pennsylvania Academy of Sciences, USA.

Mukerji, K.G., Mandeep and Varma, A.K., 1998, Mycorrhizosphere microorganisms: screening and evaluation, in: *Mycorrhiza Manual* (A. Varma, ed.), Springer-Verlag, Hiedelberg, Germany.

Mukerji, K.G., Chamola, B.P. and Sharma, M., 1997, Mycorrhiza in control of plant pathogens, in: *Management of Threatening Plant Diseases of National Importance* (Y.P. Agnihotri, A.K. Sarbhoy and D.V. Singh, eds.), Malhotra Publishing House, New Delhi.

Mukerji, K.G., Chamola, B.P., Kaushik, A., Sarwar, N. and Dixon, R., 1996, Vesicular arbuscular mycorrhiza: potential biofertilizer for nursery raised multipurpose tree species in tropical soils, *Ann. For.* 4: 12-20.

Newman, E.I., Eason, W.R., Eissenstat, D.M. and Romos, M.I.R.F., 1992, Interactions between plants: the role of mycorrhizae, *Mycorrhiza* 1: 47-53.

Noyd, R.K., Pfleger, F.L. and Norland, M.R., 1996, Field responses to added organic matter, arbuscular fungi and fertilizer in reclamation of taconite iron ore tailing, *Plant Soil* 179: 89-97.

O'Keefe, D.M. and Sylvia, D.M., 1992, The chronology and mechanism of mycorrhizal plant growth responses on sweet potato, *New Phytol.* 122: 651-659.

Osonubi, O., Mulongoy, K., Awotoye, O.O., Atayese, M.O. and Okali, D.V.V., 1991, Effect of ectomycorrhizal and vesicular arbuscular mycorrhizal fungi on drought tolerance of four leguminous woody seedlings, *Plant Soil* 136: 131-143.

Paulitz, T.C. and Linderman, R.G., 1991, Mycorrhizal interactions with soil microorganisms, in: *Handbook of Applied Mycology Vol. I: Soil and Plants* (D.K. Arora, B. Rai, K.G. Mukerji and G.R. Knudsen, eds.), Marcel Dekker, New York.

Pfeiffer, C.M. and Bless, H.E., 1988, Growth and nutrition of guayule *(Parthenium argentatum)* in a saline soil as influenced by vesicular arbuscular mycorrhiza and phosphorus fertilization, *New Phytol.* 108: 315-321.

Reinhard, S., Weber, E., Martin, P. and Marschner, H., 1994, Influence of phosphorus supply and light intensity on mycorrhizal response in *Pisum-Rhizobium-Glomus* symbiosis, EXPEA, 50: 890-896.

Rhodes, L.H. and Gerdemann, J.W., 1975, Phosphate uptake zones of mycorrhizal and non-mycorrhizal onions, *New Phytol.* 75: 555-561.

Rozema, J., Arp, W., Diggelen, J., Van Esbrock, M., Van Broekmann, R. and Punte, H., 1986, Occurrence and ecological significance of VAM in salt marsh environment, *Acta. Bot. Neerl.* 35: 457-467.

Sarwar, N., 1996, Vesicular arbuscular mycorrhiza in relation to multipurpose legumes, *Ph.D. Thesis,* University of Delhi, India.

Scannerini, S. and Bonfante-Fasolo, P., 1979, Ultrastructural, cytochemical demonstration of polysaccharides and proteins within host arbuscule interfacial matrix in an endomycorrhiza, *New Phytol.* 83: 87-94.

Sharma, M. and Mukerji, K.G., 1995, Interactions between vesicular-arbuscular mycorrhizal fungi and free-living nitrogen fixers and its effect on growth of jute, in: *Mycorrhizae - Biofertilizers for the Future* (A. Adholeya and S. Singh, eds.), TERI, New Delhi.

Sieverding, E., 1991, Vesicular-arbuscular mycorrhiza in management of tropical agrosystems, *Gesellschaft für Technische Zusammenabeit (GTZ)*, Eschborn, Germany.

Singh, R., 1997, VAM establishment in two multipurpose tree species, *M.Phil. diss.*, University of Delhi, India.

Smith, S.E., 1980, Mycorrhizas of autotrophic higher plants, *Biol. Rev.* 55: 475-510.

Smith, S.E. and Smith, F.A., 1990, Structure and function of the interfaces in biotrophic symbiosis as they relate to nutrient transport, *New Phytol.* 114: 1-38.

Smith, S.E. and Read, D.J., 1997, *Mycorrhizal Symbiosis*, Academic Press, London.

Srivastava, D., Kapoor, R., Srivastava, S.K. and Mukerji, K.G., 1996, Vesicular arbuscular mycorrhiza - an overview, in: *Concepts in Mycorrhizal Research* (K.G. Mukerji, ed.), Kluwer Academic Publ., Netherland.

Sylvia, D.M. and Williams, S.E., 1992, Vesicular-arbuscular mycorrhiza and environmental stress, in: *Mycorrhizae in Sustainable Agriculture* (G.J. Bethlenfalvay and R.G. Linderman, eds.), ASA Special Publ., Madison, WI.

Szanizzlo, P.J., Powell, R.E., Reid, C.P.P. and Cline, G.R., 1981, Production of hydroxymate siderophore iron chelators by ectomycorrhizal fungi, *Mycologia* 73: 1153-1174.

Tarafdar, J.C. and Marschner, H., 1994, Phosphatase activity in rhizosphere and hydrosphere of VA mycorrhizal wheat supplied with inorganic phosphorus, *Soil Biol. Biochem.* 26: 387-395.

Tester, M., Smith, A. and Smith, S.E., 1992, The role of ion channels in controlling solute exchange in mycorrhizal associations, in: *Mycorrhizas in Ecosystems* (D.J. Read, D.H. Lewis, A.H. Fitter and I.J. Alexander, eds.), CAB International, Cambridge.

Timmer, L.W. and Leyden, R.F., 1978, Stunting of *Citrus* seedlings on fumigated soils in Texas and its correction by phosphorus fertilization and inoculation with mycorrhizal fungi, *J. Am. Soc. Hort. Sci.*, 103: 533-537.

Van Kessel, C., Singleton, P.W. and Hobbel, H.J., 1985, Enhanced nitrogen transfer from soybean to maize by vesicular-arbuscular mycorrhizal fungi, *Plant Physiol.* 79: 562-563.

Varma, A., 1995, Ecophysiology and application of arbuscular mycorrhizal fungi in arid soils, in: *Mycorrhiza - Structure, Function and Physiology* (A. Varma and B. Hock, eds.), Springer-Verlag, Berlin.

VESICULAR ARBUSCULAR MYCORRHIZAL FUNGI: BIOFERTILIZER FOR THE FUTURE

B.P. Chamola, Bhoopander Giri and K.G. Mukerji

Applied Mycology Laboratory
Department of Botany
University of Delhi
Delhi-110 007
India

INTRODUCTION

Rapid desertification and soil erosion is a problem in the tropics due to adverse soil and climatic conditions. Marginal lands may also become barren due to deforestation. To prevent this degradation, marginal lands have to be afforested by overcoming the difficulties due to soil infertility. Sustainable agriculture and forestry have therefore taken centre stage among researchers, the public and policy makers. Afforestation or revegetation of these barren lands through vesicular arbuscular (VA) mycorrhizal technology is an immediate necessity. The overall goal of forestry research is focused on increasing food, fibre, forage, fuel wood, fertiliser (root nodule) and timber production.

Maintaining or enhancing the fertility of barren soil is one of the most important requirements for sustainable agriculture and forestry systems. Increasing soil organic matter and mineral contents, avoiding pests and pathogens and decreasing erosion are a few other areas of concern. The present paper emphasises the beneficial aspects, for example factors which influence colonisation of VA mycorrhizae and application of VAM fungi in agriculture and afforestation.

Vesicular arbuscular mycorrhizal fungi are an integral component of sustainable agriculture and tropical forestry with mutual dependency between fungus and tree host for normal functioning and survival with widest host range (Fitter, 1991, Dixon et al., 1997). VA mycorrhizae are ubiquitous soil inhabitants and are known to exist in extreme terrestrial environments. Recently much attention has been paid to the use of VAM fungi in reinstatement of forest and improving soil fertility due to their potential for increasing the growth, survival and biomass production of plants. VAM fungi are an important biotechnological tool for balancing soil nutrients and the sustainability of forest ecosystems.

Vesicular arbuscular mycorrhizal fungi are extremely widespread in their distribution (Gerdemann, 1968; Mosse, 1973; Jagpal and Mukerji, 1988). The fungal symbiont consists

From Ethnomycology to Fungal Biotechnology
Edited by Singh and Aneja, Plenum Press, New York, 1999

of aseptate hyphae belonging to the order Glomales of class Zygomycetes (Morton and Benny, 1990; Mukerji, 1996). VA mycorrhizae are included under endomycorrhiza and are characterized by specialized structures, i.e. vesicles and arbuscules produced inside the host tissue (Harley and Smith, 1983) and chlamydospores and azygospores produced in the soil (Jasper et al., 1989). VAM fungi are found in association with plant roots of arctic, temperate and tropical regions and are distributed in nearly all the families of angiosperms (Kendrick and Berch, 1985). They benefit the plants by improving the supply of nutrients especially phosphorus and other minerals such as zinc, copper, potassium and calcium (Cooper and Tinker, 1978a,b). The mycelial network of VA mycorrhizal plants enables them to obtain nutrients from beyond the zones of low concentration around roots. Exchange of nutrients and carbohydrates between the host and symbiont takes place through specific structures, the arbuscules (Bhandari and Mukerji, 1993). VAM fungi also assist plants by improving absorption of water through the roots (Safir et al., 1971). Due to the potential of VAM fungi in increasing the growth and yield of crops, the commercial use of these fungi may be an alternative to the rising costs of agriculture and fertilizer.

OCCURRENCE AND DISTRIBUTION

Vesicular arbuscular mycorrhizae (VAM) occur in almost all soils. Associations with VAM are widespread among the plant kingdom, occurring in bryophytes, pteridophytes, gymnosperms and angiosperms. However, there are a few examples where these fungi may be absent, namely in eroded soils, fumigated soils or soils disturbed by mining. In the soil these fungi are present as chlamydospores, zygospores and azygospores. There is more or less a uniform distribution of all the genera of VAM fungi in both the hemispheres and the spores have a broad ecological range (Mukerji et al., 1984). Most vascular plants show VA mycorrhizal colonization. Both cultivated and non-cultivated soils have been reported to contain VAM fungal spores. VAM fungal propagules have been isolated from forest, open woodlands, scrub, savanna, heaths, grasslands, sand dunes, semi-deserts, anthracite and bituminous coal wastes, (Mukerji et al., 1984, Pfleger et al., 1994).

Families of angiosperms that rarely form vesicular arbuscular mycorrhizal associations include Chenopodiaceae, Brassicaceae, Amaranthaceae, Polygonaceae and Fumariaceae. Some families are wholly non-mycorrhizal, such as Pinaceae, Betulaceae, Orchidaceae, Commelinaceae and Urticaceae. The members of the families Amaranthaceae, Chenopodiaceae and Brassicaceae are often non-mycorrhizal but there are some reports on occurrence of mycorrhizae in the family Chenopodiaceae and Brassicaceae (Jagpal and Mukerji, 1988). The occurrence of VAM fungi in five genera of the family Amaranthaceae in the arid and semi-arid zones are also reported (Jagpal and Mukerji, 1988; Neeraj et al., 1991).

VA mycorrhizal fungi are influenced by damage to vegetation, soil and natural processes. But in undisturbed ecosystems of tropical and temperate climates the colonization of new roots by VAM fungi occurs rapidly due to contact with a pre-existing network of infective hyphae. The number of spores in soil is correlated with soil factors such as pH, phosporus, potassium, organic matter and texture. In arid, semi-arid and waste soils, the results of studies conducted so far indicate that the number of VAM fungal spores are comparatively lower, and this may be attributed to sparse vegetation.

FACTORS WHICH INFLUENCE VA MYCORRHIZAL COLONIZATION

A range of factors influence VA mycorrhizal colonization in host roots and these include the following.

Temperature

Temperature has been shown to have a significant influence on the colonization and sporulation of VAM endophytes in nature. Higher temperature results in greater root colonization and increased sporulation of VAM fungi which also increases the rate of photosynthesis, transpiration and translocation of photosynthate in plants. In arid regions VAM fungi provide nutrition to the plant and are able to tolerate extreme temperatures. At lower temperatures (15°C) there was reduced production of shoot dry matter, as well as enhanced nodule production in tripartite associations (Glycine-Bradyrhizobium-*Glomus mosseae*). Soil temperature has an influence on all three stages of mycorrhizal development, i.e. spore germination, hyphal penetration of the root and proliferation within the cortical cells of the root.

Light

Under shade, VAM colonization, spore production and plant growth was highly reduced. This may be due to reduced spread of internal hyphae within root tissues and restricted growth of extramatrical hyphae in the soil (Srivastava et al., 1996). Day length has been shown to stimulate development of VA mycorrhizal colonization (Furlan and Fortin, 1977; Daft and El-Giahmi, 1978). The growth of VAM fungi and plants is reduced by decrease in photon irradiance (Pearson et al., 1991).

Air Pollution

Sulphur dioxide (SO_2), nitrogen oxide (NO_2) and ammonia (NH_3) are major pollutants of air. Ammonium sulphate is the most important air pollutant with high NH_3 emission. Heijne et al. (1994) reported that VAM colonization of *Allium montana* and *Helianthus pilosella* was not affected by increasing ammonium sulphate. They also suggested that colonization by VAM increases with increased ammonia concentration. The rate of colonization by VAM remains unchanged with increasing ozone levels. Increased ozone concentration results in an increase in the frequency of vesicles, hyphal coils and internal mycelium in sugar maple while there is a decrease in the number of arbuscules (Duckmanton and Wider, 1994).

Soil Characteristics

Removal of surface soil layers mainly by water erosion decreased markedly both the number of propagules of VAM fungi and the extent of mycorrhiza formation. Nadian (1996) reported that the pore size of soil affects VAM colonization, and increase in the soil bulk density due to soil compaction leads to increased mycorrhizal colonization in *Trifolium subterraneum*. In contrast, Entry et al. (1996) reported that soil compaction has no effect on mycorrhizal colonization in corn roots.

Soil moisture and soil pH also influence VAM fungal spore germination. VAM colonization in roots of *Oryzopsis hymenoides* is known to stimulate growth and increase tolerance to extreme conditions such as draught and chilling (Kothari et al., 1990). Delayed germination of *Glomus caledonium*, *Gigaspora calospora* and *Acaulospora laevis* at 2.2 MPa was reported but germination was higher at 1.4 MPa.

Rainfall

In the dry season rainfall affects the VA mycorrhizal fungi either directly or indirectly. Rainfall affects the infectivity of propagules and nutrient availability and limits the overall quantity of fungi and their colonization in roots (Braungerger et al., 1994).

Chemical Fertilizers

Agro-chemicals are extensively used to control soil- and air-borne fungal pathogens, and this has an impact on the development of VAM fungi. However, studies on the effect of different fungicides, both systemic and non-systemic, on VAM colonization do not show a consistent trend. According to some reports the population dynamics of VAM fungi are inversely affected more by fungicides than by herbicides, nematicides and insecticides (Tommerup and Briggs, 1981; Trappe et al., 1984). Addition of chemical fertilizers has been found to decrease the level of mycorrhizal colonization in a variety of crops. It is also reported that some species of VAM fungi, unlike others, are able to colonize roots with large additions of phosphate (Thompson et al., 1986). However, Anderson et al. (1987) found that adding phosphate fertilizers had no effect while Hayman et al. (1975) found a decrease in the level of mycorrhizal infection.

Crop Rotation

The effects of crop rotation on the density of VAM fungi have also been investigated. More spores have been found in continuously cropped soils than in soils maintained as clean fallows (Thompson, 1987). Similarly, greater mycorrhizal colonization was recorded in continuous crops than in crops grown after a fallow (Black and Tinker, 1979; Thompson, 1987).

BENEFICIAL ASPECTS AND APPLICATION OF VAM FUNGI

VAM in Phosphorus Uptake

Mycorrhizal colonization enhances plant growth by increasing uptake of nutrients by increasing the absorbing surface area and by mobilizing available nutrient sources (Bolan, 1991). Various forms of P present in soil have a very low solubility and the concentration of phosphorus in soluble forms is extremely low, viz. up to 0.5 micromoles/litre in forest soils and up to 1.5 micromoles/litre in cultivated soils. The organic and insoluble mineral phosphorus, which constitutes the greater part of the phosphorus in soil, is not available to plants. Since phosphorus has a very low mobility a zone of extremely low concentration is formed in the vicinity of the roots. Deficiency of phosphorus leads to a number of symptoms and inhibits survival of plants. Available reports suggest that VAM colonization increases the growth of plants significantly by improving P nutrition. Enhanced phosphorus uptake by Sorghum plants was found when inoculated with VAM fungi.

Literature reports suggest that increased concentration of phosphorus in VA mycorrhizal plants may be due to (i) increased physical exploration of the soil (Tinker, 1978), (ii) increase in movement of phosphorus into VAM fungal hyphae (Bolan et al., 1987), (iii) modified root environment by VAM fungi (Hoffland et al., 1989), (iv) increased storage of absorbed phosphorus (Martin et al., 1983), and (v) efficient utilization of phosphorus within the plant (Lambert et al., 1979).

Lambert et al. (1979) also noticed that one maize line grew better when colonized by a mycorrhizal fungus and produced more dry weight at any concentration of phosphorus in plant tissue than non-mycorrhizal plants. However, Stribley et al. (1980) found that mycorrhizal plants produce less dry weight at a given phosphorus concentration in shoots than non-mycorrhizal plants when soluble phosphate has been added. These results led to the observation that VA mycorrhizal plants may use more carbon for the growth of photosynthetic tissue.

VAM in Nitrogen Uptake

Improvement of the nitrogen status of the host is one of the ecologically significant consequences of mycorrhizal association. VAM fungal hyphae have the ability to extract nitrogen and transport it from soil to plant due to the increased absorption surface provided by the mycorrhizal fungi. Nitrogen exists in many forms, namely free nitrogen, nitrate, nitrite, ammonium ions and organic nitrogen. Ammonium constitutes a significant proportion of the inorganic nitrogen pool at low pH. Ammonium is less mobile in soil because it is fixed in the lattice of clay minerals and absorbed to negatively charged clay minerals. VA mycorrhizal hyphae transport such immobile ammonium to plant roots (Bowen, 1987).

Mycorrhizal association enhances nitrogen gain by increasing nitrogen fixation rates of the plant-N-fixing bacterial association. VA mycorrhizae make a significant contribution to the growth of nitrogen-fixing legumes by increasing the uptake of phosphorus by plants (Mosse, 1973; Rhodes and Gerdemann, 1975). The tripartite symbiosis of the leguminous plants with rhizobium and VAM fungi aids plant acquisition of two vital elements, i.e. nitrogen and phosphorus (Hayman, 1982). Dual inoculation of plants with VAM and rhizobia cause an increase in shoot dry weights and mycorrhizal colonization of roots (Sengupta and Chaudhri, 1995). Nitrogen fixed by soybeans was transported to maize through VA mycorrhizae and significantly increased growth and nitrogen status in the maize plants.

VAM in Micronutrients Absorption

Micronutrients are also very important for the proper growth and development of plants. The micronutrients include copper, zinc, magnesium, manganese, cobalt and cadmium. Mycorrhizal plants are known to accumulate greater amounts of some micronutrients, especially under conditions of low soil nutrient availability (Gerdemann, 1975; Smith, 1980; Faber et al., 1990). VA mycorrhizal plants showed higher absorption of nutrients, which was attributed to a wider exploration of the soil volume by the extended extramatricular mycelium and higher uptake of water (Safir et al., 1971). A number of workers reported increased zinc uptake due to mycorrhizal inoculation in several plants such as field bean (Kucey and Janzen, 1987); maize, (Kothari et al., 1991); pigeonpea (Wellings et al., 1991); wheat (Thompson, 1990); linseed (Thompson, 1994) and *Leucaena* sp. (Manjunath and Habte, 1988). VA mycorrhizal inoculation also increases the uptake of copper (Li et al., 1991) and zinc and copper concentration in shoots and roots of the plants (Lambert and Weidensaul, 1991; Sharina et al., 1992).

VAM in Biocontrol of Plant Pathogens

Biocontrol of plant pathogens is currently accepted as key practice in sustainable agricultural systems and forestry, because it is based on the management of natural resources which develop antagonistic activities against harmful organisms. Many workers suggest that different microbes present in the rhizosphere reduce survival of soil-borne pathogens (Hwang, 1992). Host plants previously inoculated with the VAM fungi exhibited increased resistance to several soil-borne root pathogens (Jalali and Jalali, 1991; Shukla et al., 1996; Trotta et al.,

1996). *Pyrenochaeta terristris*, which causes root disease of onion, showed reduced effects when these plants were inoculated with VAM fungi. Rice plants inoculated with VAM fungi showed increased resistance to stem and sheath blight diseases caused by *Sclerotium* and *Rhizoctonia solani* respectively. Besides these, VAM fungi are also known to reduce nematode and bacterial diseases (Morandi 1996; Garcio-Garrido and Ocampo, 1989).

VAM in Agriculture

India is one of the poorest countries with a population growing at a rate of 1.8% per year, thus enhancing the demand for food, fuel wood, fodder, fibres, timber and other agricultural and forest products. During the last five years, efforts have been made to achieve increased production through the introduction of high yielding varieties of seeds, chemical fertilizers and pesticides, but these components have been introduced without taking their side effects into account. Due to intensive use of these components the soil and groundwater become contaminated with pollutants such as nitrates and phosphates. These pollutants cause the formation of barren lands after a few years and there are frequent crop losses. Thus, scientists have begun to use organic fertilizers or traditional agricultural implements which do not harm the environment and maintain environmental balance.

Deficiency in phosphorus is one of the major factors limiting crop growth in the tropics. Phosphorus is often present at low concentrations in the soil solution in comparison with other major nutrients and it diffuses only slowly in the soil environment (Lewis and Quirk 1965; Lewis and Koide, 1990). Due to high pH of soil the fertilizers applied become unusable because of chemical reactions. It is estimated that 75% of the phosphorus applied to the crops is not utilized by plants but is converted to forms unavailable to the plants (Mosse, 1973).

VAM fungi are very beneficial microorganisms under these conditions and are often used to increase the efficiency of phosphate fertilizers in agriculture. VAM fungi enhance phosphorus uptake and compensate for the low concentration of phosphate by exploring a large volume of soil by creating a greater surface area for phosphorous uptake. VAM fungi colonize tropical forage legumes and grasses growing in both natural and cultivated environments (Safir, 1986). VAM fungi also play an important role in conservation of soil and are thus regarded as 'biological fertilizers'. Positive effects of VAM fungi on the growth of *Centrosema, Andropogon, Stylosanthes, Pueraria, Brachiaria* and *Siratro* have been reported (Crush, 1974; Lopes, 1980; Mossae, 1973).

There are several promising tropical forage species for which effects of mycorrhizal fungi have been studied. The tropical pastures programme of Centre International de Agricultura Tropical (CIAT) is screening a range of legume and grass germplasm for cultivation in well drained tropical savannahs of South America for the presence of VAM fungi. The soil of this region is characterized by acidity and infertility and here VA mycorrhizal fungi can play an important role in nutrient uptake (Safir, 1984; Salinas et al., 1985). Marginal agricultural soils can be made more productive by the addition of selective strains of mycorrhizal fungi to the soils. Several workers have demonstrated that plant species and even cultivars of the same species can differ in their dependence on mycorrhizae.

Besides agriculture, VAM fungi also give excellent results with ornamentals and flowering plants. Woody ornamentals like *Liquidamber, Ampelopsis* etc. give excellent results when inoculated with VAM fungi (Gianinazzi et al., 1990). VAM colonization also gives better performance to other plants like *Berberis, Chamaecyparis, Tegetis* etc. and vegetable crops such as Asparagus, Leek and Onion (Gianinazzi et al., 1986, 1989). Successful establishment of VAM fungi has been reported in a variety of rooting media comprising sand, gravel, peat, expanded clay, pumice, perlite, bark, vermiculite or mixtures of these materials (Menge, 1983; Dehne and Backhaus, 1986).

VAM in Afforestation

Approximately 35% of the earth's dry lands are deserts spreading over 100 nations. In India about 6% of the land biomass has attained desertification in the form of vast expanses of arid, semi-arid and waste tracts. The adverse and unfavourable soil texture of arid land results in a net deficit of forest biomass production and creates unproductive barren land areas (Mukerji et al., 1996; Dixon et al., 1997). In these regions crop yields are very low due to poor supply of nutrients (Bagyaraj and Verma, 1995). Phosphorus concentration is also very low in these areas.

Fertilizer expenditure in agriculture and forestry is a major expense to the farmers and forestry managers. With diminishing reserves of high quality raw materials and with escalating energy costs there is a need to develop sustainable agricultural and forestry technologies which will require low fertilizer inputs. This can also be made more effective by adding efficient strains of VAM fungi to the soil. VA mycorrhizal fungi are tremendously efficient in converting infertile soils to fertile soils.

The importance of mycorrhizae in the successful development of severely disturbed lands was also observed by Allen and Allen (1980). The management and use of VAM fungi is very effective for afforestation in tropical soils. Our experimental results suggest that the inoculation of seedlings at the pre-transplanted nursery stage with VAM fungi contributes to the successful establishment of these seedlings in the field (Mukerji et al., 1996). Fast-growing leguminous trees such as *Sesbania sesban, S. grandifllora* and *Leucaena lucocepllala* show a three-fold increase in biomass production with higher acid and alkaline phosphate activities by VAM inoculation (Jagpal and Mukerji, 1991 a,b; Giri, 1997). An increase in total biomass of *Albizia lebbek* on VAM colonization was also reported (Kaushik and Kaushik, 1995). Mukerji et al. (1996) found enhanced growth of inoculated juvenile seedlings of multipurpose tree species (MPTS) and their successful primary establishment in tropical soil and suggested that VAM fungi are potential 'biofertilizers' for the development of arid and semi-arid lands.

CONCLUSIONS

VA mycorrhizal colonization affects symbiosis and depends on the interaction between the mycosymbiont, VAM fungal species, edaphic factors and environment. Inoculation of tree seedlings with VA mycorrhizal fungal spores increases the growth and primary establishment of seedlings. The significant increase in biomass and percentage mycorrhizal colonization of inoculated seedlings strongly suggest the potential uses of VAM fungi. For the improvement of large tracts of arid, semi-arid, waste and barren lands in India, and the sustainable utilization of forest and agricultural products, these VAM fungi could be used as biofertilizers in the future.

ACKNOWLEDGEMENTS

This work is supported by the grant received from DBT.

REFERENCES

Allen, E.B. and Allen, M.F., 1980, Natural re-establishment of vesicular arbuscular mycorrhiza following strip mining reclamation in Wyoming, *J. Appl. Ecol.* 17: 139-147.

Anderson, EL., Miller, P.D. and Kunishi, M., 1987, Maize root length density and mycorrhizal infection as influenced by tillage and soil phosphorus, *J.Pl. Nutrit.* 10: 1349.

Bagyaraj, D.J. and Verma, A., 1995, Interaction between arbuscular mycorrhizal fungi and plants, in: *Advances in Mycorrhizal Ecology*, Vol. 14 (G.J. Jones, ed.), Plenum Press, New York.

Bhandari, N.N. and Mukerji, K.G. 1993, *The Haustorium*, Research Studies Press Ltd., Somerset, U.K; and John Wiley and Sons Inc., New York, U.S.A.

Black, R. and Tinker, P.B., 1979, The development of endomycorrhizal root systems, II, Effect of agronomicfactors and soil conditions on the development of vesicular-arbuscular mycorrhizal infection in barley and   on the endophyte spore density, *New Phytol.* 83: 401-413.

Bolan, N.S., 1991, A critical review of the role of mycorrhizal fungi in the uptake of phosphorus by plants, *Pl.Soil.* 134: 189-207.

Bolan, N.S., Robson, A.D. and Barrow, N.J., 1987, Effect of phosphorus application and mycorrhizal inoculation on root characteristics of sub-terranean clover and ryegrass in relation to phosphorus uptake, *Trans. Brit. Mycol. Soc.* 70: 443-450.

Bowen, G.D., 1987, The biology and physiology of infection and its development, in: *Ecophysiology of VA Mycorrhizal Plants* (G.R. Safir, ed.), CRC Press.

Braungerger, P.G., Abbott, L.K. and Robson, A.D., 1994, The effect of rain in the dry-season on the formation of vesicular arbuscular mycorrhizas in the growing season of annual clover-based pastures, *New Phytol.* 127: 107-114.

Cooper, K.M. and Tinker, P.B., 1978a, Translocation and transfer of nutrients in vesicular arbuscular mycorrhizas, IV, Effect of environmental variables on movement of phosphorus, *New Phytol.* 88: 327-333.

Cooper, K.M. and Tinker, P.B., 1978b, Translocation and transfer of nutrients in vesicular arbuscular mycorrhizae, II, Uptake and translocation of phosphorus, zinc and sulphur, *New Phytol.* 81: 43-47.

Crush, J.R., 1974, Plant growth responses to vesicular arbuscular mycorrhiza, VII, Growth and nodulation of some herbage legumes, *New Phytol.* 73: 743-749.

Daft, M.J. and El-Giahmi, A.A., 1978, Effect of vesicular arbuscular mycorrhiza on plant growth, VIII, Effects of defoliation and light on selected hosts, *New Phytol.* 80: 365-372.

Dehne, H.W. and Backhaus, G.F., 1986, The use of vesicular arbuscular mycorrhizal fungi in plant protection, I, Inoculation production, *Z. Pflkrankn pflschutz* 93: 415-425.

Dixon, R.K., Mukerji, K.G., Chamola, B.P. and Kaushik., A., 1997, Vesicular arbuscular mycorrhizal symbiosis in relation to afforestation in arid lands, *Ann. For.* 5(1): 1-9.

Duckmanton, L. and Wider, P., 1994, Effect of ozone on the development of vesicular arbuscular mycorrhiza in sugar maple saplings, *Mycologia* 86: 181-186.

Entry, J.A., Reeves, D.W., Mudd, E., Lee, W.J., Guertal, E. and Raper, R.L., 1996, Influence of compaction from wheel traffic and tillage on arbuscular mycorrhizal infection and nutrient uptake by *Zea mays*, *Pl. Soil* 180: 139-146.

Faber, B.A., Zasoki, R.J., Burau, R.G. and Uriu, K., 1990, Zinc uptake by corn as affected by vesicular arbuscular mycorrhiza, *Pl.Soil* 129: 121-130.

Fitter, A.H., 1991, Costs and benefits of mycorrhizas, implications for functioning under natural conditions, *Experimentia* 47: 350-355.

Furlan, V. and Fortin, J.A., 1973, Formation of endomycorrhizae by *Endogone calospora* on *Allium cepa* under three temperature regimes, *Nat. Canad.* 100: 467-477.

Furlan, V. and Fortin, J.A., 1977, Effects of light intensity on the formation of vesicular arbuscular mycorrhizas on *Allium cepa* by *Gigaspora calospora*, *New Phytol.* 79: 335-340.

Garcia-Garrido, J.M. and Ocampo, J.A., 1989, Effect of VA mycorrhizal infection of tomato on damage caused by *Pseudomonas syringae*, *Soil Biol. Biochem.* 21: 165-167.

Gerdemann, J.W., 1968, Vesicular arbuscular mycorrhizae and plant growth, *Ann. Rev. Phytopathol.* 6: 397-418.

Gerdemann, J.W., 1975, Vesicular arbuscular mycorrhizae, in: *The Development and Function of Roots* (J.G. Torrey and D.T. Clarkson, eds.), Academic Press, London.

Gianinazzi, S., Gianinazzi-Pearson, V. and Trouvelot, A., 1986, One peuton attendre des mycorrhizas dans la production des arbres fruiters, *Fruits* 41: 553-556.

Gianinazzi, S., Trouvelot, A. and Gianinazzi-Pearson, V., 1989, Conceptual approaches in agriculture for the rational use of VA endomycorrhizae in agriculture: possibilities and limitations, *Ag. Eco. Environ.* 29:153-161.

Gianinazzi, S., Trouvelot, A. and Gianinazzi-Pearson, V., 1990, Role and use of mycorrhizas in horticultural crop production, Plenary lecture at the XXIII International Horticultural Congress, I.S.H.S. Italy.

Giri, B., 1997, VAM colonization in two MPTS, *M.Phil Thesis*, Department of Botany, University of Delhi, India.

Harley, J.L. and Smith, S.E., 1983, Mycorrhizal symbiosis, Academic Press, New York.

Hayman, D.S. 1975, The occurrence of mycorrhiza in crops as affected by soil fertility, in: *Endomycorrhizas* (F.E.Sanders, B. Mosse and P.B. Thinker, eds.), Academic Press London. pp. 495-509.

Hayman, D.S., 1982, Influence of soils and fertility on activity and survival of vesicular arbuscular mycorrhizal fungi, *Phytopathol.* 72: 1119-1125.

Heijne, B., Dueck, Th.A, Van dereerden, L.J. and Heil, G.W., 1994, Effects of atmospheric ammonia and ammonium sulphate on vesicular arbuscular mycorrhizal colonization in three heathland species, *New*

Phytol., 127: 85-696.

Hoffland, E., Findenegg, G.R. and Nelemans, J.A., 1989, Solubilization of rock phosphate by rape, II, Local root exudation of organic acids as a response to P starvation, *Pl. Soil* 113: 169-176.

Hwang, S.F., 1992, Effects of vesicular arbuscular mycorrhizal fungi on the development of *Verticillium* and *Fusarium* wilts of alfalfa, *Pl. Dis.* 76: 239-243.

Jagpal, R. and Mukerji, K.G., 1991a, Reforestation in waste lands using vesicular arbuscular mycorrhizae, in: *Recent Development in Tree Plantations of Humid/Subhumid Tropics of Asia* (S.A. Abob, P.Md. Tahir, L.M. Tsai, N.A. Shukar, A.S. Sajap and D. Manikam, eds.), Univ. Pertanian Malaysis, Selonger, Malaysia pp. 488-494.

Jagpal, R. and Mukerji, K.G., 1991b, VAM fungi in reforestation, in: *Plant Roots and their Environment* (B.L. McMichael and H. Pearson, eds.), Elsevier, Amsterdam pp. 309-313.

Jagpal, R., Sharma, P.D. and Mukerji, K.G., 1988, Distribution of VAM in relation to pollution, *Proc. 1st Asian Conf. Mycorrhizae, Madras, India,* 48-49.

Jasper, D.A., Abbott, L.K. and Robson, A.D., 1989, Hyphae of vesicular arbuscular mycorrhizal fungi maintain infectivity in dry soil, except when the soil is disturbed, *New Phytol.* 112: 101-107.

Jalali, B.L. and Jalali, I., 1991, Mycorrhiza in plant disease control, in: *Handbook of Applied Mycology, Vol. 1. Soil & Plants* (D.K. Arora, B. Rai, K.G. Mukerji and G.R. Krudsen, eds.), Marcel Dekker Inc., New York, pp. 131-154.

Kaushik, J.C. and Kaushik, N., 1995, Interaction between vesicular arbuscular mycorrhizal fungi and *Rhizobium* and their influence on Albizia lebbek, in: *Mycorrhizae: Biofertilizer for Future,* (A. Adholya and S. Singh, eds.), Proc. IInd Nat. Conf. Mycorrhiza, New Delhi.

Kendrick, W.B. and Berch, S.M., 1985, in: *Comprehensive Biotechnology, Vol. III,* (M. Mooyoung, ed.), Pergamon Press, Oxford, pp. 109-152.

Kothari, S.K., Marchner, H. and Romheld, V., 1990, Direct and indirect effects of VAmycorrhizal fungi and rhizosphere microorganisms on acquisition of mineral nutrients by maize *(Zea mays* L.) in calcareous soil, *New Phytol.* 116: 637-645.

Kothari, S.K., Marshner, H. and Romheld V., 1991, Contribution of VAmycorrhizal hyphae in acquisition of P and Zn by maize grown in a calcareous soil, *Pl. Soil* 131: 177-185.

Kucey, R.M.N. and Janzen, H.H., 1987, Effects of VAM and reduced nutrient availability on growth, phosphorus and micronutrient uptake of wheat and field bean under greenhouse conditions, *Pl. Soil* 104: 71-78.

Lambert, D.H., Baker, D.E. and Cole, H., 1979, The role of mycorrhizae in the interactions of phosphorus with zinc, copper and other elements, *Soil Sci. Soc. Am. J.* 43: 976-986.

Lambert, D.H. and Weidensaul, T.C., 1991, Element uptake by mycorrhizal soybean from sewage-sludge treated soil, *Soil Sci. Soc. Am. J.* 55: 393-398.

Lewis., D.G. and Quirk, J.M., 1965, Diffusion of phosphate to wheat plant roots, Nature (London) 205: 765-766.

Lewis, J.D. and Koide, R.T., 1990, Phosphorus supply, mycorrhizal infection and plant offspring vigour, *Functional Ecol.* 4: 695-702.

Li, X.L., Marschner, H. and George, E., 1991, Acquisition of phosphorus and copper by VA mycorrhizal hyphae and root to shoot transport in white clover, *Pl. Soil* 136: 49-57.

Lopes, E.S., Oliveira, E. and Neptune, A.M.L., 1980, Efeito de especies de mycorrhizas vesicular arbuscular em siratro *(Macroplilium atropurpureum), Bargantia* 39: 241-245.

Manjunath, A. and Habte, M., 1988, Development of vesicular arbuscular mycorrhizal infection and the uptake of immobile nutrients in *Leucaena leucocephala, Pl. Soil* 106: 97-103.

Martin, F., Canet, D., Rolin, D., Muchal, J.P. and Lahrer, F., 1983, Phosphorus - 31 nuclear magnetic resonance study of polyphosphate metabolism in intact ectomycorrhizal fungi, *Pl. Soil* 71: 469-476.

Menge, J.A., 1983, Utilization of vesicular arbuscular mycorrhizal fungi in agriculture, *Can. J. Bot.* 61: 1015-1024.

Menge, J.A., Jarrell, W.M., Labmauskas, C.K., Ojala, J.C., Huszar, C., Johnson, E.L.U. and Sibert, D., 1982, Predicting mycorrhizal dependency of troyer citrange on *Glomus fasciculatus* in California citrus soils and nursery mixes, *Soil Sci. Soc. Am. J.* 46: 762-768.

Morandi, D., 1996, Occurrence of phytoalexins and phenolic compounds in endomycorrhizal interactions and their potential role in biological control, *Pl. Soil* 185: 241-251.

Mosse, B., 1973, Advances in the study of vesicular arbuscular mycorrhiza, *Ann. Rev. Phytopathol.* 11: 171-196.

Mosse, B., Hayman, D.S. and Arnold, D.J., 1973, Plant growth responses to vesicular arbuscular mycorrhiza, V, Phosphate uptake by three plant species from P-deficient soils labelled with ^{32}P, *New Phytol.* 72: 809-815.

Morton, J.B. and Benny, G.L., 1990, Revised classification of arbuscular mycorrhizal fungi (Zygomycetes): a new order Glomales, two suborders Glomineae and Gigasporineae and two new families Acaulosporaceae and Gigasporaceae with an amendation of Glomaceae, *Mycotaxon* 37: 471-491.

Mukerji, K.G., Chamola, B.P., Kaushik, A., Sarwar, N. and Dixon, R., 1996, Vesicular arbuscular mycorrhiza: Potential biofertilizer for nursery raised multipurpose tree species in tropical soils, *Ann. For.* 4(1): 12-20.

Mukerji, K.G., Sabharwal, A., Kochar, B. and Ardey, J., 1984, Vesicular arbuscular mycorrhiza: Concepts and advances, in: *Progress in Microbial Ecology* (K.G. Mukerji, V.P. Agnihotri and R.P. Singh, eds.), Print

House, India, pp. 489-525.

Nadian, H.. Smith, S.E., Alston, A.M., and Murray, R.S., 1996, The effect of soil compaction on growth and P uptake by *Trifolium subterraneum:* Interactions with mycorrhizal colonization, *Pl. Soil* 182: 39-49.

Neeraj, A., Shanker, A. and Verma, A., 1991, Occurrence of VA mycorrhizae within Indian semi desert arid soils, *Biol. Fertil. Soils* 11: 140-144.

Nelson, C.E. and Safir, G.R., 1982, Increased drought resistance of mycorrhizal onion plants caused by improved phosphorus nutrition, *Planta* 154: 407-413.

Pearson, J.N., Smith, S.E. and Smith, F.A., 1991, Effect of photon irradiance on the development and activity of VA mycorrhizal infection in *Allium porrum, Mycol. Res.* 95: 741-746.

Pfleger, F.L., Steward, E.L. and Moyel, R.K., 1994, Role of VAM fungi in mine land revegetation, in: *Mycorrhizae and Plant Health* (F.L. Pfleger and R.G. Lindermann, eds.), p. 47.

Rhodes and Gerdemann, J.W., 1975, Phosphate uptake zones of mycorrhizal and non-mycorrhizal onions, *New Phytol.* 75: 555-561.

Safir, S.R. 1984, Respuesta de las plantas forrajeras tropicales a la application de roca fosforica Y micorrhiza en un oxidol no esterilizado, in: *La Roca Fosforica. Memorias de la Conferencia Latinoamericiana de Roca Forsforica,* Oct, 1983 (V. Riealdi and S. Escalera, eds.), Cochabamga, Bolivia, pp. 300-327.

Safir, S.R., 1986, Vesicular arbuscular mycorrhizae in tropical forage species as influenced by season, soil texture, fertilizer, host species and ecotypes, *Agnew Bot.* 60: 125-139.

Safir, G.R., Boyer, J.S. and Gerdemann, J.W., 1971, Mycorrhizal enhancement of water transport in soybean, *Science* 172: 581-395.

Salinas, J.G., Sanz, J.I. and Sieverding, E., 1985, Importance of VA mycorrhizae for phosphorus supply to pasture plants in tropical oxisals, *Pl. Soil* 84: 347-360.

Sengupta, A. and Chaudhri, S., 1995, Effect of dual inoculation of rhizobium and mycorrhizae on growth response of *Sesbania grandiflora* L. in coastal saline sand dune soil, *Ind. J. Forestry* 18(1): 35-37.

Smith, S.E., 1980, Mycorrhiza of autotrophic higher plants, *Biol. Rev.* 55: 475-510.

Sharma, A.K., Johri, B.N. and Gianinazzi, S., 1992, Vesicular arbuscular mycorrhizae in relation to plant disease, *World J. Microbiol. and Biotech.* 8: 559-563.

Shukla, B.M., Khare, M.N. and Vyas, S.C., 1996, Role of vesicular arbuscular mycorrhizal fungi in controlling soil borne disease, in: *Advances in Botany* (K.G. Mukerji, B. Mathur, BP. Chamola and P. Chitralekha, eds.), Ashish Publishing House, Delhi.

Srivastava, D., Kapoor, R., Srivastava, S.K. and Mukerji, K.G., 1996, Vesicular arbuscular mycorrhiza - an overview, in: *Concepts in Mycorrhizal Research* (K.G. Mukerji, ed.), Kluwer Academic Publishers, pp. 1-39.

Stribley, D.P., Tinker, P.B. and Snellgrove, R.C., 1980, Effects of vesicular arbuscular mycorrhizal fungi on the relations of plant growth, internal phosphorus concentration and soil phosphate analyses, *J. Soil Sci.* 31: 655-672.

Thompson, B.D., Robson, A.D. and Abbott, L.K., 1996, Effects of phosphorus on the formation of mycorrhizae by *Gigaspora calospora* and *Glomus fasciculatum* in relation to root carbohydrates, *New Phytol.* 103: 751-765.

Thompson, B.D., Robson, A.D. and Abbott, L.K., 1990, Mycorrhizas formed by *Gigaspora calospora* and *Glomus fasciculatum* on subterranean clover in relation to soluble carbohydrate concentrations in roots, *New Phytol.* 114: 217-225.

Thompson, J.P., 1987, Decline in vesicular-arbuscular mycorrhizae in long fallow disorder of field crops and its expression in phosphorus deficiency of sunflower, *Aust. J. Agric. Res.* 38: 847-867.

Thompson, J.P., 1990, Soil sterilization methods to show VA mycorrhizae aid P and Zn nutrition of wheat in vertisols, *Soil Biol Biochem.* 22: 229-240.

Thompson, J.P., 1994, Inoculation with vesicular arbuscular mycorrhizal fungi from cropped soil overcomes long fallow disorder of linseed *(Linum usitatissimum* L.) by improving P and Zn uptake, *Soil. Biol. Biochem.* 26: 1133-1143.

Tinker, P.B., 1978, Effect of vesicular arbuscular mycorrhiza on plant nutrition and plant growth, *Physiol. Veg.* 16: 743-749.

Tommerup, I.C. and Briggs, G.C., 1981, Influence of agricultural chemicals on germination of vesicular arbuscular endophyte spores, *Trans. Brit. Mycol. Soc.* 76: 326-328.

Trappe, J.M., Molina, R. and Castellano, M., 1984, Reaction of mycorrhizal fungi and mycorrhizal formation to pesticides, *Ann. Rev. Phytopathol.* 22: 331-359.

Trotta, A.G., Varese, C., Gnavi, E., Fusconi, A., Sampo, S. and Berta, E., 1996, Interactions between the soil borne root pathogen *Phytophthora bicotiance* var. *parasitica* and the arbuscular mycorrhizal fungus *Glomus mosseae* in tomato plants, *Pl. Soil* 185: 199-209.

Wellings, N.P. and Thompson, J.P., 1991, Effects of VAM and P fertilizers rate on Zn fertilizers requirements of linseed, in: *Proc. Second Asian Conference on Mycorrhiza* (I. Soerianegara, Supriyanto, eds.), SEAMEO-BIOTROP, Indonesio, pp. 143-152.

MYCORRHIZAL ALLELOPATHY IN *TRIGONELLA FOENUM GRACEUM*

Rajni Gupta and K.G. Mukerji

Applied Mycology Laboratory
Department of Botany
Delhi University
Delhi-100 007
India

INTRODUCTION

Plant root exudates are important as they influence soil nutrient availability both directly and indirectly through the activity of the microbial biomass (Helal and Sauerbeck, 1984; Jones et al., 1994). Hiltner (1904) coined the term 'rhizosphere' to describe the zone of soil surrounding the roots of legumes in relation to symbiotic nitrogen-fixing bacteria. For the microbial biomass in the rhizosphere however exudates, in addition to secretions, lysates and gases, released from the plant roots provide an available substrate to support growth (Bowen and Rovira, 1991).

For several decades agricultural production has come to depend on a wide array of chemical inputs which have been developed to control a complex of weeds, insects and other disease-causing organisms. The use of these chemicals is mostly evaluated in terms of their efficacy in controlling pests, lowering yield losses and increasing output and profitability of the cropping system. However, there is growing evidence to indicate that the continuous use of these chemicals threatens the ability of agriculture to sustain production in the future. The excessive and improper use of synthetic agro-chemicals has created a number of different problems (Jackson et al., 1984). The emphasis in agriculture has been shifting recently from an objective of maximising short-term production and profit to a perspective that also considers the ability of the agricultural system to maintain productive capacity in the long term. Sustainable agriculture has been defined as agriculture that is simultaneously ecologically sound, economically viable and socially responsible.

Allelopathy

Molish (1937) coined the term 'allelopathy' to describe biochemical interactions between all types of plants including microorganisms. He defined allelopathy as any direct or indirect harmful effect of one plant on another through production of chemical compounds that escape into the environment.

Recent reviews on allelopathy have indicated several potential applications of allelopathic plants or the chemicals produced by them in the maintenance of sustainable agriculture. Natural products play important roles in plant resistance to insects, nematodes and pathogens. Numerous phytotoxic secondary products have been isolated and identified from higher plants. Introduction of these compounds into the environment occurs by exudation of volatiles from living plants, leaching of water and soluble toxins from aerial plant parts or subterranean tissues or by release of toxins from non-living plant materials.

The exploitation of allelopathic crops and the chemicals they produce has been proposed as an additional strategy for weed suppression in several agro-ecosystems (Putnam and Defrank, 1983). Some allelopathic crop plants have already been used experimentally in weed control (Rice, 1984). Leather (1983) found one of the thirteen genotypes of the cultivated sunflower to be strongly allelopathic to several weeds. In a five year field study with oats and sunflowers grown in rotation, the weed density was significantly less than in control plots with oats only. Putnam and Defrank (1979) tested the residues of several fall and spring planted crops with high levels of allelochemical production, and their use in weed control has been well demonstrated (Barnes and Putnam, 1982). The exploitation of crop plants with increased allelochemical production could limit the need for conventional herbicides for weed management.

Recently, there has been an increasing demand for natural pesticides for pest management. These natural pesticides are the secondary metabolites of plants and microorganisms. The family Meliaceae in particular has yielded the most promising chemicals with dramatic effects on the feeding, growth and development of insects. A potent insecticide, azadirachtin has been isolated from the leaves and fruits of the neem tree in India (*Azadirachta indica*) and from species of *Melia* in China (Jacobson, 1988).

Allelopathy is separate from competition, which involves the removal or reduction of some factors from the environment that are required by other plants sharing the habitat. Factors which may be reduced include water, minerals, food and light. Muller (1969) suggested the term 'interference' to refer to the overall influence of one plant or microorganism on another. Interference would thus encompass both allelopathy and competition. Evidence indicates that allelopathic compounds come from plants by volatilization, exudation from roots, leachates, and plants residues by rain or by decomposition of residues. Their toxic effects on the surrounding plants have been reported as stunting of growth, inhibition of primary roots, increase of secondary roots, inadequate nutrient absorption, chlorosis, premature leaf abscission, slow maturation, delay or failure of reproduction and inhibition of seed germination.

Microorganisms compete with each other for food and essential elements in the soil and around the rhizosphere (Baker, 1981). Vesicular arbuscular mycorrhizal fungi are widespread in occurrence and because of their potential for crop improvement have been investigated extensively (Mukerji, 1995). Most soils contain spores or propagules of VAM fungi which affect the rhizosphere mycoflora of the plant (Bansal and Mukerji, 1994a,b; Mukerji et al., 1997). In the present investigation efforts have been made to study the allelopathic effect of VAM on the rhizosphere of *Trigonella foenum graceum*. *T. foenum graceum* belongs to the family Leguminaceae and grows as a weed in gardens, fields, pastures and road sides. It is also a good fodder for animals. It is an annual erect weed, with toothed leaflets, and pale yellow flowers. Pods are 5-8 cm long, with a long persistent beak. Dried leaves of the plant have a peculiar odour.

For this purpose, the *T. foenum graceum* was collected randomly from two different sites: (i) growing as a weed in the Botanical garden and (ii) growing in association with VAM fungi (e.g. *Glomus macrocarpum*) in the flower beds. Association of vesicular arbuscular mycorrhizal fungi in the roots of this weed was studied following the method of Philips and

Hayman (1970). Percent VAM colonization was calculated using Nicolson's simple formula (1905).

% colonization =   No. of segments colonized with VAM   x 100
                      Total no. of segments observed

    The VAM fungal spores from the rhizosphere soil of each plant were isolated by Gerdemann and Nicolson's (1963) wet sieving and decanting technique (Table 1). Spores were mounted in lactophenol and identified using standard keys (Hall, 1983; Schenck and Perez, 1990) (Figure 1). Soil dilution and plate count method (Timonin, 1941) was used for qualitative and quantitative analysis of the microflora in the rhizosphere (Table 2). For this purpose 100 g each of soil samples were collected from different sites and dilutions were prepared with D.W., i.e. 10^{-1}, 10^{-2}, 10^{-3} and 10^{-4}. Five replicates were prepared for each dilution then the plates were incubated at $25 \pm 2°C$ for seven days. After this period the fungi present in the petriplates were identified.

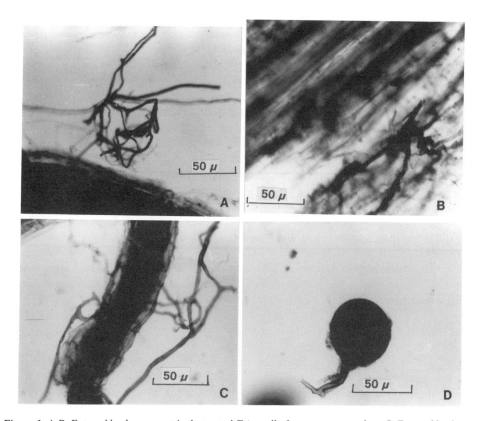

Figure 1. A-B: External hyphae present in the treated *Trigonella foenum graceum* plant. C: External hyphae entering in the root of control *Trigonella-foenum graceum* plant. D: Spore of *Glomus macrocarpum*.

    From Tables 1 and 2 it is clear that mycoflora in the rhizosphere varied in the treated and control plants. *Trigonella* is a good partner for VAM fungi and the control plants were densely associated with VAM fungi as extramatrical hyphae, arbuscules, and about 80 spores in 100 g of rhizosphere soil. By comparison, however, in the *Trigonella foenum graceum* plants inoculated with *Glomus macrocarpum*, the roots had 100% arbuscules, densely present internal hyphae and about 20 vesicles in 1 cm root portions. Furthermore, the number of spores present was considerably higher at 450 spores in 100 g of rhizosphere soil. From Table 2 it is clear that *Trichoderma viride* was present frequently in the *Trigonella foenum graceum* inoculated with *Glomus macrocarpum*. Relative abundance increased with increase in the concentration i.e. 0.10, 0.17, 0.30 and 0.49 at 1 x 10^{-1}, 1 x 10^{-2}, 1 x 10^{-3} and 1 x 10^{-4} concentrations respectively. In the control plants *T. viride* was totally absent at the 1 x 10^{-1} and 1 x 10^{-2} concentrations but was present at the 1 x 10^{-3} and 1 x 10^{-4} concentrations.

Cephalosporium acremonium was also present in the rhizosphere of the treated plants and the relative abundance also increased with increase in concentration, i.e. 0.20, 0.25 and 0.35 at 10^{-2}, 10^{-3} and 10^{-4} concentrations respectively, whilst it was totally absent in the control plants. Various species of *Penicillium* were absent in the treated plants, while they were frequently present in the control plants. This was also the case with various species of *Aspergillus* which was abundantly present in the control plants and absent in the treated plants.

Trichoderma viride and *Verticillium albo atrum* were frequently present in the rhizosphere of treated *Trigonella foenum graceum*. These species are good biocontrol agents, inhibiting the growth of other fungi and bacteria. Along with mycorrhizal fungi, plant exudates also affect the flora of the rhizosphere. Mallik and Zhu (1995) studied the effect of leaf extract of kalmia with mycorrhizal fungi on the germination of Black Spruce seeds. They also reported the failure and reduction in mycorrhizal formation by some isolates due to the toxicity of the extracts affecting mycorrhizal colonization by restricting host fungus contact. Blum (1995) studied the plant microbe soil system in relation to root exudates. Phenols present in the root also affect the presence of microflora in the soil.

External hyphae of VAM fungi provide a physical or nutritional substrate for bacteria. Analysis of the rhizosphere soil of VAM (*Glomus fasciculatum*) and non-VAM Tomato plants (*Lycopersicon esculentum*) by Bagyaraj and Menge (1978) showed that greater populations of bacteria and actinomycetes occurred in the mycorrhizosphere, compared with the non-inoculated control. The microbial effect was clearly time-dependent and dynamic, changing as the plants developed. McAllister et al. (1995), while studying the interaction between saprophytic fungi (*Aspergillus niger, Trichoderma koningii, Fusarium solani* and *Glomus mossae*), observed that *G. mossae* decreased the fungal population through its effect on the plant. Mycorrhizal association significantly changes the physiology and morphology of roots and plants in general which in turn causes a new microbial equilibrium to be established. These changes presumably involve the same type of organisms as were involved in the rhizosphere before formation of mycorrhizae but quantitative changes occur within these types as a result of direct metabolic interaction with the mycorrhizal fungal hyphae or spores, or the indirect effect mediated by the host (Lindermann, 1988).

Microbial Substrates in the Rhizosphere

Rovira et al. (1983) listed five main sources of rhizosphere substrates ranging from exudates and secretions from intact cells to mucilage of plant origin to lysates from senescing cells. Microbial growth in the rhizosphere is stimulated by the continual input of readily assimilable organic substrates from the root. In ecological terms, the substrate from the roots is the product of photosynthesis and is thus primary productivity and the utilization of these organic substrates by rhizosphere microorganisms results in secondary productivity. In turn, these microorganisms can influence the plant as primary producer. Lynch (1990) classified the four groups of chemical components coming from roots depending on their mode of arrival.

Table 1. Presence of different structure of VAM fungi in *Trigonella foenum graceum* with different treatment.

Treatment	Ex. hyphae	Int. hyphae	Appressoria formation	Arbuscules	Vesicles Number	No. of spores per 100 g rhizosphere
Control	DP	LP	A	40%	02	80
Glomus macrocarpum	LP	DP	P	100%	20	450

A	=	Absent
P	=	Present
DP	=	Densely present
LP	=	Less present

Table 2. Relative abundance and percentage frequency of various fungi present in the rhizosphere of both treated and control plants.

Fungal species	Mycorrhizal (Concentration of Rhizosphere Soil)								Non-mycorrhizal (Concentration of Rhizosphere Soil)							
	10^{-1}		10^{-2}		10^{-3}		10^{-4}		10^{-1}		10^{-2}		10^{-3}		10^{-4}	
	RA	F	RA	F	RA	F	RA	F	RA	F	RA	F	RA	F	RA	F
Aspergillus flavus	-	0	0.08	100	-	-	-	-	0.07	100	0.09	100	0.09	100	-	-
A. niger	0.06	100	0.09	100	0.10	100	0.10	100	0.49	100	0.11	100	0.04	100	-	-
A. paradoxsus	-	-	-	-	-	-	-	-	0.06	100	0.04	100	-	-	-	-
A. sydowii	-	-	-	-	-	-	-	-	-	-	-	-	0.11	100	-	-
Alternaria alternata	-	-	-	-	0.125	100	0.34	100	-	-	-	-	-	-	-	-
Cladosporium herbarum	0.23	100	0.24	100	0.25	100	-	-	0.14	100	0.22	100	0.30	100	-	-
Cephalosporium acremonium	-	-	-	-	-	-	-	-	-	-	0.20	100	0.25	100	0.35	100
Mucor racemosus	0.13	100	0.13	100	0.10	100	-	-	0.15	100	0.18	100	0.09	100	-	-
Penicillium nigricans	-	-	-	-	-	-	0.13	25	0.03	50	0.04	50	-	-	-	-
P. funiculosum	-	-	-	-	-	-	-	-	0.03	50	0.09	50	0.09	100	0.05	50
Trichoderma viride	0.10	100	0.17	50	0.30	100	0.49	100	-	-	-	-	0.23	50	0.29	50
Xanthomonas	0.13	100	-	-	-	-	-	-	-	-	-	-	-	-	-	-

RA - Relative abundance
F - Percentage frequency

(i) Water soluble exudates, such as sugars, amino acids, organic acids, hormones and vitamins which leak out from the root without the involvement of metabolic energy.

(ii) Secretions, such as polymeric carbohydrates and enzymes which depend upon metabolic processes for their release.

(iii) Lysates, released when cells autolyse including cell walls and, with time, whole roots.

(iv) Gases, such as ethylene and carbon dioxide.

These groupings cover all stages of plant growth and development with the balance of these various processes changing with the age of plant. This can range from the release of simple sugars during membranae re-organization of the embryo during seed germination (Hale and Moore, 1979) through to root cortical death in the mature plant followed by whole root senescence. Addition of microorganisms generally increases the carbon loss. These types of rhizodeposition have been analysed quantitatively and qualitatively by chemical analytical techniques (Lee and Gaskins, 1982).

Another aspect of rhizodeposition concerns the involvement of symbiotic root inhabitants such as rhizobium and mycorrhizae. Legumes benefit from *Rhizobium* infection by improved nitrogen nutrition from nitrogen-fixation and the majority of plants benefit from mycorrhizae by improved mineral nutrition, particularly of phosphorus, and by drought tolerance. In return the heterotrophs gain organic material from the endorhizosphere. Studies on vesicular arbuscular mycorrhizal infections indicate that conditions which increase permeability of roots or exudation lead to increased infection (Graham et al., 1981).

Once synthesized, the cost of maintaining VA mycorrhizae in terms of extra carbohydrate moved to the roots is approximately 6-11% net fixed carbon (Koch and Johnson, 1984). In experiments with *Vicia faba* plants, it was shown that nodulated root systems of VA mycorrhizal plants fixed more nitrogen than nodulated root systems of non-mycorrhizal plants. The plants may compensate for the cost of these symbionts by increasing rates of photosynthesis or by changing morphological characteristics to increase specific leaf area (Harris et al., 1985). Greater amounts of carbon are transferred to roots with ectomycorrhizae compared with non-mycorrhizal plants (12-22 % total fixed carbon in Pine) and this is also partially compensated for by increased rate of photosynthesis (Ried et al., 1983). Such large inputs of carbon to the roots may seem wasteful but in ecological terms they may enable plants to grow where they otherwise could not. Photosynthetic compensation can occur and in situations where lack of nutrients or environmental conditions limit growth, plants may have excess source potential (Herold, 1980).

Uptake and Translocation of Nutrients by Roots

Rhizosphere microorganisms influence the uptake and translocation of plant nutrients by the roots. Barber (1966, 1967) found that microbes competed with the roots for phosphate, resulting in reduced transport to the tops of barley seedlings. Krasilinikov (1958) and Bowen and Rovira (1966, 1968) reported that more phosphate was taken up, incorporated into DNA phosphate and RNA phosphate and translocated to the tops by non-sterile wheat roots than by sterile wheat roots. Barber and Martin (1976) found that decreased or increased uptake by non-sterile roots depends upon age, nutrition and species of plants, the origin of microflora, the concentrations of phosphate and the duration of the experiment. The rhizosphere microflora has also been demonstrated to increase the uptake of manganese and potassium by cereal roots (Barber and Lee, 1974; Williamson and Wyn Jones, 1973). These effects of microorganisms on root growth, root morphology, root hair development, uptake and translocation of nutrients to plant tops highlight the importance of the rhizosphere microflora in plant nutrition and the need for multidisciplinary research.

Plant Growth Promotion

Reports from the USSR of large increases in yield from the treatment of seed with *Azotobacter chroococcum* led to the assumption that these increases were due to non-symbiotic nitrogen-fixation in the rhizosphere. By 1958, 10 million hectares were sown with seed treated either with *Azotobacter* or *Phosphobacteriam*. The results were positive in greenhouse crops: the wheat and tomato responded to seed treatment by flowering earlier and increasing yields (Rovira, 1963). Elklund (1970) showed that plant growth-promoting compounds, probably gibberellic acid and indole acetic acid, are produced by a range of bacteria including species of *Azotobacter, Bacillus* and *Pseudomonas* and are responsible for better development. Schroth and coworkers have demonstrated the widespread nature of plant growth-promoting rhizobacteria (PGPR's) (Kloepper et al., 1980), emergence promoting rhizobacteria and their effects on seed germination, plant growth and yield (Kloepper et al., 1986). Mycophagous protozoa, nematodes and microarthropods have been demonstrated to interact with and in some cases control root pathogens, e.g. *Rhizoctonia solani* (Curl, 1979) and *Gaeumannomyces graminus* (Chakraborty and Warcup, 1983).

In agricultural soil, nutrient availability is a limiting factor for microbial growth and activity. Consequently, most microorganisms, including pathogenic fungi, exist in a state of exogenous dormancy of fungistasis (Lockwood, 1977). Propagules of soil-borne fungi that arise from fungal thalli developing on and in infected plant tissues during pathogenesis, as well as those propagules released into soil upon decomposition of plant residues, are susceptible to fungistasis. However fungistasis suppression can be readily overcome once propagules are supplied with appropriate organic and inorganic stimuli (Lockwood, 1988).

The stimulants most commonly introduced into the soil environment that alleviate fungistasis are components of seed and root exudates released during seed germination and root development. The majority of host-parasite interactions in soil do not occur without the release of such stimuli (Curl and Truelove, 1986). Therefore, knowledge of the molecules involved in the activation of quiescent fungal propagules by seed and root exudates and understanding of the initiation and control of seed and root infections by soil-borne pathogens is of prime importance. Inoculation of decomposing roots with *Trichoderma viride* has been found to reduce the release of their allelopathic chemicals (Burgosleon et al., 1980). There are several studies which indicate that the release of allelochemicals from higher plants is inhibitory to root rot pathogens. Root exudates of some plants favour the establishment of rhizosphere microorganisms which are antibiotic to the soil-borne pathogens (Jackson 1965). P-courmaric, syringic and ferulic acids found in the soil of *Alnus rubra* are inhibitory to *Poria weirii*, the root rot pathogen. Such non-host plants which release adequate quantities of allelochemicals may be exploited as inter-cropping plants for the control of plant diseases.

Most of the allelochemicals involved in the resistance of hosts to pathogens fall in the category of phytoncides. These are generally divided into two categories: (i) secondary compounds generally present in the host but which may increase subsequent to infection; and (ii) phytoalexins, new compounds formed only after infection (Kuc, 1966; Ingham, 1972). Such compounds render the plants resistant to disease.

Using modern biotechnological techniques the allelochemicals of the resistant varieties should be incorporated into the susceptible ones through genetic manipulation and this will certainly decrease the demand for chemical pesticides.

CONCLUSIONS

Formation of mycorrhizae is the rule, not the exception. When the symbiosis is established, very significant morphological and physiological changes take place in the plant, and significant changes occur in the mycorrhizosphere soil. The microbial composition in the soil reaches a new equilibrium as a result of the selective pressure induced by mycorrhizae. For this reason, mycorrhizae should be considered a fundamental component in rhizosphere studies. Mycorrhizal fungi are present in most soils, but their establishment may be delayed due to cultural practices or to distribution of scarce inoculum in the soil. Other beneficial microbes could be added to the inoculum to enhance further the growth and health of plants due to increased nitrogen fixation, biological control of disease, nutrient acquisition potential and drought tolerance.

REFERENCES

Bagyaraj, D.J. and Menge, J.A., 1978, Interactions between a VA mycorrhizae and *Azotobacter* and their effects on the rhizosphere microflora and plant growth, *New Phytol.* 80 : 567-573.

Baker, R., 1981, Eradication of plant pathogens by adding organic amendments to soil in: *Handbook of Pest Management in Agriculture, Vol.2* (D. Pimental, ed.), pp.137-157, Chemical Rubber Company Press, Boca Raton, Florida.

Bansal, M. and Mukerji, K.G., 1994a, Efficacy of root litter as a biofertilizer, *Biol. Fertil. Soils* 18 : 228-230.

Bansal, M. and Mukerji, K.G., 1994b, Positive correlation between VAM induced changes in root exudation and mycorrhizosphere mycoflora, *Mycorrhizae* 5 : 39-44.

Barber, D.A. and Martin, J.K., 1976, The release of organic substances by cereal roots into the soil, *New Phytol.* 76 : 69-80.

Barber, D.A., 1967, Microorganisms and the inorganic nutrition of higher plants, *Ann. Rev. Pl. Physiol.* 19: 71-88.

Barber, D.A. and Lee, R.B., 1974, The effect of microorganisms on the absorption of manganese by plants, *New Phytol.* 73: 97-106.

Barber, D.A., 1966, Effect of microorganisms on nutrient absorption by plants, *Nature* (London) 212: 638-640.

Barnes, J.P. and Putnam, A.R., 1982, Annual Meeting, *Weed Sci. Soc. America*, pp. 67.

Blum, Udo, 1995, The value of model plant microbe soil systems for understanding processes associated with allelopathic interaction, in: *Allelopathy: Organisms, Processes and Applications* (K.M.M. Inderjit, F. Dakshini and A. Einhelling, eds.), pp.39-58, American Chemical Society, Washington D.C.

Bowen, G.D. and Rovira, A.D., 1968, *The Influence of Microorganisms on Growth* (W.J. Whittington, ed.), pp.170-201, Butterworths, London, England.

Bowen, G.D. and Rovira, A.D., 1966, Microbial factor in short term phosphate uptake studies with plant roots, *Nature* 211: 666-668.

Bowen, G.D. and Rovira, A.D., 1991, The rhizosphere of the hidden half, in: *Plant Roots, the Hidden Half* (Y. Waisel, A. Eshel and U. Kafkafi, eds.), pp. 641-669, Marcel Dekker, New York, USA.

Burgosleon, W., Gangy, F., Nicou, R., Chopart, J.L. and Dommergues, Y., 1980, in: *Recent Advances in Phytochemistry - Chemically Mediated Interactions between Plants and Other Organisms,* (Cooper-Driver, Swain and Conn), Plenum Press, New York, pp. 546.

Chakraborty, S. and Warcup, J.H., 1983, Soil amoebae and saprophytic survival of *Gaeumannomyces graminis* var. *tritici* in a suppressive pasture soil, *Soil. Biol. Biochem.* 14: 241-245.

Curl, E.A., 1979, Effects of mycophagous collembola on *Rhizoctonia* and cotton seedling disease, in: *Soil Borne Plant Pathogens* (B. Schippens and W. Gams, eds.), pp.253-269, Academic Press, London, England.

Curl, E.A. and Truelove, B., 1986, *The Rhizosphere,* Springer-Verlag, Berlin, pp. 288.

Elklund, E., 1970, Secondary effects of some *Pseudomonas* in the rhizoplane of peat grown cucumber plants, *Acta. Agric. Scand. Suppt.* 17, Stockholm, pp. 57.

Gerdemann, J.W. and Nicolson, T.H., 1963, Spores of mycorrhizal endogone species extracted from soil by wet sieving and decanting, *Trans. Br. Mycol. Soc.* 46: 235-244.

Graham, J.T., Leonard, R.T. and Menge, J.A., 1981, Membrane mediated decrease in root exudation responsible for phosphorus inhibition of vesicular arbuscular mycorrhiza formation, *Plant Physiol.* 68: 548-552.

Hale, M.G. and Moore, L.D., 1979, Factors affecting root exudation, II. 1970-1978. *Adv. Agron.* 31: 93-124.

Hall, I.R., 1983, A summary of the features of endogonaceous taxa, *Tech. Rep. No.8.* Agricultural Research Centre. Mosigel, New Zealand.

Harris, D., Pacovsky, R.S. and Paul, E.A. 1985, Carbon economy of soybean. *Rhizobium - Glomus* association, *New Phytol.* 101: 427-440.

Helal, H.M. and Sauerbeck, D.R., 1984, Influence of plant roots on C and P metabolism in soil, *Plant and Soil* 76: 175-182.

Herold, A., 1980, Regulation of photosynthesis by sink activity - the missing link, *New Phytol.* 86: 131-144.

Hiltner, L., 1904, Uber neuere Erfahrungen and probleme auf dem Gebiet der Bodem bakteriologie und unter besonderer Berucksi chtigung der Grundungung and Brache, *Arb. Dtsch. Landwirt. Ges.* 98: 59-78.

Ingham, J.C., 1972, Phytoalexins and other natural products as factors, *Bot. Rev.* 38: 343-424.

Jackson, R.M., 1965, in: *Ecology of Soil Borne Plant Pathogens* (K.E. Baker and W.C. Synder , eds.), pp. 363-369, University of California Press, Berkeley.

Jackson, W., Berry, W. and Colman, B., 1984, *Meeting the Expectations of the Land*, North Point Press, Berkeley, pp. 247.

Jacobson, M. 1988, *Focus on Phytochemical Pesticides: The Neem Tree,* C.R.C. Press Inc., Boca Raton, Florida, pp. 178.

Jones, D.L., Edwards, A.C., Donachie, K. and Darrah, P.R., 1994, Role of proteinaceous amino acids released in root exudates in nutrient acquisition from the rhizosphere, *Plant and Soil* 158: 183-192.

Kloepper, J.W., Schroth, M.N. and Miller, T.D., 1980, Effects of rhizosphere colonization by plant growth promoting rhizobacteria on potato plant development and yield, *Phytopathol.* 709: 1078-1082.

Kloepper, J.W., Scher, F.M., Leliberte, M. and Tipping, B., 1986, Emergence promoting rhizobacteria. Description and implications for agriculture, in: *Iron, Siderophores and Plant Disease* (T.R. Swinburne, ed.), pp.155-164, Plenum, New York.

Koch, K.E. and Johnson, C.R., 1984, Photosynthate partitioning in split root citrus seedlings with mycorrhizal and non-mycorrhizal root systems, *Plant Physiol.* 75: 26-30.

Krasilnikov, N.A., 1958, *Soil Microorganisms and Higher Plants*, Acad. Sci. USSR, Moscow, 474 pp., Translated in 1961 for the NSF, USDA and the Israel Program for Scientific Translation.

Kuc, J., 1972, Phytoalexins, *Ann. Rev. Phytopathol.* 10: 207-232.

Leather, G.R., 1983, Sunflowers (*Helianthus annus*) are allelopathic to weeds, *Weed Sci* 31: 37-42.

Lee, K.J. and Gaskins, M.H., 1982, Increased root exudation of ^{14}C compounds by *Sorghum* seedlings inoculated with nitrogen fixing bacteria, *Plant and Soil* 69: 391-399.

Linderman, R.G., 1988, Mycorrhizal interactions with the rhizosphere microflora: the mycorrhizosphere effect, *Phytopathol.* 78: 366-371.

Lockwood, J.L., 1977, Fungistasis in soil, *Biol. Rev.* 52: 1-43.

Lockwood, J.L., 1988, Evolution of concepts associated with soil borne plant pathogens, *Ann. Rev. Phytopathol.* 26: 93-121.

Lynch, J.M. (ed.), 1990, *The Rhizosphere,* John Wiley and Sons, New York, 458 pp.

Mallik, A.V. and Zhu, H., 1995, Overcoming allelopathic growth inhibition by mycorrhizal inoculation, in: *Allelopathy: Organisms, Processes and Applications* (K.M.M. Inderjit, F. Dakshini and A. Einhelling, eds.), pp. 39-58, American Chemical Society, Washington D.C.

McAllister, C.B., Garcia-Romera, I., Martein, J., Crodeas, A. and Ocampo, J.A., 1995, Interactions between *Aspergillus niger* van Tiegh and *Glomus mossae*, (Nical & Gerd.) Gerd. & Trappe, *New Phytol.* 129: 309-316.

Molisch, H., 1937, *Der Einfluss einer Pflanze auf die andere - Allelopathie*, Fischer, Jena.

Mukerji, K.G., 1995, Taxonomy of endomycorrhizal fungi, in: *Advances in Botany* (K.G. Mukerji, B. Mathur, B.P. Chamola and P. Chitralekha, eds.), pp. 212-218, APH Publishing Corporation, New Delhi.

Mukerji, K.G., Mandeep and Verma, A., 1997, Mycorrhizosphere microorganisms - screening and evaluation, in: *Mycorrhiza Manual* (A.K. Verma, ed.), pp.85-97, Springer, Heidelberg, Germany.

Muller, C.H., 1969, Allelopathy as a factor in ecological process, *Vegetatio* 18: 348-357.

Nicolson, T.H., 1905, The mycotrophic habit in grass, *Thesis*, University of Nottingham, Nottingham, pp. 66-82.

Philips, J.N. and Hayman, D.S., 1970, Improved procedures for clearing roots and staining parasitic and vesicular arbuscular mycorrhizal fungi for rapid assessment of infection, *Trans. Br. Mycol. Soc.* 55: 158-161.

Putnam, A.R. and DeFrank, J., 1983, Use of phytotoxic plant residues for selective weed control, *Crop Protection* 2: 173-181.

Putnam, A.R. and Defrank, J., 1979, Use of cover crops to inhibit weeds, *Proc. IX Int. Cong. Plant Protection* pp. 580-582.

Putnam, A.R. and Duke, W.B., 1978, Allelopathy in agroecosystems, *Ann. Rev. Phytopathol.* 16: 431-451.

Rice, E., 1984, *Allelopathy (2nd Edition)*, Academic Press, New York, 385 pp.

Ried, C.P.P., Kidd, F.A. and Ekwebelam, S.A., 1983, Nitrogen nutrition, photosynthesis and carbon allocation in ectomycorrhizal pine, *Plant and Soil* 71: 415-431.

Rovira, A.D., 1963, Microbial inoculation of plants. I. Establishment of free living nitrogen fixers in the rhizosphere and their effects on maize, tomato and wheat, *Plant and Soil* 19: 304-314.

Rovira, A.D., Bowen, G.D. and Foster, R.C., 1983, The significance of rhizosphere microflora and mycorrhizas on plant nutrition, in: *Encyclopedia of Plant Physiology, New series Vol. 15*. (A. Lauchi and R.L. Bieleski, eds.), pp. 61-93, Springer-Verlag, Berlin, Heidelberg.

Schenck, N.C. and Perez, Y., 1990, *Manual for the Identification of VA Mycorrhizal Fungi, VAM, 3rd Edn.,* Gainsville, Florida, University of Florida.

Timonin, M.I., 1941, The interaction of higher plants and soil microorganisms III. Effects of by products of plant growth on activity of fungi and actinomycetes, *Soil Sci.* 52 : 395-413.

Williamson, F.A. and Wyn Jones, R.G., 1973, The influence of soil microorganisms on growth of cereal seedlings and on potassium uptake, *Soil. Biol. Biochem.* 5: 569-575.

STORAGE FUNGI IN EDIBLE AGRICULTURAL COMMODITIES

Anjula Pandey and K.G. Mukerji

Applied Mycology Laboratory
Department of Botany
University of Delhi
Delhi-110 007
India

INTRODUCTION

Agriculture plays an important role in the national economy of a country. A great emphasis on pre-and post-harvest technologies in recent years has gained much attention and has resulted in a steady increase in agricultural production. However, materials under storage frequently undergo deterioration due to infestation by various pests and pathogens.

The deterioration of stored materials is a complex process which is influenced by several biotic and abiotic environmental factors as well as the physico-chemical nature of the plant produce. The population of microorganisms in storage is greatly influenced by abiotic factors such as temperature, relative humidity, substrate moisture content and pH. Temperature is an important factor which controls the rate of metabolism, growth, development and reproduction of microorganisms; very low temperatures are fatal to them.

Among the major biotic factors in the storage of edible commodities, fungal contamination appears to be responsible for most of the problems related to quality deterioration. However, if those conditions which are favourable for fungal growth are avoided, their proliferation can be controlled and the quality of the stored material maintained.

STORAGE FACTORS

The hot and humid climate of tropical countries provides favourable conditions for the growth of a wide variety of microorganisms which cause rapid deterioration of materials. The fungi play an important role in the spoilage of a wide range of economically useful products (Allsopp and Allsopp, 1983; Eggins and Allsop, 1975; Garg et al., 1993).

High temperature and humidity seem to favour the incidence of basal rot in onion bulbs (Dwivedi et al., 1995). Chilli fruits stored under humid conditions are attacked by a number of storage fungi, for example *Aspergillus flavus, A. terreus, A. candidus, A. niger, A.*

sclerotium, Paecilomyces variotii and *P. corylophilum* which were isolated from most deteriorated fruits (Prasad, 1997).

Abiotic factors such as temperature, relative humidity and substrate moisture are responsible for competition of *A. flavus* with other storage fungi in different seasons. Choudhary and Sinha (1993) studied such competition and established a correlation between these toxigenic isolates with that of aflatoxin fungal incidence: 63% in the monsoon season, and 52% in winter which corresponded to maximum aflatoxin B_1 accumulation levels in monsoon (1360 $\mu g/kg$) with a 33% reduction in winter. There exists a direct correlation between storage mycoflora and toxin contents in the stored material. The major storage fungi in cashew nut and quince apple (dry fruit slices) are *Aspergillus, Penicillium, Fusarium, Chaetomium* and *Cladosporium* which under humid conditions are responsible for deterioration and toxin accumulation (Giridhar and Reddy, 1997; Sharma and Sumbali, 1997). In general more acidic tissues are attacked by fungi while those having a pH above 4.5 are more commonly infected by bacteria (Desai et al., 1986).

FUNGAL COLONIZATION AND INVASION UNDER STORAGE

The storage fungi not only remain on the surface of infected tissue but also invade the outer layers and internal tissues, causing damage in cereals, legumes, fruits and vegetables, oilseeds, dry fruits and nuts, spices and condiments, masticatories and other products.

In the case of cereals, storage fungi may even invade the seed before harvest and cause embryo damage which subsequently leads to low germination or poor seedling growth (Fields and King, 1962). In pepper (*Piper nigrum*) most of these fungi can apparently invade the outer layer of the fruit before harvest despite its pulpy flesh and heavy skin. Christensen et al. (1967) suggested that the nature of the pepper fruit might not allow the invasion of storage fungi before harvest. However, it seems probable that aflatoxin-producing aspergilli can colonize the fruits during their development. Faulty practices such as incorrect harvesting, improper drying and storage conditions and poor quality processing may also result in such contamination in storage (Banerjee et al., 1993).

The onion is an important bulb crop which is affected by a number of fungal diseases (Neergaard, 1977; Richardson, 1990; Walker, 1952). Of the important pathogens, *Fusarium oxysporum* causes bottom or basal rot under storage (Abawi and Lorbeer, 1971; Dwivedi et al., 1995; Mishra and Rath, 1986; Sumnar and Gay, 1984). The 'Black Mould' of onion is yet another major storage problem associated with heavy colonization by *Aspergillus niger.* The growth and sporulation of this fungus is pronounced on the stem plate (?) and in between the outer scales of the first foliage. Fungal colonization results in wet rot of the bulb (Dwivedi et al., 1995).

The causal storage fungus *F. oxysporum,* responsible for basal rot or soft rot, becomes associated with the bulbs in the field as a result of direct infection and this continues even further during storage (Abawi and Lorbeer, 1971). The heavily infected bulbs become shrivelled and pulpy with basal rot. Cottony fungal growth in the basal and upper parts, rarely on the outer dry scales, is indicative of the vertical spread of the fungus (Dwivedi et al., 1995).

Several cultivars of *Cucurbita maxima* and *C. moschata* have been found to be invaded by fungi during storage. Among these, *Fusarium culmorum, F. solani* and *Didymella bryoniae* are the major infectants (Hawthorne, 1988). Owing to their high moisture content and tender nature, vegetables and fruits pose a characteristic post-harvest problem. High moisture content makes them difficult and expensive to conserve as dry products, and they are metabolically more active than dry products.

Like any other food crop, vegetables and fruits are prone to microbial spoilage due to fungi, bacteria, yeast and moulds. The succulent nature of these commodities make them easily invaded by the microorganisms. It is estimated that 30 to 40% of vegetable decay is caused by soft rot bacteria. The sources of infection are soil in the field, water used for cleaning, surface contact with equipment and the storage environment. The most common pathogens causing deterioration of vegetables are fungi such as *Alternaria, Botrytis, Diplodia, Monilia, Phomopsis, Penicillium, Rhizopus and Fusarium,* and bacteria such as *Erwinia, Ceratocystis* and *Pseudomonas.* While most of the pathogens can invade only damaged tissue, a few such as *Colletotrichum* are able to penetrate the skin of healthy storage tissue.

The injured and uninjured rhizomes of *Curcuma longer,* the turmeric of commerce, are known to harbour several rhizome rot fungi which become prominent in monsoons and reach a minimum in summer months in samples collected from different markets (Kumar and Roy, 1990). Similarly the seed rhizome of ginger *(Zingiber officinale)* seriously suffers from storage rot caused by pathogens such as *Fusarium* and *Rhizoctonia* (Beena et al., 1997). The handling of propagating materials after harvest and storage conditions may prove vital in restricting losses caused by these fungi.

EFFECT OF FUNGAL GROWTH ON QUALITY

The quality of the seed is the major deciding factor in achieving set production targets (Neergaard, 1977). Deterioration of stored rice seed under seasonal fluctuations has caused considerable variation in seed viability, mycoflora and seed moisture levels (Mallick and Nandi, 1982).

Changes in Nutritional Levels

The biochemical changes leading to seed deterioration due to colonization of storage fungi led to a decrease in carbohydrate contents in most cases (Ghosh and Nandi, 1986). In wheat, embryo and endosperm damage due to unfavourable storage conditions was observed by Gajapathy and Kalyanasundaram (1986). Among the cereals stored in bulk in warehouses wheat is generally more susceptible to damage by storage fungi than milled rice. In wheat, the protein constitutes 11.1% and lipid 1.7% as against 7.6% and 0.3% in polished rice. Brown rice (hulled) is more resistant to storage fungi than the polished rice.

The selective preference of fungi in cereals under storage for different nutrients such as lipids, proteins and carbohydrates is responsible for quality deterioration. Qualitatively, the pattern of occurrence of some fungal species, for example *A. flavus, A. nidulans* and *A. niger* on the peripheral layers and *A. glaucus, A. candidus* and *A. terreus* in deeper layers is indicative of this fact. The penicillia were less commonly occurring: only *P. citrinum* was found in contaminated materials. It is possible that in the aleurone layer, the nutrient-rich membrane, species such as *A. flavus* are the most successful competitors.

Deo and Gupta (1989) observed biochemical changes associated with loss of germinability. Pea seeds stored at variable temperatures and relative humidity gradually showed reduction in total nitrogen, proteins, sugars and non-reducing sugars with passage of storage period; the reducing sugars however increased in quantity. Similar results were observed by Roy and Chourasia (1990) in stored seeds of *Mucuna pruriens* under different relative humidity conditions. Among the major constituents, proteins, phenols and alkaloids were greatly influenced by *Aspergillus flavus* and *Penicillium citrinum.*

Differential capability of storage moulds for secretion of protein hydrolysing enzymes and invasion in the seeds showed varying patterns of declined protein levels in *Sesamum indicum* and sunflower *(Helianthus annuus).* These enzymes were responsible for protein breakdown into oligopeptides which further hydrolyse into free amino acids (Pathak, 1988).

Toxins and Mould Contamination

Mould contamination and deterioration of foodstuffs and animal feeds are constantly occurring phenomena (Christensen, 1957, 1972). Association of storage fungi is usually accompanied by simultaneous production of highly toxic compounds called mycotoxins which are lethal to man and animals (Bilgrami, 1983; Christensen and Kaufmann, 1969; Dickens and Jones, 1965; Goldlatt, 1969). Among the most important fungi causing high mammalian toxicity, *A. flavus* and *A. parasiticus* cause deterioration in many agricultural commodities (Saxena et al., 1988, 1989). Various species of *Fusarium* are known to cause storage damage and toxigenic effects in seeds of sorghum and many other cereals (Gupta, 1996, 1996a). These highly toxic compounds, mycotoxins, reduce the value of food and make it unsafe for human consumption.

There are several reports of mould infestation in cereals, legumes, spices and other edible products where toxins have been reported in storage (Bilgrami et al., 1985; Biswas et al. 1989). Among the spices, pepper (*Piper nigrum*) and cardamom (*Elettaria cardamomum*) are particularly known for their contamination by the storage fungus *A. flavus* (Banerjee et al., 1993; Christensen, 1967; Flanningan and Hui, 1976). Fennel, largely consumed for chewing purposes, is not free from toxins such as Aflatoxin B_1 which is known to cause carcinogenic effects on humans (Bilgrami et al., 1985; Rani and Singh, 1989). The natural occurrence of patulin in dry fruit slices of quince consumed by local people of Jammu and Kashmir was reported to be as high as 0.25 to 1.425 mg/kg dry fruit weight (Singh and Sumbali, 1997).

As well as materials for human consumption, animal feeds are also not free from storage mould contamination. Sastry et al. (1967) reported on groundnut toxicity in murrah buffaloes in Andhra Pradesh (India). Oilcakes, which constitute a major part of the animal feed, are generally contaminated by toxigenic strains of *A. flavus* and are known to contain very high levels of aflatoxins (Agarwal et al., 1990; Kumar and Singh, 1989). These aflatoxins not only cause mortality in cattle but also make their milk unsafe for human consumption (Agarwal et al., 1990; Anonymous, 1976).

USE OF ALLELOCHEMICALS IN THE CONTROL OF STORAGE FUNGI

Extracts from a number of higher plants are known to have potent inhibitory effects on plant pathogenic microorganisms. The 'chemicals' present in many plants contain a spectrum of secondary metabolites which act as natural inhibitors and protect the plant from microbial invasions. The importance of these substances as antimicrobial agents and their role as allelochemicals has been stressed by several workers (Mahadevan, 1970, 1979; Roychoudhury et al., 1997; Thapliyal and Nene, 1967; Whittakar and Feeny, 1971).

In view of the fact that disease resistance in some plants is due to the presence of these chemical substances in host tissue to specific pathogens, several plants have been screened for control of plant diseases when applied in combination or singly. Some examples of such plants are *Argemone mexicana, Zingiber officinale, Azadirachta indica, Calotropis procera, Ocimum sanctum, O. basilicum, Allium cepa, A. sativum, Curcuma longa* and several spices and condiments (Amer et al., 1980; Nanir and Kadu, 1987; Prasad and Simlot, 1982).

In countries such as India, where wide climatic variations and improper conditions in grain storage management exist, several pests and pathogens are reported to cause extensive damage in stored agricultural commodities (Mishra, 1985). Annual loss due to spoilage of cereals is estimated to be 10 to 15 percent of total production (Girish and Goyal, 1986). Decomposers of food grains such as fungi and bacteria exist on the surface of the grain in dormant form and only invade under favourable conditions.

The use of chemical controls to combat storage associated problems and microbial spoilage of food grain is now commanding greater attention by research workers and environmentalists (Bankole, 1996; Castro et al.,1995; Sholberg and Gaunce, 1996). The reasons are their high cost, residual and environmental deterioration problems, toxicity in plants and animals and adverse impacts on ecological balance, as well as their adverse effect on flavours (Deshpande et al., 1986; Prakash and Kaurav, 1983; Tyagi et al., 1986).

The use of allelochemicals as antimicrobial agents is an old concept but it has received considerable attention in recent years. There are several examples in which plant extracts have been known to control or reduce the growth of microorganisms. Leaf extracts of *Pyrus communis* (at 1:250 dilution) have been found to influence the conidial germination of *Venturia inequalis* which causes apple scab disease. Similarly Tripathi et al. (1978) have reported an antifungal factor against *Helminthosporium oryzae* in leaves of *Lawsonia inermis.* Crude alcohol extracts of *Mimosa hamata* have been found to possess an antibacterial property (Hussain et al., 1979). Several species of *Croton*, particularly *C. lacifers,* are known to have antifungal activity against *Cladosporium* species (Ramesh et al., 1995). *Croton sparsiflorus, Azadirachta indica* and *Lawsonia inermis* have been tested for their antiseptic and anti-microbial activity (Acharya et al., 1964).

Using this concept, Maraghy (1995) investigated the potential of the spices Chinese cassia *(Cinnamomum aromaticum),* clove *(Syzygium aromaticum)* and *thyme (Thymus vulgaris)* to control storage mould *A. flavus* and to inhibit aflatoxin production in seed during storage when used separately or in combination. In earlier works, Rao and Ratnasudhakar (1992) used the rhizome powder of *Acorus calamus* and leaf powder of *Azadirachta indica* separately and in combination (1:2 ratio by w/w). The fungal and bacterial population during the storage of rice was found to be reduced; moreover, there were no adverse effects on the quality of the cooked rice nor did it have any undesirable odour.

Information on the usefulness of indigenous plant materials in minimising or controlling the microflora of stored commodities and consequent effects thereon remains scanty. Several attempts have been made by various workers to explore such plants which could be exploited (either as crude extracts or in powdered form) to reduce the incidence of microflora contamination on different commodities during storage. Their easy availability, low cost and non-toxic nature has prompted workers to attempt to discover the ideal source of antifungal material from nature.

IMPACT OF POST-HARVEST METHODS ON STORAGE OF BULBS AND RHIZOMATOUS CROPS

The effect of humidity on the development of storage fungi is associated with that of temperature and it operates indirectly through the moisture content of the material.

In recent years Bartali et al. (1990) and Sinha et al. (1991) have made some attempts to find an ideal storage system for cereals. They compared storage pits lined with polythene and with the traditional straw layer. The mean temperature of the straw-lined pits was consistently higher than that of the plastic-lined pits. Results in the straw-lined with respect to *A. flavus* infection, loss of germinability etc. were found to be lower as compared with those of the plastic-lined pits. Similar experiments were carried out by Sinha et al. (1991) with wheat kept in ventilated conditions during storage. Moist wheat was protected from damage by *Penicillium* species, *Aspergillus* species and bacteria by ventilation which reduced germination loss and carbon dioxide production.

While investigating the impact of different storage methods, the development of storage mycoflora and quality deterioration in rhizomatous and bulbous crops was studied. The study was conducted in experimental plots in the Botanical Gardens, University of Delhi where turmeric (*Curcuma longa*) and garlic (*Allium sativum*) were grown. On maturity the crops were harvested for the edible rhizomes and bulbs respectively, which were stored after post-harvest processing.

The turmeric rhizomes were given three treatments: storage in plastic containers (P), in earthen pots (E) and in cardboard boxes (T). In each treatment, freshly harvested samples (F) were stored besides commercially available market samples (C). The samples were drawn at regular intervals and screened for presence of storage fungi. Qualitative and quantitative assessments of the mycoflora were made using the soil dilution and plate methods (Timonin, 1941) using Czapek Dox medium. Similarly for garlic bulbs, samples stored in well aerated places with leaf base remains on the bulbs (T) were compared to the controls (C) and the ones without leaf bases.

Earlier, the storage fungi in rhizomes of *Curcuma longa* from different sources in Delhi markets were studied by Kumar and Roy (1990). In the storage of *Allium cepa* (onion), several storage pathogens have been reported to cause damage and bulb rot (Bhadraiah and Surya Teja, 1997; Pathak, 1988; Ranjan et al., 1992; Singh and Saha, 1994).

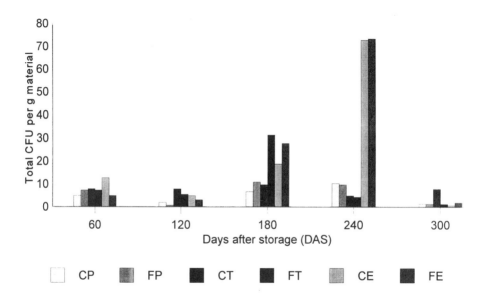

Figure 1. Isolation of mycoflora from turmeric *(Curcuma longa)* rhizomes under storage (1: 1000 dilution). C: commercial (market) samples; F: freshly harvested samples; P: plastic containers; T: cardboard boxes; E: earthen pots.

The turmeric material stored in earthen pots had maximum storage fungi (represented as colony forming units, CFU) as compared to the plastic containers. The porous and absorbent walls of the earthen pots allowed the atmospheric humidity to percolate and influence the fungal growth to a great extent. In contrast, the plastic containers maintained a constantly low humidity in the samples, so were least infected. In comparison to the freshly harvested samples in all treatments, the commercial samples were very highly infected until the end of the summer. This may be due to the smoother and milled rhizome surface of market samples that was easily attacked by storage fungi. Moreover, the presence of thick rhizome skin (the peel) in fresh samples protected them from easy invasion by the fungi. However, during monsoon the relative humidity allowed the fungal spores present in soil-contaminated rhizomes to germinate and set more infection in freshly harvested samples as compared to fresh ones (Figure 1). A similar trend was observed by Kumar and Roy (1990) in turmeric and

Wallace et al. (1976) and Singh and Saha (1994) in other stored products.

The garlic bulbs with leaf bases stored in a well aerated place showed a consistently uniform pattern of occurrence of storage fungi on bulbs as compared to the controls (Figure 2). In general the samples with the leaf bases had low infectivity by storage fungi as compared to those without them. This confirms the role of leaf bases in the protection of cloves/bulbs from fungal invasion. The post-harvest practice of leaving the leaf bases intact is widely followed by the farmers for storing garlic bulbs for long-term uses or for seed purposes.

Initially before storing, turmeric was contaminated by various fungi, for example *Penicillium chrysogenum, P. funiculosum, Gliocladium fimbriatum, Alternaria alternaria, Aspergillus niger, Fusarium oxysporum.* However, after one year of storage, a remarkable increase in contamination was evidenced by the appearance of *Aspergillus flavus, A. nidulans, A. fumigatus, Fusarium sporotrichioides, Cephalosporium achrimonium, Alternaria solani, Macrophomina phaseolina, Cladosporium herbarum* and *Paecilomyces lilacinus.* In garlic a similar trend was observed, although differences were evident both qualitatively and quantitatively in the storage mycoflora of turmeric and garlic in different storage conditions.

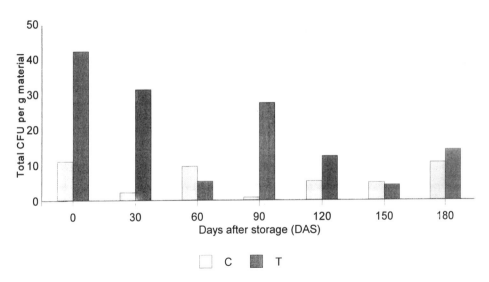

Figure 2. Isolation of mycoflora from garlic (*Allium sativum*) bulbs under storage (1:1000 dilution). C: control (surface sterilized) samples; T : unsterilized samples.

CONCLUSIONS

Deterioration caused by storage fungi leads to poor quality of stored agricultural commodities and hance poses a great threat to the national economy. The extent of infection in different materials, their causal organisms and their effects on human health are critical issues which demand great concern and therefore this information needs to be well documented to increase public awareness. Much emphasis needs to be laid on manipulations of the storage environment to combat this problem through effective pre- and post-harvest operations. The losses can be minimised by adopting the necessary cultural operations, careful handling and effective storage management practices. The use of appropriate chemicals at pre- or post-harvest storage may reduce the loss and increase the storage life. Keeping in view the importance of some antifungal plants in the control of storage fungi, further research needs to be carried out on these with respect to various groups of edible commodities. This would not only reduce the financial burden of excessive use of expensive chemical fungicides but the user-friendly, non-residual, low-cost, renewable nature of these antifungal plant materials would also contribute to environmental protection.

REFERENCES

Abawi, G.S. and Lorbeer, J.W., 1971, Reaction of selection onion varieties infected by *F. oxysporum* f. sp. *cepae, Pl. Dis. Repr.* 55: 1000-1004.

Acharaya, B.K., Modi, M.L. and Sinha, S.N., 1964, The antibacterial activity *(in vitro)* of the alkaloid of *Croton sparsiflorus* Morong on some pathogenic organisms, *J. India Med. Assoc.* 43: 592-595.

Agarwal, Renu, Kumar, A. and Singh, S., 1990, *Geobios* 17: 234-238.

Allsopp, C. and Allsopp, D., 1983, An updated survey of commercial products used to protect material against biodeterioration, *Biodet. Bull.* 19: 99-146.

Amer, M., Taha, M. and Tossong, S., 1980, The effect of aqueous garlic extract on growth of dermatophytes, *Int. J. Dermatology* 19: 285-287.

Anonymous, 1976, IARC Monograph, WHO Publications, vol. 10: 51-72.

Aspinall, D. and Paleg, L.G., 1971, Deterioration of wheat embryo endosperm in function with age, *J. Exp. Botany* 22: 925-935.

Banerjee, A., Mathews, R.P., Prakash, H.S. and Shetty, H.S., 1993, *Mycol. Res.* 97(11): 1403-1406.

Bankole, S.A., 1996, Effect of ethylene oxide and methyl formate fumigation on seed mycoflora and germination of some stored oil seeds in Nigeria, *Crop Res.* 11 (2): 224-27.

Bartali, H., Dunkel, F.V., Said, A. and Sterling, R.L.F., 1990, Performance of plastic lining for storage of barley in traditional underground structures (Matmora) in Morocco, *J. Agric. Eng. Res.* 47(4): 297-314.

Beena, N., Rajan, P.P., Sharma, Y.R. and Anandraj, M., 1997, Control of ginger storage rot by agrochemicals and biocontrol, in: *Indian Phytopathological Society - Golden Jubilee International Conference on Integrated Plant Diseases Management for Sustainable Agriculture,* 10-15 Nov 1997, New Delhi, India.

Bhadraiah, B. and Surya Teja, K.G.V.N., 1997, Mycological and biochemical changes in stored ragi *Eleusine coracana* L. Gaertn. seeds, in: *Indian Phytopathological Society - Golden Jubilee International Conference on Integrated Plant Diseases Management for Sustainable Agriculture,* 10-15 Nov., 1997, New Delhi.

Bilgrami, K.S., 1983, Mycotoxin problem in food and feed - some social obligations of strategy for future, in: *Proc. Symp. Mycotoxin in Food and Feed.* pp. 1-33, Allied Press, Bhagalpur, India.

Bilgrami, K.S., Prasad, T., Sinha, K.K., Singh, A. and Ranjan, K.S. (eds.), 1985, *Mycotoxin in Dry Fruit and Spices,* Narendra Publishing House, Delhi.

Biswas, G., Raj, H.G. and Mukerji, K.G., 1989, Glutathione levels and Y Glutamyl transpeptidase activities in aflatoxigenic and non-aflatoxigenic strains of *Aspergillus flavus. J. Toxicol. Toxin Reviews* 8: 329-338.

Castro, M.F., De, P.P.M. and Pacheco, M., 1995, Utilization of Phosphine fumigant for control of fungi naturally present in stored paddy rice *(Oryza sativa* L.)*, Revista de Microbiologia* 26(93): 230-235.

Choudhary, A.K. and Sinha, K.K., 1993, Competition between a toxigenic *Aspergillus flavus* strain and other fungi on stored maize kernels, *Stored Pro. Res.* 29(1): 75-80.

Christensen, C.M., 1957, Deterioration of stored grains by fungi, *Bat. Rev.* 23: 108-134.

Christensen, C.M., 1972, Microflora and seed deterioration, in: *Viability of Seed* (E.H. Roberts, ed.), pp. 59-93, Chapman and Hall Ltd., London.

Christensen, C.M., Fanse, H.A., Nelson, G.H., Bales, F. and Mirocha, C.J., 1967, Microflora of black and red pepper, *Appl. Microbiol.* 15: 622-629.

Christensen, C.M. and Kaufmann, H.H. (eds.), 1969, *Grain Storage, the Role of Fungi in Quality Loss,* University of Minnesota Press, Minneapolis.

Deo, P.P. and Gupta, J.S., 1989, Fungal deterioration of gram seeds during storage under various control conditions, *Int. J. Trop. Pl. Dis.* 6(1): 115-128.

Desai, U.T., Kadam, S.S. and Salunkhe, D.K., 1986, Post-harvest handling, storage and processing of vegetables, in: *Vegetable Crops in India* (T.K. Bose and M.G. Sum, eds.), Naya Prokash, Calcutta.

Deshpande, G.D., Choulwar, S.B., and Salwe, P.T., 1986, Carbandazim spray on seed health parameters in *Sorghum, Indian Phytopath.* 39: 143.

Dickens, F. and Jones, H.E.H., 1965, Carcinogenic activity of a series of reaction lactone and related substances, *British J. Cancer* 15: 85-88.

Dwivedi, Arvind, Singh, T.K. and Singh, D., 1995, Basal rot of onion bulbs caused by *Fusarium oxysporum, Acta Botanica Indica* 23: 55-58.

Eggins, H.O. and Allsopp, D., 1975, Biodeterioration and biodegradation by fungi, in: *The Filamentous Fungi* (J.E.Smith and B.R. Berry, eds.), pp.301-319, Edward Arnold, London.

Fields, R.W. and King, T.H., 1962, Influence of storage fungi in deterioration of stored pea seeds, *Phytopath.* 52: 336-39.

Flannigan, B. and Hui, S.C., 1976, The occurrence of aflatoxin producing strains of *A. flavus* in the moulds of groundnut species, *J. Appl. Bacteriol.* 41: 41-18.

Forgacs, J., 1962, Mycotoxicosis, the neglected disease, *Feed Stuffs* 34: 124-34.

Gajapathy, M.K. and Kalyanasundaram, Indira, 1986, Distribution of rice seed mycoflora with special reference to storage fungi, *Indian Phytopath.* 39: 288-292.

Garg, K.L., Garg, N. and Mukerji, K.G., eds., 1993, *Recent Advances in Biodeterioration and Biodegradation,* Naya Prokash, Calcutta.

Ghosh, J. and Nandi, B., 1986, Deteriorative abilities of some common storage fungi of wheat, *Seed Sci. & Technol.* 14: 141-49.

Giridhar, C.P. and Reddy, S.M., 1997, Mycoflora and mycotoxins contamination of cashew nut, in: *Proceedings of the 84th Indian Science Congress Part III (Adv. Abs.).*

Girish, G.K. and Goyal, R.K., 1986, Presence of mycoflora and mycotoxin in different food grains with particular reference to paddy (rice), *Bull. Grain Tech.* 24: 157-177.

Goldlatt, L.A. (ed.), 1969, *Aflatoxin Scientific Background and Control and Implication,* Academic Press, New York.

Gupta, Rajni, 1996, Screening of *Sorghum* cultivars for the elaboration of Zearolenone by two toxigenic isolates of *Fusarium, Proc. Indian Natn. Sci. Acad,* B 62(4): 277-280.

Gupta, Rajni, 1996a, Seed borne fungi in *Sorghum, Plant Genetic Research Newsletter* 106 (1): 1-2.

Halloin, J.M., 1986, Microorganism and seed deterioration physiology of seed destruction, *CSSA, Spec. Pub.* 11: 89-99.

Hawthorne, B.T., 1988, Fungi causing storage rot on fruits of *Cucurbita sp., NZ J. Exp. Agric. Auckland, New Zealand,* 16(2): 151-158.

Hesseltine, C.W., 1968, Flour and wheat research on their microbiological flora, *Baker's Digest* 42(3): 40-42.

Hussain, N., Madan, M.H., Shabbir, S. and Zaidi, A.A.H., 1979, Anti-microbial principles in *Mimosa hamata, J. Nat. Prod.* 87: 125-128.

Krishna Rao, V. and Ratnasudhakar, T., 1992, Effect of some plant powders on grain microflora of paddy during storage, *Indian Phytopath.* 45(1): 55-58.

Kumar., A. and Singh, S., 1989, National occurrence of rnycoflora and aflatoxins in *Linum usitatisimum* seeds, *Geobios,* 49-53.

Kumar, Harish and Roy, A.N., 1990, Occurrence of fungal rot of turmeric *(Curcuma longa)* rhizomes in Delhi in markets, *Indian J. Agric. Sci.* 60(3): 189-191.

Mahadevan, A., 1970, Prohibitions and disease resistance, *Phytopath. Z.* 68: 73-80.

Mahadevan, A., 1979, Biochemical aspects of plant diseases resistance, *Biochem Rev.* 449: 51-66.

Mallick, A.K. and Nandi, E., 1982, Deterioration of stored rough rice: Grain in private storage, *Seed Sci & Technology* 10: 527-533.

Maraghy, S.S.M., 1995, Effect of some spices as preservatives for storage of lentil *(Lens esculenta* L.) seeds, *Folia Microbiologica* 40 (5): 490-492.

Mishra, D. and Rath, G.C., 1986, Survey of post harvest decay of vegetable caused by *Fusarium, Indian Phytopath.,* 39: 273-274.

Mishra, P.L. (ed.), 1985, *Environment Influence on Agriculture,* Ritu Prakashan, Meerut.

Nanir, S.P. and Kadu, B.B., 1987, Effect of some medicinal plant exhibits on some fungi, *Acta Botanica Indica* 15: 170-75.

Neergaard, P. (ed.), 1977, *Seed Pathology,* The MacMillan Press Ltd., London.

Pathak, Shyam Sunder, 1988, Lipid and protein degradation to mustard seeds by fungal activity in different storage systems, *PhD Thesis,* Bhagalpur University, Bhagalpur.

Prakash, A. and Kaurav, L.P., 1983, Compatibilities between certain pesticides used for control of insect pests and seed-borne fungi in stored paddy, *Pesticides* 17: 21-22.

Prasad, B. and Simlot, M.M., 1982, Antifungal activity of fruit of lemru *(Diospyros cordifolia), Sci. and Cult.* 48: 290-291.

Prasad, B.K., 1997, Decay of chill fruits in India during storage, *Proc. 84th Ind. Sci. Cong. Part III (Adv. Abst.).*

Ramadevi, P., Subramanyam, K., Krishna Rao, V. and Chiranjeevi, V., 1988, Effect of some dried leaf powders on grain mycoflora and viability of rice during storage, *Bull. Grain Tech.* 26: 138-145.

Ramesh, V.M., Hilda, A. and Manjula, V.K., 1995, Fungitoxic effect of leaf extract of *Croton sparsiflorus* Morong on phytopathogenic fungi, *Acta Botanica Indica* 23: 63-66.

Rani, Neeta and Singh, S., 1989, Natural occurrence of mycoflora and aflatoxins in *Foeniculum vulgare* Mill. seeds, *Nat. Acad. Sci. Letters.*

Ranjan, K.S., Sahay, S.S. and Singh, A.K., 1992, The influence of storage structure on aflatoxin contamination in wheat and mustard, *Stored Prod. Res.* 28(3): 221-224.

Rao, V.K. and Ratnasudhakar, T., 1992, Effect of some plant powders on grain microflora of paddy during storage, *Indian Phytopath.* 45(1): 55-58.

Richardson, M.J., 1990, *An Annotated List of Seed-Borne Disease*, Seed Test Assoc., IInd Jurisch, Switzerland.

Roy, A.K. and Chourasia, H., 1990, Fungal association and deterioration of active principles of *Mucuna pruriens* seeds under different relative humidity, *Acta Botanica Indica* 18(2): 235-39.

Roychoudhury, R., Vadhanayam, S.M., Bhatt, K. and Sinha, P., 1997, Management of Yellow Vein mosaic disease in bhindi *Abelmoschus esculentus* by sowing dates and with seem producers, in: *Indian Phytopathological Society - Golden Jubilee International Conference on Integrated Plant Disease Management for Sustainable Agriculture*, 10-15 Nov. 1997, New Delhi.

Sastry, G.A., Rao, Narayan, Christopher, P.R. and Hill, K.P., 1967, Groundnut toxicity in murrah buffaloes in Andhra Pradesh (India), *Indian Vet. J.* 42: 79-82.

Saxena, M., Allameh, A., Mukerji, K.G. and Raj, H.G., 1989, Studies on glutathione S-Transferases of *Aspergillus flavus* group in relation to aflatoxin production, *J. Toxicol. - Toxin Reviews*, 8: 319-328.

Saxena, M., Mukerji, K.G. and Raj, H.G., 1988, Positive correlation exists between glutathione S-Transferase activity and aflatoxin formation in *Aspergillus flavus*, *Biochem. J.* 254: 567-570.

Sharma, Y.P. and Sumbali, G., 1997, Association of Patulin with market samples of dry fruit slices of quinces *(Cydonia oblonga* Mill.) from J & K, in: *From Ethnomycology to Fungal Biotechnology - Exploiting fungi from natural resources for novel products*, Dec. 15-16, 1997, Simla, India.

Sholberg, P.L., Gaunce, A.P., 1996, Fumigation of high moisture seed with acetic acid to control storage mold, *Can. J. Pl. Sci.* 76(3): 551-555.

Singh, B.K. and Saha, N.K., 1994, Impact of storage systems on mycoflora of *Maduca longifolia* and seeds, *Indian Phytopath.* 47(3): 266-269.

Singh, P.L. and Gupta, M.N., 1989, Mycoflora associated with the surface of fresh market sample of *Pinus geradiana* seeds, *J. Indian Bot. Soc.* 68: 105.

Sinha, R.N., Muir, W.E., Sanderson, D.B. and Tuma, D., 1991, Ventilation of bin-stored moist wheat for quality preservation, *Can. Agric. Eng.* 33(1): 55-65.

Sumnar, D.R. and Gay, J.D., 1984, Basal rot of onion caused by *F. oxysporum* f. sp. *cepae, Georgia Pl. Dis.* 68: 450.

Thapliyal, S.C. and Nene, N.L., 1967, Inhibition of plant pathogens by higher plant substances, *J. Sci. Ind. Res.* 26: 2899-2910.

Timonin, M.I., 1941, The interactions of higher plants and soil microorganisms III: Effects of by products of plant growth on activity of fungi and actinomycetes, *Soil Sci.* 52: 395-413.

Tripathi, S.C., Srivastava, H.S. and Dixit, S.N., 1978, A fungitoxic principle from the leaves of *Lawsonia inermis, Experimentia* 34: 51-52.

Tyagi, R.P.S., Singh, T. and Srivastava, S.K., 1986, Efficacy of some newer fungicides against storage fungi, *Bull. Grain Tech.* 24: 178-179.

Ullah, M.W., 1990, Effect of moisture on rough rice stored in open and closed container, in: *Proceedings of the International Agricultural Engineering Conference and Exhibition* (V.M. Salokhe and S.G. Ilangantileke, eds.), pp. 97-605, Bangkok, Thailand, Dec. 3-6, 1990.

Usha, C.M., Patkar, K.L., Shetty, H.S., Kennedy, R. and Lacey, 1993, Fungal colonization and mycotoxin contamination of developing rice grain, *Mycol. Res.* 97: 795-798.

Walker, J.C., 1952 (ed.), *Disease of Vegetable Crops,* McGraw Hill Book Co., New York, USA.

Wallace, H.A.H., Sinha, R.N. and Mills, J.T., 1976, The fungal mycoflora and its influence by age, *Can. J. Bot.* 54: 132-143.

Whittakar, R.H. and Feeny, P.P., 1971, Allelochemics chemical interactions between species, *Science* 171: 757-770.

256

SEED-BORNE MYCOFLORA OF TWO UNDER-EXPLOITED LEGUMES: *VIGNA UMBELLATA* AND *PSOPHOCARPUS TETRAGONOLOBUS* FROM NORTH-EASTERN PARTS OF INDIA

Vandana Joshi and K.G. Mukerji

Applied Mycology Laboratory
Department of Botany
University of Delhi
Delhi-110 007
India

INTRODUCTION

About ninety percent of all the food crops grown on earth are propagated through seed (Neergaard, 1977). Seeds thus play a vital role in the total biological yield. However, fungal deterioration of crop seed has been an established fact for many years (Christensen, 1957; Christensen and Kaufmann, 1965). Seeds are known to be colonised by various types of fungi amongst which many are plant pathogens.

The International Seed Testing Association (ISTA) acknowledges the importance of seed-borne pathogens and makes efforts to improve planting material through testing and certification of seed. The ISTA published its first list of seed-borne pathogens in 1958 (Noble et al., 1958) as part of its Handbook on Seed Health Testing. The latest edition of this list (Richardson, 1990) includes approximately 1300 organisms which are pathogenic and known to be seed-borne. The great majority of them are fungi and seed-borne pathogenic fungi have been reported on 306 host genera in 96 plant families.

Seed-borne fungi can be grouped into: (i) obligate parasites, (ii) facultative saprophytes and (iii) facultative parasites. For obligate parasites, parasitism is an essential part of their life-cycles. Facultative saprophytes normally function as parasites but are able to complete their life-cycles as saprophytes. Of the facultative parasites, many of the fungal genera are known to cause considerable damage to seeds of food crops. They exist as saprophytes but have the ability to survive as parasites. These can be divided into field fungi and storage fungi (Christensen and Kaufmann, 1969; Singh and Mukerji, 1994). Field fungi contaminate or colonise the seed in the field during ripening and harvesting operations. Species of *Fusarium, Alternaria, Drechslera, Phoma, Curvularia, Epicoccum, Nigrospora, Stemphylium* and *Cladosporium* are included under the field fungi. Field fungi usually cause seed rot, damping-off, blights, discolouration and mycotoxin production that damages the seed. Some mycotoxins also cause diseases in plants that grow from the infected seeds.

Species of *Aspergillus* and *Penicillium* are the important storage fungi which develop on seeds during storage. These fungi occur individually or together in seeds. *Aspergillus* species can occur as major pre-harvest pathogens on seeds of many crop species. The predominant species of storage fungi are *Aspergillus restrictus, A. galucus, A. candidus, A. flavus* and *A. ochraceus*, and *Penicillium brevicompactum, P. cyclopium* and *P. viridicatum*. Other genera of fungi that invade stored seeds are *Absidia, Chaetomium, Mucor, Paecilomyces, Rhizopus* and *Scopularia*.

Among the seed-borne organisms, fungi cause maximum seed damage, for example seed abortion, shrunken seeds, seed rot, sclerotisation, seed necrosis, seed discolouration, reduced germination and reduced vigour. A large number of seed-borne fungi produce toxic metabolites (mycotoxins) which often kill the embryo. Various studies have been carried out on the production, structure and biological activities of mycotoxins from seed-borne fungi. These mycotoxins are non-enzymatic substances of low molecular weight and are responsible for destructive diseases in many food crops (Upadhyay and Mukerji, 1997).

Alternaria, Aspergillus, Fusarium and *Penicillium* are some of the common genera which produce mycotoxins (Bokhary and Naguib, 1983). The reduction in germination of seeds, pre- and post-emergence death and discolouration of seeds are other undesirable effects of seed-borne fungi (Abou-heilah, 1984; Ashokhan et al., 1979; Baird et al., 1996, Bokhary, 1986; Handoo and Aulakh, 1979; Shafie and Webster, 1981).

Reddy and Reddy (1990) studied the mycotoxins produced by seed-borne fungi of maize including species of *Aspergillus* and *Fusarium, Trichothecium roseum, Trichoderma viride, Myrothecium roridum* and *Penicillium griseofulvum*. The toxigenic potential of individual fungi varied. Tseng et al. (1995) made a comparative investigation of mycoflora and mycotoxins in dry bean (*Phaseolus vulgaris*). When the seed-borne fungi were isolated and characterised, *Fusarium* and *Aspergillus* were identified as the most probable mycotoxin-producing fungi (Gupta, 1996).

EFFECT OF STORAGE AND AGEING

After harvesting, seeds are usually stored for different periods depending upon consumption and transportation. Seeds in storage are subjected to environmental conditions different from those in the field. This leads to invasion of the seeds by an entirely different set of organisms which are also largely responsible for the deterioration of seeds in storage.

Deterioration of stored grains by fungi was first investigated by Duvel (1909) and Shanahan et al. (1910). Since then, a substantial amount of work has been done (Christensen, 1951, 1955a,b, 1957, 1962, 1973; Christensen and Drechsler, 1954; Christensen and Kaufmann, 1965, 1969; Gilman and Barron, 1930; Semeniuk and Barre, 1944). Legume seeds have been shown to harbour a number of fungi (Siddiqui et al., 1974; Suryanarayan and Bhombe, 1961).

One very important factor of storage is the heating which almost always occurs spontaneously in stored grains and causes grain damage. The biochemical changes taking place in the seeds subsequent to fungal invasion have been reviewed by Christensen (1957). According to her, the more obvious alterations were increases in fatty acids and reducing sugars, and decreases in non-reducing sugars. There is found to be loss in organic matter, increase in fat acidity and decrease in total nitrogen as a result of mould development (Milner, et al., 1947; Semeniuk and Barre, 1944).

Maheshwari and Mathur (1987) studied biochemical changes (protein nitrogen, protein and reducing and non-reducing sugars) in lobia seeds due to certain *Aspergillus* species (*A. nidulans* and *A. terreus*) under different temperatures (20, 28 or 36°C) and relative humidities

(75, 85 or 95%) and observed that deterioration was increased at higher temperature and relative humidity. Infection by *A. nidulans* was more deleterious than infection by *A. terreus.*

Khairnar and Bhoknal (1990) studied biodeterioration in pearl millet by eleven seed moulds on surface-sterilised, unautoclaved and autoclaved seeds. They observed that in unautoclaved seeds, reduction of protein content was caused by *Rhizoctonia solani, Curvularia lunata* and *Fusarium moniliforme*, whilst it was increased by *Aspergillus niger* and *Rhizopus nigricans*. In autoclaved seeds, most of the fungi expect *Pythium* sp. and *Gibberella fujikuroi* caused protein increases. Maximum starch reduction was caused by *C. lunata, A. flavus, Drechslera tetramera* and *G. fujikuroi.*

Gupta et al. (1993) investigated the effect of *A. niger* and *A. glaucus* on seed deterioration during accelerated ageing. Both fungi reduced germination in rolled paper towel and modified sand tests. *A. niger* reduced germination more than *A. glaucus*. Regina and Tulasi (1992) investigated biochemical changes in stored caraway seeds due to *Aspergillus flavus, A. niger* and *Fusarium moniliforme*. All three fungi reduced protein, carbohydrates and total oil in the seeds and increased fatty acids. Nwaiwu et al. (1995) analysed biodeterioration of African bread fruit (*Treculia africana*) pods and seeds in south-eastern Nigeria and its control. The microorganisms associated with biodeterioration of pods were largely *A. niger, Rhizopus stolonifer* and *Botryodiplodia theobromae* (72%). *R. stolonifer* caused the highest degree of seed spoilage. High correlations were found between the microorganisms from pods and seeds. Storage at 5°C and pre-storage treatment with a 15% sodium chloride solution reduced spoilage from 100 to 0% and from 100 to 5% respectively.

SEED-BORNE PLANT PATHOGENIC MICROORGANISMS

The incidence and spread of seed-borne diseases have increased in recent years due to the introduction of new varieties into territories where they are not indigenous and also by growing the same crop over wide areas to facilitate its handling and harvesting to get maximum yields per hectare by incentive cropping. An enormous amount of literature is available on seed-borne fungi and the diseases they cause (Agarwal, 1981; Alcock, 1931; Doyer, 1938; Malone and Muskett, 1964; Neergaard, 1977, 1979; Neergaard and Saad, 1962; Noble et al., 1958; Noble and Richardson, 1968; Richardson 1979, 1990; Sao et al., 1989 and Suryanarayan 1978). Several investigators isolated and characterised the seed mycoflora in wheat; Orsi et al. (1994) in durum wheat kernels; Jayaweera et al. (1988) and Ahmad et al. (1989) in paddy; Moreno-Martinez et al. (1994) in maize and barley; Kanapathipillai and Derris (1988) in *Psophocarpus tetragonolobus*; Ramadoss and Sivaprakasan (1989) from two months stored cowpea seeds; Sinha and Prasad (1989) in fenugreek; Sereme (1991) in seventeen samples of bambara groundnut; Balardin (1992) in bean seeds; Tseng et al. (1995) in dry bean (*Phaseolus vulgaris*); Khamees and Schlösser (1990) from sesame seeds; Zad (1990) from 16 samples of sunflower; Shah and Jain (1993) in mustard; Basak et al. (1989) in different cultivars of brinzal; Kumkum et al. (1989) in okra; Hashmi (1989) and Liang (1990) in chilli; Chandi Ram and Maheshwari (1992) in sponge gourd and bottle gourd; Kowalik (1989) in beetroot; Gowda et al. (1989) in tobacco; Mercer (1994) in linseed; Patil et al. (1994) in sugar-cane; and Singh and Saha (1994) in *Madhuca longifolia.*

Kanapathipillai and Derris (1988) isolated thirty-three fungal species from seeds of *Psophocarpus tetragonolobus* using three media: blotter, malt agar and malt agar + sodium chloride. More fungi were isolated on malt agar and malt agar + NaCl. Soil inoculation tests on emerging seedlings using 12 selected species indicated that *Botryodiplodia theobromae* and *Diaporthe phaseolorum* were pathogenic to seeds and seedlings of *P. tetragonolobus*. Disease symptoms were observed on leaves and stems of young plants inoculated with each of the 12 species. Sereme (1991) isolated fungi from 17 seed samples of bambara groundnut (*Vigna*

Figure 1. *Vigna umbellata* and *Psophocarpus tetragonolobus*, the under-exploited protein-rich legumes grown in experimental field. 1A: Dwarf variety of *P. tetragonolobus* (Dwarf mutant); High yield cultivars of *Vigna umbellata:* 1B: RBL-1; and 1C: RBL-6.

subterranea). Pathogenicity was confirmed for *Macrophomina phaseolina, Botryodiplodia theobromae, Colletotrichum dematium, Fusarium moniliforme, F. solani, Aspergillus niger* and *A. flavus*. Balardin et al. (1992) studied the incidence of *Colletotrichum lindemuthianum, Fusarium spp., Rhizoctonia solani* and storage fungi (*Aspergillus* spp. and *Penicillium* spp.) in seed samples of *Phaseolus vulgaris*. Tseng et al. (1995) isolated and characterised seed-borne fungi in dry seeds of *Phaseolus vulgaris*. The fungi most frequently isolated were *Alternaria, Fusarium, Rhizoctonia, Penicillium, Rhizopus, Sclerotinia, Gliocladium* and *Mucor*, from diseased ontario beans. The fungi most frequently isolated from diseased Taiwan beans were *Aspergillus, Penicillium, Eurotium, Rhizopus* and *Curvularia*.

DISEASES OF *VIGNA UMBELLATA* AND *PSOPHOCARPUS TETRAGONOLOBUS*

The literature on disease status of crops, particularly seed-borne diseases, is meagre. Therefore, in view of the importance of storage and seed deterioration fungi, an attempt was made to study the mycoflora invading seeds of *Vigna umbellata*, rice-bean (Figures 1B and 1C) and *Psophocarpus tetragonolobus*, winged bean (Figure 1A).

Vigna umbellata (Thunb.) Ohwi & Ohashi (syn. *Phaseolus calcaratus* Roxb.), a native of South and South-East Asia is an important multi-purpose crop grown for food, fodder, green manure and cover crop in the north-eastern and north-western hills of India. Its seeds are usually boiled and eaten with rice. *Psophocarpus tetragonolobus* (L.)DC., native of Papua New Guinea, is grown primarily for its immature pods which are cooked like french beans. The leaves, young sprouts, flowers and fruits are also used as vegetables and in soups and the plant is confined to humid sub-tropical parts of the north-eastern region. The cultivation of these minor under-utilized pulse crops is greatly handicapped by a lack of information on pathological and agronomical aspects.

Winged beans are attacked by *Cercospora arantea, C. canescens, C. cruenta* and *C. psophocarpi, Corticum solani, Corynespora cassicola, Erysiphe cichoracearum, Meliola erythrinae, Myrothecium roridum, Oidium* sp., *Periconia byssoides, Pythium debaryanum, Sporidesmium bakeri* and *Synchytrium psophocarpi*. Mosaic of *Crotalaria hirsuta* and yellow mosaic of *Cajanus cajan* are viruses that attack winged bean.

Diseases in rice bean are caused by *Corticum solani, Myrothecium roridum* and *Woroninella umbilicata*. In the Philippines, powdery mildew and rust occur sparingly. Cucumber mosaic virus also attacks the plant. However, relatively speaking, the rice bean is a pest-free crop and is even comparatively immune to most storage insects.

SEED-BORNE MYCOFLORA OF STORED SEEDS OF *PSOPHOCARPUS TETRAGONOLOBUS* AND *VIGNA UMBELLATA*: A CASE STUDY

The interaction between the seed and its mycoflora is complex. The seed mycoflora varies with different seeds, its moisture content, temperature, season, method of harvesting, threshing and storing.

Screening of seed mycoflora associated with *Vigna umbellata* and *Psophocarpus tetragonolobus* was characterised over a period of two years in post-harvest seeds during their storage at room temperature. The mycoflora associated with these samples has been studied in relation to fungal succession and its effect on seed viability. Non-surface sterilised (NSS) seeds were plated in sterilised petriplates (8.5 cm diameter) containing four layers of white blotter, soaked in sterilized water. These were incubated at 22±2°C under alternating cycles of 12 hours of near ultraviolet (NUV) light and darkness for seven days. Different fungi appearing on seeds were observed under stereobinocular microscope (40x) and compound microscope for identification and confirmation. For the detection of storage fungi, surface sterilized (SS) seeds with 2% chlorine solution were directly plated in moist chambers. Identification of the fungal species was carried out using monographs by Barnett (1960) and Singh et al. (1991). Isolations of various seed-borne fungi were made on Czapek-Dox agar medium.

Seed-borne mycoflora of *Vigna umbellata* showed the presence of a total of 41 species belonging to 19 genera in moist chamber plates. The number of fungi recorded in both the cultivars were more or less equal (34 fungal species belonging to 18 genera in cultivar RBL-1 and 35 fungal species belonging to 16 genera in cultivar RBL-6). However, the composition of fungi differed between the cultivars. Vaidehi (1997) reported the seed mycoflora of two maize cultivars.

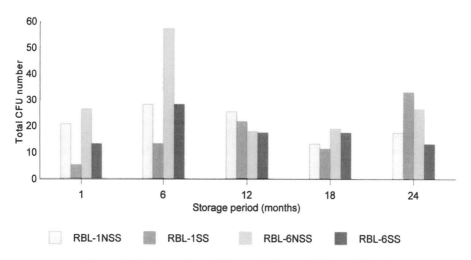

Figure 2A. Seed mycoflora of *Vigna umbellata* cv. RBL-1 and RBL-6.

The seed mycoflora superficially associated with the seeds of rice bean (cultivars RBL-1 and RBL-6) predominantly consisted of field fungi, namely *Aspergillus niger, A. flavus, Chaetomium globosum* and *Rhizopus nigricans*. The seed mycoflora of *Psophocarpus tetragonolobus*, winged bean (cultivar Dwarf mutant) showed the presence of a total of 38 fungal species belonging to 22 genera and predominantly consisted of *A. niger, A. flavus, A. nidulans, Macrophomina phaseolina* and *Fusarium oxysporum*. Dominant, moderately frequent and rarely frequent fungal species identified in each crop cultivars are given in Table 1. Misra (1989) studied mycoflora of an oilseed plant, *Buchanan lanzan* by blotter and agar plate methods and recorded more species of *Aspergillus*.

The mycoflora gradually increased in quality and quantity from the time of harvest to the time they were kept in storage. The numbers of fungi isolated in colony-forming units (CFU) were higher in seeds from kharif crop than those from rabi crop. Seasonal variation was noticed in the prevalence of different fungi on seed samples due to variation in weather conditions (Fig. 2A and 2B). *Alternaria alternata* was predominant in the rabi season and *Curvularia lunata* was prevalent in kharif season. Fungal populations increased in post-harvest seeds reaching a maximum in stored seeds. In general, some fungi (*Alternaria alternata, Curvularia lunata, Cladosporium herbarum, Penicillium* spp. and *Mucor* spp.) were absent in the latter part of the storage. This has also been reported by earlier workers (Conca et al., 1996; Hegde and Hiremath, 1987; Jain et al., 1982; Kumari and Karan, 1981; Mathur, 1954; Pattnaik and Narain, 1994; Purohit and Jamaluddin, 1993 and Vaidehi, 1992). The increase in population of microorganisms may be due to the sporulation of storage fungi contaminating through air and dust around the seeds. It is well established that the deterioration of seeds in storage is due to the associated mycoflora (Gupta and Saxena, 1984).

Succession of fungal species in storage depends on the spectrum of fungi originally present. The preponderance of certain fungi in stored seed depends on the interspecific fungal interaction in addition to the conditions prevailing in storage (Vaidehi, 1997). Significant among them are the moisture content, relative humidity, temperature and duration of storage.

Dutta (1989) studied seasonal variation of mycoflora with *Strychnos nux-vomica* seeds under storage and isolated eighteen fungal species. The maximum numbers of fungi were recorded during August and the minimum in November and this was related to high temperature and relative humidity. The dominant organisms were *Aspergillus flavus* and *A. niger*.

Lokesh and Hiremath (1993) studied the effect of relative humidity (RH) on seed mycoflora and nutritive value of red gram (*Cajanus cajan*). A higher seed moisture content increased fungal infestation, particularly by *Aspergillus* spp., of which *A. flavus* and *A. niger* were predominant at higher and lower moisture contents respectively. Higher relative humidity had a detrimental effect on the nutritive value of the seed. However, the incidence of *Fusarium moniliforme* decreased with time and increase in RH values. It is claimed that the colonisation of field fungi and intermediate flora is simultaneous before the grains are dry enough for the interface of storage fungi (Pelhate, 1981).

In the surface-sterilized seeds, treatment with chlorine solution reduced the number of saprophytic fungi (Figures 2A and 2B). Species of *Aspergillus* and *Penicillium*, commonly regarded as strong fungi, occurred with high incidence on unsterilized samples at the time of seed setting, however, they did not occur frequently on surface-sterilized ones. An increase in the number of pathogenic fungi was observed in surface-sterilized seeds.

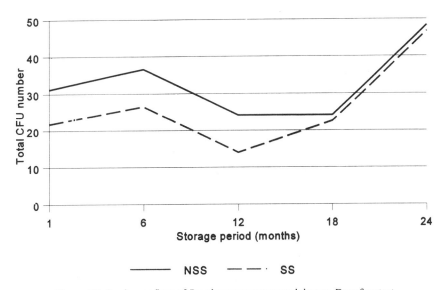

Figure 2B. Seed mycoflora of *Psophocarpus tetragonolobus* cv. Dwarf mutant.

Table 1: Isolation of mycoflora from seeds of different crop cultivars using blotter method.

Crop Cultivar	Fungal Frequency		
	Dominant	Moderately Frequent	Rarely Frequent
Rice bean RBL-1	*Aspergillus flavus* *A. niger, A. nidulans* *Chaetomium globosum*	*Rhizopus nigricans* *Mucor, A. fumigatus* *Penicillium citrinum* *P. chrysogenum* *Fusarium oxysporum* *Alternaria alternata* *Cladosporium herbarum* *Curvularia lunata*	*Chaetomium apiculata* *Epicoccum nigrum* *Macrophomina* *phaseolina* *Drechslera tetramera* *Stachybotrys atra* *Acremonium indicum* *Actinomycetes*
Rice bean RBL-6	*A. niger, A. flavus* *A. nidulans,* *Chaetomium globosum*	*A. niveus, A. aeneus,* *A. flavus* (yellow strain) *A. terreus* *Cladosporium herbarum* *Rhizopus nigricans* *Drechslera tetramera* *Penicillium chrysogenum* *Macrophomina* *phaseolina* *Chaetomium* *bostrychodes*	*Verticillium albo-atrum* *A. fumigatus* *Acremonium indicum* *A. candidus* *A. parasiticus* *Penicillium chrysogenum*
Winged bean Dwarf Mutant	*A. niger, A. flavus* *A. nidulans* (yellow, grey strain) *Sterile mycelium* *Macrophomina* *phaseolina* *Fusarium* *oxysporum*	*Curvularia lunata* *Curvularia pallescens* *Phoma lingam* *Alternaria alternaria* *Penicillium citrinum* *Penicillium chrysogenum* *A. aeneus, A. terreus* *Chaetomium globosum*	*Verticillium albo-atrum* *Thelavia sepedonium* *Stachybotrys atra* *Paecilomyces fusisporus* *A. sydowii* *F. sporotrichoides* *Mucor racemosus* *A. candidus* *Rhizopus nigricans* *Actinomycetes*

Germination in seeds infested with pathogenic fungi viz. *Fusarium oxysporum, F. solani, Macrophomina phaseolina* and *Phoma hibernica* was very poor and seeds were covered by profuse growth of the mycelium concerned. It is quite probable that the location of pathogen and intensity of seed infection may have a more direct bearing on seed germination and survival of seedlings than mere presence of pathogen on the seeds. Moreover, failure to germinate may also be due to unknown physical, physiological or genetic factors. Interactions with other fungi such as the saprophytes present on the seed surface as well as in the soil may also have an influence on germination and survival of seedlings. With a few exceptions, the percentage germination recorded by the blotter method was higher than percentage germination recorded in soil. There was negative correlation between germination percentage and infection percentage of seeds. However, surface-sterilized seeds exhibited better seed germination as compared with untreated seeds in all the experiments. Anwar et al. (1995) reported that high incidence of storage fungi (*Aspergillus, Penicillium* and *Rhizopus* spp.) in soyabean reduced seed germination potential *in vitro*.

CONCLUSIONS

In recent years, seed-borne diseases have adversely affected worldwide trading of seeds as well as research programmes of international research and germplasm centres. Seed health could be improved by producing seed in areas less affected by seed transmissible diseases, and by application of certification schemes to prevent the introduction of important pathogens through seed. Cataloguing of seed-borne pathogens is necessary and particularly to those pathogens that meet the criteria of limited distribution and of potential economic importance. Information on local pathogenic races/strains of seed-borne pathogens and their infection capacity to cultivars grown in other countries is also of great importance.

REFERENCES

Abou-Heilah, A.N., 1984, Seed-borne fungi of wheat and their control by seed treatments, *Indian Phytopathol.* 37: 656-659.

Agarwal, V.K.and Srivastava, A.K., 1981, A simpler technique for routine examination of rice seed lots for rice bunt, *Seed Tech. News* 11: 1-2.

Ahmad, S.I., Siddiqui, N.U. and Khan, M.A., 1989, Seed-borne fungi associated with seed lots of different paddy cultivars in Pakistan, *Pakistan J. Bot.* 21(2): 309-312.

Alcock, N.L., 1931, Notes on common diseases sometimes seed-borne, *Trans. Bot. Soc. Edinburgh* 30: 332-337.

Anwar, S.A., Abbas, S.F., Gill, M.M., Rauf, C.A., Mahmood, S. and Bhutta, A.R., 1995, Seed-borne fungi of soyabean and their effect on seed germination, *Pak. J. Phytopath.* 7(2): 184-190.

Ashokhan, A., Rambandan, R. and Emayavrambran, N., 1979, Influence of seed-borne fungi on germination and post emergence mortality of rice (AOT) and ragi (C07) seeds, *Indian J. Microbiol.* 19: 232-234.

Baird, R.E., Huber, D.M. and Mullinix, B.G., 1996, The mycobiota from seeds of Shrunken-2 (Sh 2) Sweet Corn, *Mycopathologia* 132(3): 147-154.

Balardin, R.S., Piva, C.A. Dal and Ogliari, P.J., 1992, Sanitation of bean seeds in Santa Catarina. Preliminary results, *Ciencia Rural* 22(2): 151-155.

Barnett, H.L. and Hunter, Barry B., 1972, *Illustrated genera of imperfect fungi*, p.241, Burgers Publishing Company, USA.

Basak, A.B., Mridha, M.A.U., Shamima, S., 1989, Mycoflora associated with different cultivars of brinzal seeds collected from the chittagong district, *Seed Res.* 17(1): 93-95.

Bokhary, H.A. and Naguib, K., 1983, Production of mycotoxins by seed-borne fungi from Saudi Arabia, *Egyptian J. Phytopathol.* 15: 55-64.

Bokhary, H.A., 1986, Seed mycoflora of wheat in Saudi Arabia, *Int. J. Tropical Plant Dis.* 4: 31-39.

Chandi Ram and Maheshwari, S.K., 1992, Seed-borne fungi of sponge gourd and their control, *Agricultural Science Digest* 12(2): 62-64.

Chandi Ram and Maheshwari, S.K., 1992, Seed mycoflora of bottle gourd and their control, *Agricultural Science Digest* 12(2): 79-81.

Christensen, C.M., 1951, Fungi on and in wheat seed, *Cereal Chem.* 28: 408-415.

Christensen, C.M., 1955a, Grain storage studies XVIII. Mold invasion of wheat stored for sixteen months at moisture contents below 15%, *Cereal Chem.* 32: 107-116.

Christensen, C.M., 1955b, Grain storage studies XXI. Viability and moldiness of commercial wheat in relation to the incidence of germ damage, *Cereal Chem.* 32: 507-518.

Christensen, C.M., 1957, Deterioration of stored grain by fungi, *Bot. Rev.* 23: 108-134.

Christensen, C.M., 1962, Invasion of stored wheat by *Aspergillus ochraceous*, *Cereal Chem.* 39: 100-106.

Christensen, C.M., 1963, Longevity of fungi in barley kernels, *Pl. Dis. Reptr.* 47: 639-642.

Christensen, C.M., 1973, Loss of viability in storage: microflora, *Seed Sci. & Technol.* 1: 547-562.

Christensen, C.M. and Drescher, R.F., 1954, Grain storage studies XIV. Changes in moisture content, germination percentage and moldiness of wheat samples stored in different portions of bulk wheat in commercial bins, *Cereal Chem.* 31: 206-216.

Christensen, C.M. and Kaufmann, H.H., 1965, Deterioration of stored grains by fungi, *Ann. Rev. Phytopath.* 3: 69-84.

Christensen, C.M. and Kaufmann, H.H., 1969, *Grain storage. The role of fungi in quality loss*, University of Minnesota, Minneapolis, pp. 7, 153.

Conca, G., Golinelli, F. and Porta-Puglia, A., 1996, Health of soyabean seed samples used in Italy, *Sementi Elette.* 42 (1): 3-6.

Doyer, L.C., 1938, *Manual for the determination of seed-borne diseases*, Wageningen, International Seed Testing

Association, 59 pp.

Dutta, G.R., 1989, Seasonal variation of mycoflora with *Strychnos nux-vomica* L. seeds under storage, *Indian J. Appl.and Pure Biol.* 4(2): 133-135.

Duval, J.W.J., 1909, The deterioration of Corn in storage, *USDA Bur. Plant Ind. Clr.* 43.

Gilman, J.C. and Barron, D.H., 1930, Effect of molds on temperature of storage grain, *Pl. Physiol.* 5: 565-573.

Gowda, D.N., Swamy, H.N., Setty, M.V.N. and Gowda, C.K., 1989, Studies on seed mycoflora of tobacco, *Farming Systems* 5 (1-2): 46-48.

Gupta, Rajni, 1996a, Screening of *Sorghum* cultivars for the elaboration of zearalenone by two toxigenic isolates of *Fusarium, Proc. Indian Natn. Sci. Acad* B62, (4): 277-280.

Gupta, Rajni, 1996b, Seed-borne fungi in *Sorghum, Plant Genetic Research Newsletter* 106 (1): 1-2.

Gupta, R.C. and Saxena, A., 1984, Efficacy of some fungicides on incidence of seed-borne fungi of *Eruca sativa* grown in Tarai region of Naintal, *Madras Agric. J.* 71: 512-515.

Gupta, I.J., Schmitthenner, A.F. and McDonald, M.B., 1993, Effect of storage fungi on seed vigour of soyabean, *Seed Sci.Technol.* 21(3): 581-591.

Handoo, M.L. and Aulakh, K.S., 1979, Control of seed-borne fungi by maize by coating seeds with antagonistic ones, *Seed Res.* 7: 151-156.

Hashmi, M.H., 1989, Seedborne mycoflora of *Capsicum annuum* L., *Pakistan J. Bot.* 21(2): 302-308.

Hegde, D.G. and Hiremath, R.V., 1987, Seed mycoflora of cowpea and its control by fungicides, *Seed Res.* 15: 60-65.

Jain, S.C., Singh, R.C., Sharma, R.C. and Mathur, J.R., 1982, Seed mycoflora of moth bean (*Vigna aconitifolia*) its pathogenicity and control, *Indian J. Mycol. & Pl. Pathol.* 12: 137-141.

Jayaweera, K.P., Wuesundera, R.L.C. and Medis, S.A. 1988. Seed-borne fungi of *Oryza sativa*, *Indian Phytopathol.* 41 (3): 355-358.

Kanapathipillai, V.S., Derris, N.M., 1988, The seed microflora of *Psophocarpus tetragonolobus*, four-angled bean and their pathogenic activity, in: *Movement of Pests and Control Strategies* (K.G. Singh, P.L. Manalo, S.S. Sastrrontemo, K.C. Chan, L.G. Lin, A.N. Ganapathi, P.S.S. Rahim, Durai and M.C. Doss, eds.), Selangor, Malaysia, Asean Plant Quarantine Centre and Training Institute, pp 311-319.

Khairnar, D.N. and Bhoknal, B.D., 1990, Biodeterioration in seeds by some seed moulds of pearl millet, *Geobios New Reports* 9(2): 197-198.

Khamees, M.A.F. and Schlösser, E., 1990, Seed-borne fungi on sesame in the Sudan, *Rifksuniversiteit Gent.* 55 (3a): 877-887.

Kowalik, M., 1989, Mycoflora of seed material of the selected breeding lines of red beetroot, *Roczniki Akademii Rolniczej W Poznaniu, Orgodnictwo* 194 (16): 137-148.

Kumari, V. and Karan, D., 1981, Seed mycoflora of *Abelmoschus esculentus* (L.) Moench: 1. Survey and enumeration, *Acta Botanica Indica* 17(2): 200-206.

Liang, L.Z., 1990, Seed-borne *Fusarium* of chilli and their pathogenic significance, *Acta Phytopathologica Sinic.* 20(2): 117-121.

Lokesh, M.S. and Hiremath, R.U., 1993, Effect of relative humidity on seed mycoflora and nutritive value of red gram (*Cajanus cajan* (Linn.) Millsp.), *Mysore J. Agric. Sci.* 27 (3): 268-271.

Maheshwari, R.K. and Mathur, S.K., 1987, Changes in lobia (*Vigna sinensis* Savi) seeds due to some Aspergilli, *Alexandria J. Agric. Res.* 32 (2): 289-293.

Malone, J.P. and Murkett, A.E., 1964, Seed-borne fungi. Description of 77 fungus species, *Proc. Int. Seed Test. Ass.* 29: 176-384.

Mathur, R.S., 1954, Diseases of pulse crops in Uttar Pradesh, *Agric. Anum. Husb.* 5: 24-28.

Mercer, P.C., 1994, Seed-borne pathogens of linseed in the UK in seed treatment: progress and prospects, in: *Proceedings, Symposium held at the University of Kent, Canterbury, 5-7 January, 1994* (T. Martin, ed.), Franham, UK.

Milner, M., Christensen, C.M. and Geddes, W.F., 1947, Grain storage studies VII. Influence of certain mold inhibitors on respiration of moist wheat, *Cereal Chem.* 24: 507-517.

Misra, N., 1989, Studies on the mycoflora of *Buchanania lanzan* spreng, *Agricultural Science Digest* 9(4): 214-216.

Moreno-Martinez, E., Vazquez-Badillo, M.E., Navarrete, R. and Ramirez-Gonzalez, J., 1994, Effect of fungi and chemical treatment on viability of maize and barley seeds with different storage characteristics., *Seed Sci. and Tech.* 22(3): 541-549.

Neergaard, P., 1977, *Seed Pathology,* The Macmillan Press Ltd., London and Basingstoke.

Neergaard, P., 1979, *Seed Pathology,* The Macmillan Press Ltd., London and Basingstoke.

Neergaard, P. and Saad, A., 1962, Seed health testing of rice, a contribution to development of laboratory routine testing methods, *Indian Phytopath.* 15: 85-111.

Noble, M., De Tempe, J. and Neergaard, P., 1958, *An Annotated List of Seed-borne Diseases*, Kew-CMI.

Noble, M. and Richardson, M.J., 1968, An annotated list of seed-borne diseases, *Proc. Inst. Seed Test. Assoc.* 33: 1-19.

Nwaiawu, M.Y., Efuivweb-were, B.J.O. and Princewill, P.J.T., 1995, Biodeterioration of African breadfruit

(*Treculia africana*) in south-eastern Nigeria and its control, *Trop. Sci.* 35 (1): 22-29.

Orsi, C., Chiusa, G. and Rossi, V., 1994, Further investigations on the relationship between the mycoflora of durum wheat kernels and the incidence of black point, *Petria* 4 (3): 225-235.

Patil, A.S., Pawar, B.H., Hapase, R.S. and Hapase, D.G., 1994, Note on seed mycoflora and seedling mortality in sugarcane, *Pl. Dis. Res.* 9 (1): 76-79.

Pattnaik, Mousumi and Narain, A., 1994, Studies on fungi associated with seeds of maize, wheat and ragi and their effect on seed germination, *Orissa. J. Agric. Res.* 7: 82-84.

Pelhate, J., 1981, Mycoflora of maize seeds. II Ear storage cryptogamie, *Mycologie* 2 (1): 61-84.

Pilt, J.I., Hocking, A.D., Kanjana Bhudhasmal, Miscamble, B.F., Wheeler, K.A. and Tanboon-Ek, P., 1994, The normal mycoflora of commodities from Thailand. 2. Beans, rice, small grains and other commodities, *Intl. J. Food Microbiol.* 23(1): 35-53.

Purohit, Mamta and Jamaluddin, 1993, Seed mycoflora of Palas (*Butea monosperma* (Lam) Taub.) during storage, *Seed Res.* 21(2): 126-127.

Ramadoss, S. and Sivaprakasan, K., 1989, Seed borne fungi of *Vigna unguiculata* and their control by seed treatments, *Madras Agric. J.* 76(1): 55-57.

Reddy, V.K. and Reddy, S.M., 1990, Elaboration of mycotoxins by seed borne fungi of maize, *Nat. Acad. Sci. Letters* 13 (1): 11-13.

Regina, M. and Tulasi, Raman, 1992, Biochemical changes in stored Caraway seeds due to fungi, *Indian Phytopathol.* 45(3): 384.

Richardson, M.J., 1979, *An Annotated List of Seed-Borne Diseases (Third Edition)*, Proc. Int. Seed Test. Ass., Wageningen, Netherlands.

Richardson, M.J., 1990, *An Annotated List of Seed-Borne Diseases (Fourth Edition)*, Proc. Int. Seed Test. Ass., Zurich, Switzerland.

Sao, R.N., Singh, R.N., Narayan, N., Kumar, S. and Prasad, B.K., 1989, Seed-borne fungi of vegetables belonging to Brassicaceae, *Indian Phytopathol.* 42: 538-543.

Semeniuk, G. and Barre, H.L., 1944, *Pathology and Mycology of Corn,* Iowa Agr. Exp. Sta. Rept. Part II, pp. 52-59.

Senapati, A.K. and Narain, A., 1990, A note on seed-borne diseases of wheat in Orissa *Orissa J. Agr. Res.* 3(1): 78.

Sereme, P., 1991, Diseases transmitted by bambarra groundnut seed in Burkina Faso, *Sahel PV Info* 32: 2-5.

Shafie, A.A. and Webster, E.J., 1981, A survey of seed-borne fungi of *Sorghum bicolor* from the Sudan, *Trans. Br. Mycol. Soc.* 77: 339-342.

Shah Rakesh and Jain, J.P., 1993, Seed mycoflora of mustard and its control, *Indian J. Mycol. Pl. Pathol.* 23(3): 291-295.

Shanaan, J.D., Leighly, C.E. and Boerner, E.G., 1910, *American Export Corn (Maize) in Europe,* USDA Bureau of Plant Industry, Circular 55.

Siddiqui, M.R., Mazumdar, A. and Gaur, A., 1974, Fungal flora associated with the seeds of cereals and vegetables in India, *Seed Res.* 2: 46-50.

Singh, B.K. and Saha, N.K., 1994, Impact of storage systems on mycoflora of *Madhuca longifolia* seeds, *Indian Phytopath.* 47(3): 266-269.

Singh, Kulwant, Frisvad, Jens C., Thrane ulf and Mathur, S.B., 1991, *An Illustrated Manual on Identification of some Seed-borne Aspergilli, Fusaria, Penicillia and their Mycotoxins,* Danish Govt Inst. of Seed Pathology for Developing Countries, Denmark.

Singh, Kulwant and Mukerji, K.G., 1994, Recent trends in classification of seed-borne fungi, in: *Vistas in Seed Biology Vol. 1* (Tribhuvan Singh and Pravin Chandra Trivedi, eds.), pp. 1-8, Printwell, Jaipur, India.

Sinha, D.C. and Prasad, R.K., 1989, Seed mycoflora of fenugreek and its control, *Indian Phytopath.* 43(1): 177-179.

Suryanarayan, D., 1978, *Seed Pathology (Seed-borne Diseases of some Important Crop Plants, their Identification and Control),* Vikas Publishing House, New Delhi.

Suryanarayan, D. and Bhombe, B., 1961, Studies on the fungal flora of some vegetable seeds, *Indian Phytopath.* 14: 30-31.

Suryanarayan, T.S. and Suryanarayan, C.S., 1990, Fungi associated with stored sunflower seeds, *J. Econ. Tax. Bot.* 14(1): 174-176.

Tseng, T.C., Tu, J.C. and Tzean, S., 1995, Mycoflora and mycotoxins in dry bean (*Phaseolus vulgaris*) produced in Taiwan and in Ontario, Canada, *Bot. Bull. Academia Sinica* 36(4): 229-234.

Upadhyay, R.K. and Mukerji, K.G., 1997, *Toxins in Plant Disease Development and Evolving Biotechnology,* Science Publishers Inc., New Hampshire, USA.

Vaidehi, B.K., 1992, Studies on fungi from pre-harvest, post-harvest and stored groundnut kernels, *Indian J. Microbiol. Ecol.* 2: 51-55.

Vaidehi, B.K., 1997, Seed mycoflora of maize: an appraisal, in: *New Approaches in Microbial Ecology* (J.P. Tewari, G. Saxena, N. Mittal, I. Tewari and B.P. Chamola, eds.), pp. 337-357, Aditya Books Private Limited, New Delhi, India.

Webster, N., 1962, *Webster's New Twentieth Century Dictionary,* The World Pub. Co., Cleveland and New York.
Zad, J., 1990, Mycoflora of sunflower seeds, *Rijksuniversiteit Gent* 55 (2a): 235-238.

ASSOCIATION OF PATULIN WITH MARKET SAMPLES OF DRY FRUIT SLICES OF QUINCES (*CYDONIA OBLONGA* MILL.) FROM JAMMU AND KASHMIR, INDIA

Yash Pal Sharma and Geeta Sumbali

Department of Biosciences
University of Jammu
Jammu-180 004
India

INTRODUCTION

Patulin, also known as expansin, penicidin, leucopin, tercinin, clavacin, claviformin and clavatin, is a mycotoxin that was first isolated by Chain et al. (1942) and was originally investigated because of its potential as an antibiotic. All the mycotoxins are, in general, important environmental pollutants synthesised by moulds as secondary metabolites possessing low molecular weight, non-antigenic properties and capable of eliciting a toxic response in man and animals. It is known that most of the antibiotics are in practice mycotoxins, the difference being one of degree rather than of kind. Although patulin does possess some desirable antimicrobial properties, it has also been found to be toxic to a wide range of other biological systems (rats, cats, mice and rabbits), on account of which its use as an antibiotic was abandoned (Stott and Bullerman, 1975).

Patulin is produced by some toxigenic moulds only under favourable environmental conditions which include suitable substrate, optimal temperature, water activity (a_w) and hydrogen ion concentration. Unfortunately, under current agronomic practices most of these parameters are met during pre-harvest, storage and processing operations of agricultural commodities which result in the production of this toxin. Several species of *Penicillium* and *Aspergillus* are known to produce patulin but the most active toxin producing strains belong to *Penicillium expansum, P. chrysogenum, P. roqueforti, P. aurantiogriseum, Eupenicillium brefeldianum* and *Aspergillus amstelodami* (Steiman et al., 1989). In addition, numerous other fungi imperfecti, a few mucorales, ascomycetes and basidiomycetes are also reported to synthesise this molecule.

PHYSICO-CHEMICAL CHARACTERISTICS OF PATULIN

Patulin, 4-hydroxy-4H-furo[3,2-C]pyran-2(6H)-one is a lactone of fungal origin with an empirical formula $C_7H_6O_4$ and molecular weight of 154. It is a colourless to white crystalline compound that melts at 110.5°C. Patulin is optically inactive and has an ultraviolet absorption peak at 276nm in alcohol (Katzman et al., 1944). It is soluble in water and polar organic solvents but is insoluble in petroleum ether. It is unstable in alkaline solution and loses its biological activity but retains stability in acid (Chain et al., 1942). Another notable feature about the chemical stability of patulin is that it is not destroyed by exposure to a temperature of 80°C for 10-20 minutes (Scott and Sommers, 1968). It is also reported to be resistant to thermal destruction at a pH range of 3.5-5.5 when heated up to 125°C (Lovett and Peeler, 1973).

Maximum production of patulin occurs at 20-25°C (Roland and Beuchat, 1984b). *Penicillium griseofulvum* is recorded to produce maximum patulin at 20-30°C and pH 3.5 with submerged and aerated conditions by orbital shaking (Sanchis et al., 1992). In the case of *P. expansum,* Podgorskar (1993a) reported maximum patulin production at 25°C and pH 6.0 on Czapek's Dox agar medium containing fructose as a carbon source. Water activity of 0.95 is also necessary for patulin production (Northolt et al., 1978).

Another interesting property of patulin is its reactivity with sulphydryl and amino compounds to form non-toxic adducts. Therefore foodstuffs rich in sulphydryl groups (cysteine, glutathione etc.) limit or reduce the amount of patulin by reacting with it and thus inactivating or destroying it (Ashoor and Chu, 1973a,b). Alcoholic fermentation of fruit juices and wine-making also destroy more than 99% of the patulin (Stinson et al., 1978). Addition of ascorbic acid and saccharose in the medium has also been reported to reduce patulin production by *P. expansum* in pears and apples at 25°C and modified atmosphere of 3% $CO_2/2.7\%$ O_2. Recently, Al-Garni and Al-Fassi (1987) have reported that in orange fruits infected with *Penicillium digitatum* and treated with a combination of heat (50°C for 10 minutes) and 0.2 M rad gamma irradiation, production of patulin was inhibited.

TOXICITY AND BIOLOGICAL EFFECTS OF PATULIN

Patulin is a molecule with cytostatic, antibiotic and antifungal properties, but its toxicity precludes its therapeutic use. It is toxic to a wide range of bacteria, fungi and protozoa (Ciegler et al., 1971). Besides, patulin possesses moderately high toxicity to mice, rats and chicks (Lovett, 1972). It is regarded as exhibiting mutagenic, neurotoxic and gastrointestinal effects in laboratory animals (Stott and Bullerman, 1975) and the most important target organs are the liver, spleen, kidney, skin, nervous system and reproductive system (Figure 1). The LD_{50} values for mice vary with the route of administration (Mckinnen and Carlton, 1980) whereas, in rats, the LD_{50} doses are slightly higher (15 to 25 mg/kg). Patulin exhibited teratogenicity when low levels (1-2 μg) were injected into chick embryos (Ciegler et al., 1976). Malformed feet, ankles, beaks, exencephaly and exopthalmia were the major abnormalities encountered. Choudhary et al. (1992) studies the effect of patulin on reproduction in pregnant albino rats and found that its oral administration showed anti-implantational and significant abortifacient activity.

Evidence for carcinogenicity of patulin came from the study of Dickens and Jones (1961) who found that localised tumours developed in rats when they were repeatedly injected with sublethal doses of this toxin. However, in view of the route of entry from contaminated food, oral administration of patulin, which is a more valid test, did not produce tumours (Osswald et al., 1978). Patulin is also known to inhibit various enzymes. It is a highly toxic inhibitor of RNA polymerase (Tashiro et al., 1979). Arafat and Musa (1995) reported that patulin

Table 1: Fungal species reported to be patulin producers.

Fungal Genera	Species	Source
Zygomycetes		
Cunninghamella	*C. bainieri*	Steiman et al. (1989)
Mortierella	*M. bainieri*	
Mucor	*M. hiemalis, M. racemosus* var. *globosus*	
Basidiomycetes		
Trametes	*T. squalens*	
Ascomycetes		
Byssochlamys	*B. nivea, B. fulva*	Frisvad and Samson (1991)
Gymnoascus	*G. reesii*	Steiman et al. (1989)
Eupenicillium	*E. brefeldianum*	
Sporormiella	*S. minimoides*	
Deuteromycetes		
Aspergillus	*A. clavatus, A. giganteus*	Gorst-Allman and Steyn (1979)
	A. terreus	
	A. amstelodami, A. echinulatus	Steiman et al. (1989)
	A. fumigatus, A. manginii	
	A. parasiticus, A. repens	
	A. variecolor, A. versicolor	
Penicillium	*P. digitatum*	Al-Garni and Al-Fassi (1997)
	P. patulum, P. roqueforti	
	P. expansum, P. variable	Gorst-Allman and Steyn (1979)
	P. claviforme, P. lapidosum	
	P. melinii, P. rugulosum	
	P. equinum, P. divergens	
	P. griseofulvum, P. leucopus	
	P. novae-zeelandiae, P. cyclopium	
	P. chrysogenum	
	P. aurantiogriseum, P. canescens	Steiman et al. (1989)
	P. citreonigrum, P. funiculosum	
	P. corylophilum, P. fellutanum	Vismer et al. (1996)
	P. coprobium, P. glandicola	Frisvad and Samson (1991)
	P. viridicatum	
Acremonium	*A. zeae*	Steiman et al. (1989)
Alternaria	*A. alternata, A. papaveris*	
Aschochyta	*A. imperfecta*	
Aurobasidium	*A. pullulans var. pullulans*	
Botrytis	*B. allii*	
Calcarisporium	*C. arbuscula*	
Chrysosporium	*C. pannorum*	
Cladobotryum	*C. verticillatum, C. varium*	
Cladorrhinum sp.		
Colletotrichum	*C. musae*	
Curvularia	*C. lunata*	
Fusarium	*F. culmorum, F. oxysporum*	
	F. proliferatum	
Oidiodendron	*O. echinulatum, O. tenuissimtum*	
Pseudodiplodia sp.		
Rhinocladiella	*R. atrovirens*	
Scopulariopsis sp.		
Sporotherix	*S. schenckii*	
Trichoderma	*T. pseudokoningii, T. polysproum*	
Trichophyton	*T. mentogrophytes*	
Paecilomyces	*P. lilacinus*	
	P. variotii	Frisvad and Samson (1991)

induced inhibition of protein synthesis in hepatoma tissue culture by inhibition of amino acid uptake into the cell and their incorporation into the proteins.

DETECTION OF PATULIN IN FOOD PRODUCTS

Analysis for patulin in food products can be accomplished primarily by thin layer gas and liquid chromatography. This toxin can be extracted from food by using different organic solvents especially ethyl acetate and ethyl acetate-water mixtures (Pohland and Allen, 1970).

Thin-layer chromatography (TLC) and high pressure liquid chromatography (HPLC) are the most commonly used methods to quantify patulin after extraction (Crosby, 1984). A review of TLC developing solvents and spray reagents for visualisation have been given by Scott (1974). The most sensitive reagents which can detect patulin at levels as low as 0.02-0.05 μg are ammonia and phenyl hydrazine hydrochloride (Scott and Kennedy, 1973). Other analytical methods include the use of gas-liquid chromatography, colorimetric methods and bioassay methods (Stott and Bullerman, 1975). Diphasic dialysis membrane procedure has also been adopted for detection of patulin (Prieta et al., 1994a).

NATURAL OCCURRENCE OF PATULIN IN DRY QUINCES (*CYDONIA OBLONGA*)

Almost all plant products can serve as substrate for fungal growth and subsequent mycotoxin formation under favourable environmental conditions. Realising the general importance of patulin as a mycotoxin being produced by *Penicillium, Aspergillus* and some other moulds which can occur in fruit products, an investigation was undertaken to elucidate the association of patulin with market samples of dry fruit slices of quinces from Jammu and Kashmir state of India.

Quince is an important rosaceous pome fruit cultivated in temperate regions of the Jammu and Kashmir state of India for its juicy, fragrant, astringent, expectorant and nutritionally rich fruits. Like most other fruits, quinces have a short shelf-life and perish on account of the growth of a variety of microorganisms. Therefore, in order to minimise post-harvest wastage, these fruits are usually sliced and sun-dried to approximately 12% moisture level. These dehydrated quince slices are cooked or eaten as such and also constitute a major part of the 'Prasad' of the holy shrine of Vaishno Devi (Katra) visited by pilgrims from all over India and abroad. However, due to the prevailing warm and humid conditions, many storage fungi cause extensive deterioration including discolouration, production of bad odour, off flavour, loss of nutritive value, various biochemical changes and production of diverse types of secondary metabolites (Sharma, 1997). Investigations on the surface mycoflora revealed many fungi harbouring on the dried slices both in the market (Sumbali and Sharma, 1997) and in storage (Sharma and Sumbali, 1996). Among the various fungi isolated, species of *Aspergillus* and *Penicillium* were found to be most abundant and fairly tolerant to extreme environments, growing at water activity levels that preclude most other fungi (Sharma, 1997). These fungal species are notorious for the production of various types of mycotoxins in general which have been implicated in eliciting chronic and sub-chronic toxicological effects in humans and susceptible animals.

Natural contamination of dry quince slices with patulin was analysed by following standard techniques developed by Stoloff et al. (1971) and 28% of the investigated samples were found to be positive for this mycotoxin. The amount of patulin detected in these samples varied from 0.250 mg/kg to 1.425 mg/kg of dry fruit which is far above the maximum permissible concentration (MPC) of 20-50 μg/L fixed by health authorities (Anonymous, 1980) and regulated by countries such as Sweden, Norway and Switzerland (Scott, 1985).

Aspergillus terreus isolates recovered from dry quince slices were also screened for patulin production *in vitro* following the method of Subramanian (1982) and using phenyl hydrazine hydrochloride as spray reagent for its detection and confirmation (Scott et al., 1972). 45% of the investigated *A. terreus* isolates were found to be patulin producers and the range varied from 1.625 mg to 2.3 mg/L.

NATURAL OCCURRENCE OF PATULIN IN VARIOUS OTHER FOOD SUBSTRATES

Substantial data is available worldwide on the natural occurrence of patulin in apple, orange, pears, banana, guava, tomato, commercial fruit juices, other fruit products, wines, vegetables, arecanut and pulse products (Table 2). The occurrence of patulin was first reported by Brian et al. (1956) from apples infected with *Penicillium expansum*. Wilson and Nuovo

Table 2: Occurrence of patulin in various foods.

Year	Place	Food Substrate	Conc. Range (μg/L / μg/kg)	Reference
1970	Canada	Rotten apple	1000	Harwig et al. (1973)
1971	USA	Apple	45000 max.	Wilson and Nuovo (1973)
1976-77	USA	Apple juice	10-350	Brackett and Marth (1979)
1978	Finland	Apple juice	5-72	Lindroth & Niskanen (1978)
1979	Spain	Apple and pears	1000-250000 900-10000	Burdaspal & Pinilla (1979)
1982	South Africa	Cereals & animal feedstuffs	-	Dutton & Westlake (985)
1982	India	Scented supari	-	Neelakanthan et al. (1982)
1982	-	Grape juice & wines	1-230	Altmayer et al. (1982)
1982	Italy	Fruit juices	5-15	Cavaliaro & Carreri (1982)
1983	-	Apples and pears	-	Gimeno & Martins (1983)
1984	-	Apple juice	-	Roland & Beuchat (1984a,b)
1984	Poland	Fruit wines	25-100	Czerwiecki (1985)
1984	Georgia	Apple cider	244-3990	Wheeler et al. (1987)
1985	-	Apple and grape juice	5-56	Mortimer et al. (1985)
1987	Spain	Cereals	75-127	Miguel & Andres (1987)
1988	India	Pearl millet	20-60	Girisham & Reddy (1992)
1990	Australia	Apple juice	5-629	Watkins et al. (1990)
1992	India	Guava	-	Madhukar & Redy (1992)
1992	-	Apple	-	Taniwaki et al. (1992)
1992	Australia	Apple, pear & mixed fruits	5-1130	Burda (1992)
1993	-	Grape juice	50000	Benkhemmar et al. (1993)
1994	-	Apple juice	100	Durakovic et al. (1993)
1994	Spain	Apple juice	10-170	Prieta et al. (1994a,b)
1995	-	Pears and apples	-	Paster et al. (1995)
1996	France	Apple cider	-	Herry & Lemetayer (1996)
1996	South Brazil	Apple juice	6.4-77	Machinsky & Midio (1996)
1996	India	Pulse products	130-800	Neeta Rani & Singh (1996)
1997	-	-	-	Al-Garni and Al-Fassi (1997)
1997	-	-	-	Lopez-Diaz & Flannigan (1997)

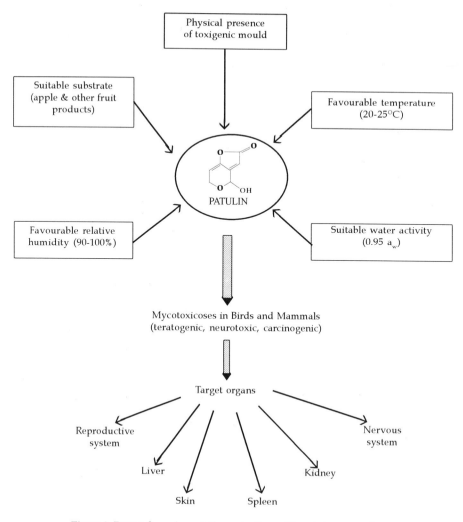

Figure 1. Factors favouring patulin production and its role in mycotoxicoses.

(1973) detected up to 45 mg of patulin per litre of organic apple cider samples that were obtained from processors where decayed apples were not sorted out prior to processing.

Patulin has also been reported in spontaneously mouldy bread and baked goods (Reiss, 1973). Buchanan et al. (1974) found patulin in the lesions of pears and stone fruits decayed by *Penicillium expansum.* Brackett and Marth (1979) surveyed apple juice from roadside stands in Wisconsin and found 10-350 μg of patulin per litre in 23 out of a total of 40 samples. Burdaspal and Pinilla (1979) observed patulin levels ranging from 1-250 mg/kg in apples and pears. Eller et al. (1985) analysed apple powder, moulded strawberry and moulded mandarin oranges and found 80 mg, 32 μg and 10-80 μg of patulin per kg respectively. Patulin was also detected in mixed feeds, fodder beets, soyabean meals, corn, barley, wheat, apples and apple products by Wizniewska and Piskorka-Pliszczynska (1983). They found 17 out of 30 fodder beet samples and 9 out of 20 apple product samples positive for patulin and the contamination level ranged from 20-6188 μg/kg and 60-1500 μg/kg respectively in these two commodities. Czerwiecki (1985) noted 6 out of 71 samples of Polish fruit wines positive for patulin and the detection level was 25-100 μg/L. Burda (1992) analysed as many as 328 apple, pear and mixed fruit products including jams, sauces, purees, jellies, diced apples and apple pulps from 38 Australian producers and found that 75 out of 258 juice concentrate samples contained patulin ranging from 5-50 μg/L and 73 samples were contaminated with 51-1130 μg of patulin per litre. However, of the 70 other remaining samples, patulin was detected in 18 at levels well below 50 μg. Machinsky and Midio (1996) found an average of 21% of the total samples of commercial apple juice from South Brazil to contain patulin in the range of 6.4-77 μg/L. During malting of barley and wheat, Lopez-Diaz and Flannigan (1997) reported patulin in all samples analysed with the equivalent of 22.4 mg/kg in one sample at 16 °C. It is evident that incidence of patulin contaminated is fairly high worldwide, but the level of contamination is usually low in substrates other than apple and apple products.

Patulin has also been detected from some food commodities of India where the tropical climatic conditions are very congenial for both fungal deterioration of food and consequent contamination with mycotoxins. Neelakanthan et al. (1982) analysed 147 samples of different arecanut items and found only one sample of scented supari to be contaminated with patulin. Girisham and Reddy (1988) observed a significant increase in the patulin content of pearl millet seeds stored at 90-100% relative humidity. Madhukar and Reddy (1992) detected patulin from *Aspergillus terreus* infected guava fruits. Recently, Neeta Rani and Singh (1996) screened 66 samples of pulse products and found seven of them to be positive for patulin (130-880 μg/kg).

Although there is currently no evidence to prove that patulin has the potential to produce adverse human health effects (Hopkins, 1993; Machinsky and Midio, 1995), the findings in animals emphasise the need for more research concerning this mycotoxin and its presence in various other raw fruits and their products.

ACKNOWLEDGEMENTS

The authors are grateful to Dr. S.P. McMormick, National Centre for Agricultural Utilization Research, USDA, Illinois, USA for generously providing a gift sample of patulin standard.

REFERENCES

Al-Garni, S.M. and Al-Fassi, F.A., 1997, Influence of gamma irradiation and heat on patulin production from orange fruits infected with *Penicillium digitatum, Adv. Food Sci.* 19: 59-62.

Altmayer, B., Eichhorn, K.W. and Plapp, R., 1982, Analysis of patulin in grape juices and wine, *Z. Lebensum. Unters. Forsch.* 175: 172-174.

Anonymous, 1980, Environmental Health Criteria, WHO publications, Document 11: Mycotoxins.

Arafat, W. and Musa, M.N., 1995, Patulin-induced inhibition of protein synthesis in hepatoma tissue culture, *Communication Mol. Pathol. Pharmacol.* 87: 177-186.

Ashoor, S.H. and Chu, F.S., 1973a, Inhibition of alcohol and lactic dehydrogenases by patulin and penicillic acid in vitro, *Food Cosmet. Toxicol.* 11: 617-624.

Benkhemmar, O., Fremy, J.M., Lahlou, H., Bompeix, G., El-Mniai, H. and Boubekri, C., 1993, Production of patulin by *Penicillium expansum* in table grape juice, *Sci. Aliments.* 13: 149-154.

Brackett, R.E. and Marth, E.H., 1979, Patulin in apple juice from roadside stands in Wisconsin, *J. Food Prot.* 42: 862-863.

Brian, P.W., Elson, G.W. and Lowe, D., 1956, Production of patulin in apple fruits by *Penicillium expansum, Nature* 178: 263-264.

Buchanan, J.R., Somer, N.F., Frotlage, R.J., Maxie, E.C., Mitchell, F.G. and Hsieh, D.P.H., 1974, Patulin from *Penicillium expansum* in stone fruits and pears, *J. Amer. Soc. Hort. Sci.* 99: 262-265.

Burda, K., 1992, Incidence of patulin in apple, pear and mixed fruit products marketed in New South Wales, *J. Food Prot.* 55: 796-798.

Burdaspal, P.A. and Pinilla, I., 1979, Naturally occurring mycotoxin contamination of apples and other projects: Application of a new analytical technique, *Alimentaria* 107: 35-37.

Cavallaro, A. and Carreri, D., 1982, Evaluation of patulin content in some Italian commercial fruit juices and homogenates. *Boll. Chim. Unione Ital. Lab. Prov.* 33: 527-534.

Chain, E., Florey, H.W. and Jennings, M.A., 1942, An antibacterial substance produced by *Penicillium claviforme, Brit. J. Exp. Pathol.* 29: 202-205.

Choudhary, D.N., Sahay, G.R. and Singh, J.N., 1992, Effect of some mycotoxins on reproduction in pregnant albino rats, *J. Food Sci. Technol.* 29: 264-265.

Ciegler, A., Detroy, R.W. and Lillehoj, E.B., 1971, Patulin, penicillic acid and other carcinogenic lactones, in: *Microbial Toxins, Vol. VI: Fungal Toxins* (A. Ciegler, S. Kadis and S.J. Ajl, eds.), Academic Press, New York, pp. 409-434.

Ciegler, A., Bechwith, A.C. and Jackson, L.K., 1976, Teratogenicity of patulin and patulin adducts formed with cysteine, *Appl. Environ. Microbiol.* 31: 664-667.

Crosby, N.T., 1984, Review of current and future analytical methods for the determination of mycotoxins, *Food Addit. Contaminant.* 1: 39-44.

Czerwiecki, L., 1985, Detection and determination of patulin in Polish fruit wines, *Rocz. Panstw. Zakl. Hig.* 35: 347-349.

Dickens, F. and Jones, H.E.H., 1961, Carcinogenic activity of a series of reactive lactones and related substances, *Br. J. Cancer* 15: 85-100.

Durakovic, S., Radic, B., Golem, F.V., Durakovic, Z., Beritic, T. and Lalic, L.M., 1993, The determination of patulin in apple juice, *Arhiv. Za Higijenu Rada i Toksikologiju* 44: 263-268.

Dutton, M.F. and Westlake, K., 1985, Occurrence of mycotoxins in cereals and animal feedstuffs in Natal, South Africa, *J. Assoc. Off. Anal. Chem.* 68: 839-842.

Eller, K.L., Masimenko, L.V., Sobolev, V.S. and Tutelyan, V.A., 1985, Procedure for evaluation of the patulin content of foods, *Zh. Anal. Khim.* 40: 542-544.

Frisvad, J.C. and Samson, R.A., 1991, Filamentous fungi in foods and feeds: ecology, spoilage and mycotoxins production, in: *Handbook of Applied Mycology, Foods and Feeds Vol. 3* (D.K. Arora, K.G. Mukerji and E.H. Marth, eds.), Marcel Decker Inc., New York.

Gimeno, A. and Martins, M.L., 1983, Rapid thin layer chromatographic determination of patulin, citrinin and aflatoxins in apples and pears and their juices, *J. Assoc. Off. Anal. Chem.* 66: 85-91.

Girisham, S. and Reddy, S.M., 1988, Influence of humidity on biodeterioration and patulin and terreic acid production by *Aspergillus terreus* in pearl millet, *J. Ind. Bot. Soc.* 67: 59-60.

Gorst-Allman, C.P. and Steyn, P.S., 1979, Screening methods for the detection of thirteen common mycotoxins, *J. Chromatography* 175: 325-331.

Harwig, J., Chen, Y.K., Kennedy, B.P.C. and Scott, P.M., 1973, Occurrence of patulin and patulin producing strains of *Penicillium expansum* in natural rots of apple in Canada, *Can. Inst. Food Sci. Technol. J.* 6: 22-25.

Herry, M.P. and Lemetayer, N., 1996, Liquid chromatography determination of patulin in French apple ciders, *J.A.O.A.C. Intl.* 79: 1107-1110.

Hopkins, J., 1993, The toxicological hazards of patulin, *Food Chem. Toxicol.* 31: 455-456.

Katzman, P.A., Hays, E.E., Cain, C.K., van Wyk, J.J., Reithel, F.J., Thayer, S.A., Doisy, E.A., Gaby, W.L., Carroll, C.J., Muir, R.D., Jones, L.R. and Wade, N.D., 1944, Clavicin, an antibiotic substance from *Aspergillus clavatus, J. Biol. Chem.* 154: 475-486.

Lindroth, S. and Niskanen, A., 1978, Comparison of potential patulin hazard in home-made and commercial apple products, *J. Food Sci.* 43: 1427-1432.

Lopez-Diaz, T.M. and Flannigan, B., 1997, Production of patulin and cytochalasin E by *Aspergillus clavatus* during malting of barley and wheat, *Intl. J. Food Microbiol.* 35: 129-136.

Lovett, J., 1972, Patulin toxicosus in poultry, *Poult. Sci.* 51: 2097-2098.

Lovett, J. and Peeler, J.T., 1973, Effect of pH on the thermal destruction kinetics of patulin in aqueous solution, *J. Food Sci.* 38: 1094-1095.

Machinsky, M. Jr. and Midio, A.F., 1995, Toxicological and analytical aspects of patulin in foods, *Rev. Farm Bioquim Univ. S. Paulo* 31: 1-19.

Machinsky, M. Jr. and Midio, A.F., 1996, Incidence of patulin in industrialized apple juice, *Alimentaria* 34: 61-

64.

Madhukar, J. and Reddy, S.M., 1992, Incidence of mycotoxin elaboration in guava fruits, *Geobios New Reports* 11: 188-189.

McKinnen, E.R. and Carlton, W.W., 1980, Patulin mycotoxicosis in Swiss ICR mice, *Food Cosmet. Toxicol.* 18: 181-189.

Miguel, J.A. and Andres, V., 1987, Metodo rapido de romatografia en capa fina de alta eficacia para la determinacion de patulina y aciso penicilico en granos, *Invest. Agrar. Prod. Prot. Veg.* 2: 225-235.

Mortimer, D.N., Parker, I., Shepherd, M.J. and Gilbert, J., 1985, A limited survey of retail apple and grape juices for the mycotoxin patulin, *Food Addit. Contaminant* 2: 165-170.

Neelakanthan, S., Balasaraswathi, R., Balasubramanian, T. and Indirejasmin, G., 1982, Incidence of patulin in scented supari, *Madras Afric. J.* 69: 326-328.

Neeta Rani and Singh, S., 1996, Natural occurrence of mycoflora and mycotoxins in some pulse products of human use, *J. Ind. Bot. Soc.* 75: 283-285.

Northolt, M.D., van Egmond, H.P. and Paulsch, W.E., 1978, Patulin production by some fungal species in relation to water activity and temperature, *J. Food Prot.* 41: 885-890.

Osswald, H., Frank, H., Komitowski, D. and Winter, H., 1978, Long term testing of patulin administered orally to Sprague Daulley rats and Swiss mice, *Food Cosmet. Toxicol.* 16: 243-247.

Paster, N., Huppert, D. and Barkai-Golan, R., 1995, Production of patulin by different strains of *Penicillium expansum* in pear and apple cultivars stored at different room temperatures and modified atmospheres, *Food Addit. Contaminant.* 12: 51-58.

Podgorskar, E, 1993a, Effect of *Penicillium expansum* culture conditions on patulin production, *Acta Microbiol. Pol.* 41: 89-95.

Podgorskar, E, 1993b, Effect of preservatives on patulin production by *Penicillium expansum, Acta Microbiol. Pol.* 41: 97-107.

Pohland, A.E. and Allen, R., 1970, Analysis and chemical confirmation of patulin in grains, *J. Assoc. Off. Anal. Chem.* 53: 686-687.

Pohland, A.E., Sanders, K. and Thorpe, C.W., 1970, Determination of patulin in apple juice, *J. Assoc. Off. Anal. Chem.* 53: 692-695.

Prieta, J., Moreno, M.A., Diaz, S., Suarez, G. and Dominguez, L., 1994a, Survey of patulin in apple juice and children's apple food by the diphasic dialysis membrane procedure, *J. Afric. Food Chem.* 42: 1701-1703.

Prieta, J., Moreno, M.A., Diaz, S., Suarez, G. and Dominguez, L., 1994b, Patulin as quality indicator in apple foods, *Alimentaria* 31: 75-80.

Reiss, J., 1973, Mycotoxine in Nahrungsmittein III Bildung von patulin auf verschieden Schnittbrotarten durch, *Penicillium expansum, Chem. Microbiol. Technol. Lebensm.* 2: 171-173.

Roland, J.O. and Beuchat, L.R., 1984a, Biomass and patulin production by *Byssochlamys nivea* in apple juice as affected by sorbate, benzoate, SO_2 and temperature, *J. Food Sci.* 49: 402-406.

Roland, J.O. and Beuchat, L.R., 1984b, Influence of temperature and water activity on growth and patulin production by *Byssochlamys nivea* in apple juice, *Appl. Environ. Microbiol.* 47: 205-207.

Sanchis, V., Lafuente, F., Vinas, I., Torres, M. and Canela, R., 1992, Influence of incubation conditions in the patulin production by *Penicillium griseofulvum* Dierks., *Rev. Iberoam. Micol.* 9: 88-90.

Scott, P.M., 1974, Collaborative study of a chromatographic method for determination of patulin in apple juice, *J. Assoc. Off. Anal. Chem.* 57: 621-625.

Scott, P.M., 1985, Background of patulin, in: *Mycotoxins: A Canadian Perspective* (P.M. Scott, H.L. Trenholm and M.D. Sutton, eds.), NRCC/CNRC Publ. Canada.

Scott, P.M. and Kennedy, B.P.C., 1973, Improved method for thin layer chromatograpic determination of patulin in apple juice, *J. Assoc. Off. Anal. Chem.* 56: 813-816.

Scott, P.M. and Sommers, E., 1968, Stability of patulin and penicillic acid in fruit and flour, *J. Afric. Food Chem.* 16: 483-485.

Scott, P.M., Miles, W.F., Toft, P. and Dube, J.G., 1972, Occurrence of patulin in apple juice, *J. Afric. Food Chem.* 20: 450-451.

Sharma, Y.P., 1997, Studies on the mycoflora and mycotoxins associated with cut dried fruits of quinces (*Cydonia oblonga* Mill.) from Jammu province, *Ph.D. Thesis*, University of Jammu, Jammu.

Sharma, Y.P. and Sumbali, G., 1996, Impact of different storage conditions, systems and associated mycoflora on the ascorbic acid content of dried fruit slices of quinces (*Cydonia oblonga* Mill.), *J. Ind. Bot. Soc.* 75: 1-3.

Steiman, R., Seigle-Murandi, F., Sage, L. and Krivobok, S., 1989, Production of patulin by micromycetes, *Mycopathologia* 105: 129-133.

Stinson, E.E., Osman, S.F., Huhtanen, C.N. and Bills, D.D., 1978, Disappearance of patulin during alcoholic fermentation of apple juice, *Appl. Environ. Microbiol.* 36: 620-622.

Stoloff, L., Nesheim, S., Yin, L., Rodricks, J.V., Stack, M. and Campbell, A.D., 1971, A multimycotoxin detection method for aflatoxins, ochratoxins, zearalenone, sterigmatocystin and patulin, *J. Assoc. Off. Anal. Chem.* 54: 91-97.

Stott, W.T. and Bullerman, L.B., 1975, Patulin: a mycotoxin of potential concern in foods, *J. Milk Food Tech.* 38: 695-705.

Subramanian, T., 1982, Colorimetric determination of patulin produced by *Penicillium patulum, J. Assoc. Off. Anal. Chem.* 65: 5-7.

Sumbali, G. and Sharma, Y.P., 1997, Mycoflora associated with cut dried fruits of quinces (*Cydonia oblonga* Mill.) from Jammu province, in: *Microbial Biotechnology* (S.M. Reddy, H.P. Srivastava, D.K. Purohit and S.R. Reddy, eds.), Scientific Publishers, Jodhpur.

Taniwaki, M.H., Hoenderboom, C.J.M., Vitali, A.D.A. and Eiroa, M.N.U., 1992, Migration of patulin in apples, *J. Food. Prot.* 55: 902-904.

Tashiro, F., Hirai, K. and Ueno, Y., 1979, Inhibitory effects of carcinogenic mycotoxins on deoxyribonucleic acid dependent ribonucleic acid polymerase and ribonuclease, *H. Appl. Environ. Microbiol.* 38: 191-196.

Vismer, H.F., Sydenham, E.W., Schlechter, M., Brown, N.L., Hocking, A.D., Rheeder, J.P. and Marasas, W.F.O., 1996, Patulin producing *Penicillium* species isolated from naturally infected apples in South Africa, *South African J. Sci.* 92: 530-534.

Watkins, K.L., Fazekas, G. and Palmer, M.V., 1990, Patulin in Australian apple juice, *Food Aust.* 42: 438-439.

Wheeler, J.L., Harrison, M.A. and Koehler, P.E., 1987, Presence and stability of patulin in pasteurized apple cider, *J. Food Sci.* 53: 479-480.

Wilson, D.M. and Nuovo, G.J., 1973, Patulin production in apples decayed by *Penicillium expansum, Appl. Microbiol.* 26: 124-125.

Wizniewska, H. and Piskorka-Piliszczynska, J., 1983, Patulin: methods for determination and occurrence, *Prezegl. Lek.* 40: 591-592.

COMPARISON BETWEEN *SERPULA LACRYMANS* FOUND IN THE INDIAN HIMALAYAS AND MOUNT SHASTA, CALIFORNIA

Jørgen Bech-Andersen

Hussvamp Laboratoriet
Bygstubben 7
DK-2950 Vedbæk
Denmark

INTRODUCTION

True Dry Rot Fungus in Denmark

The true dry rot fungus produces a white to greyish mycelium on wood. The decayed wood breaks up into a greyish brown cuboidal rot with cross-cracks every 7 cm. The mycelium may form hyphal strands which connect to sources of moisture and which may spread over mortar, insulation materials or soil if living in the wild. Fruiting bodies are more than 2 mm thick and are easily separated from the substrate. Size varies from 100 to 1000 cm². Brown spores measuring about 10 x 8 μm are formed on the surface.

A survey of the occurence of dry rot in houses showed that every second block of flats in Copenhagen built before 1920 was infected (Bech-Andersen, 1985) (Figures 1 and 2). The timbers infected were mainly of Scotch Pine (*Pinus silvestris*) and Norway Spruce (*Picea abies*).

The infected areas measured from a single square metre to several hundred square metres. The survey also showed that dry rot always occurs within 1 m from sources of chalk (calcium) or iron, which are found in building materials such as mortar, bricks, rock wool etc. (Bech-Andersen, 1987a). The fungus is able to transport water up to 6 m from the source of moisture. An earlier investigation had already shown that dry rot needs calcium to neutralize its production of oxalic acid (Bech-Andersen, 1987).

Iron is a component of the enzymes used to oxidize cellulose during the so-called Fenton reaction (Ritschkoff and Viitanen, 1989). During this process hydrogen peroxide (H_2O_2) is produced, which together with iron (Fe^{2+}) releases hydroxyl ions (OH^-), which in turn breaks down the cellulose.

The true dry rot fungus was reported to be ocurring naturally on Sweet Chestnut (*Castanea sativa*) in Denmark one hundred years ago (Rostrup and Weismann, 1898). However, an analysis of the specimens still held at the Royal Veterinary and Agricultural

University in Copenhagen has since proved them to belong to another species, *Serpula himantioides* (Bech-Andersen, 1992). There is at presence no evidence for the occurrence of *Serpula lacrymans* in the wild in Europe.

Figure 1. A staircase tower in Copenhagen after the stairs have fallen apart due to attack by *Serpula lacrymans*.

Figure 2. Fruiting bodies of *Serpula lacrymans* on wooden boards in a cellar in Copenhagen.

Wild Dry Rot in the Himalayas

Fifty years ago Dr K.A. Bagchee from Forest Research Institute (FRI) in Dehra Dun, India found specimens of *Serpula lacrymans* in the wild at Narkanda, Pulga and Kulu in the foothills of the Himalayas, about 400 km north of Delhi (Bagchee, 1954). The timber of houses in Simla infected by *Serpula lacrymans* originated from the forests around Narkanda. A survey of the area showed that dry rot occurred on stumps and fallen logs of West Himalayan Spruce (*Picea smithiana*), Himalayan Blue Pine (*Pinus griffithii*) and *Abies pindrow*. Fruiting bodies were seen during the period July to October at an altitude of about 3000 m.

In the following decades there were no records of dry rot from the area. However, it turned out that the scientists involved had turned their research area to another field, viz. leaf pathogens in nurseries.

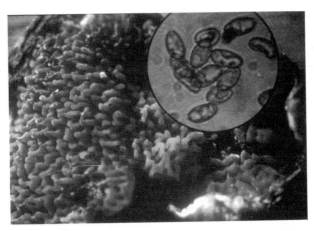

Figure 3. Fruiting body and spores (inserted) of *Serpula lacrymans* found at Narkanda in the Indian Himalayas in 1992.

Figure 4. Wood with a coarse cuboidal rot due to attack by *Serpula lacrymans* found at Narkanda in the Indian Himalayas in 1997.

At the museum in the FRI, Dehra Dun we found ten collections of dry rot from the period 1940 to 1955. Hussvamp Laboratoriet organized expeditions to the area in 1983, 1987 and 1990 without finding the fungus. However in 1992, in company with Jagjit Singh, a single fruiting body was found at Narkanda (Figures 3 and 4). This might indicate that *Serpula lacrymans* is not very common in its natural habitat where competitive species are more abundant than in houses.

Wild Dry Rot in America

In 1936 to 1942 Dr William Bridge Cooke was a ranger at Mt Shasta in California, where he stayed at Horse Camp Alpine Lodge, at an altitude of 2500 metres from May to October. During that period he surveyed the trees and fungi of the area. He found among others *Serpula americana* (Burt.) W.B. Cooke on White Fir (*Abies concolor*), Shasta Fir (*Abies magnifica* var. *shastensis*) and Mountain Hemlock (*Tsuga mertensiana*). Both *Serpula americana* and *Merulius vastator* were found on Ponderosa Pine (*Pinus ponderosa*). *Merulius vastator* is an old synonym of *Serpula lacrymans*, used together with *Serpula americana*. Later Cooke used the name *Serpula lacrymans* var. *himantioides*, which is a synonym of *Serpula himantioides*.

In 1994 representatives from Hussvamp Laboratoriet went to Mt Shasta in search of *Serpula lacrymans* var. *shastensis* and after a week succeeded in finding one infected log (Figures 5 and 6).

Figure 5. Fruiting body of *Serpula lacrymans* var. *shastensis* found on Mt Shasta in California in 1994.

Figure 6. Strand mycelium of *Serpula lacrymans* var. *shastensis* on decayed log of *Abies magnifica* var. *shastensis* on Mt Shasta. California.

Genetic Experiments

Louis Harmsen from the Danish Technological Institute in Copenhagen tried crossing Danish, Indian and American isolates to establish their relations (Harmsen, 1960). It turned out that they were all compatible and formed a dicaryotic mycelium with clamp connections. The American specimens however showed thinner fruiting bodies with more distant folds in the hymenium. In culture the mycelium was also more gracile than in the other isolates. Based on these facts Harmsen proposed the new variety *Serpula lacrymans* var. *shastensis*.

At the same time Harmsen showed that isolates of *Serpula himantioides* from Denmark, India and America were compatible with each other but not with *Serpula lacrymans*. These two species are therefore distinct.

Figure 7. Ships sailing from England to India carried 500 tons of limestone as ballast. The calcium of limestone is necessary for *Serpula lacrymans* to thrive.

Chemical Analysis

Analysis of soil samples from Narkanda in India and Mt Shasta in the USA showed the presence of calcium and iron in both places. However building materials such as mortar, concrete and rockwool have ten times as much of these substances.

Moisture Conditions

The optimal wood moisture content is as low as 20%. In the Himalayas decayed wood was often found with a moisture content above 100%. The optimal wood moisture content of 20% corresponds to a wood density of 0.5 in fresh wood. However the wood density of decayed wood with 100% moisture content was reduced to 0.1. This means that sufficient amounts of oxygen were available for the fungus despite the high moisture content because the decayed cellulose had been replaced by water and air.

Temperature

The optimal temperature for *Serpula lacrymans* is about 20°C. Both in the Himalayas and on Mt Shasta huge masses of snow fall before the frost becomes dominant, thus the temperature at the growth zone of the fungus is about 0°C the whole winter through.

Outdoors in Denmark the ground often freezes before snowfall, the latter often being very sparse. Dry rot mycelium exposed at ground surface in Denmark died at -10°C, while mycelium buried at frost-proof depth survived. The temperature in Danish houses is always above 0°C both in cellars and battlements where dry rot is found.

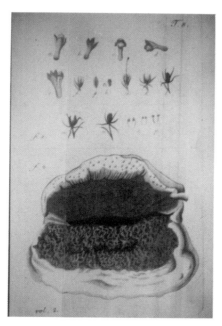

Figure 8. Drawing of a fruiting body of *Serpula lacrymans* from the publication in which it was first formally described by Wulfen from Austria in 1781.

Dispersal of *Serpula lacrymans*

The oldest information about dry rot in Europe comes from England, dating back to the time of James the First, 1566-1625, but the descriptions are rather uncertain (Coggins, 1980). After the fire of London in 1616 import of timber from India was initiated. It has turned out that the British ships carried 500 tons of limestone from Dover as ballast (Figure 7). The living conditions onboard for dry rot were thus excellent, namely the presence of wood, chalk and moisture.

The fungus was first validly described as *Boletus lacrymans* by Wulfen from Austria in 1781 (Figure 8). Later it was also described from Germany under the name *Merulius vastator* by Tode (1783).

In 1798 a ship was described from Woolwich which sunk by the weight of one man and where many brown fruiting bodies hung in festoons from deck to deck (Coggins, 1980).

Persoon described the species again in 1801 as *Merulius destruens,* probably based on finds from the Netherlands. In the same publication he also cited a find from Gotha in the former East Germany.

In 1812 James Sowerby reported on how *Boletus lacrymans* had attacked the ship Queen Charlotte, which had been launched in 1810 but needed a main repair only two years later (Coggins, 1980).

The presently used name *Serpula lacrymans* (Wulf.: Fr.) Schroet. was given by Schroeter in 1889 from Germany. Patouillard introduced the new combination *Gyrophana lacrymans* (Wulf.: Fr.) Pat. in 1897 probably based on finds from Tunisia. In 1912 Falk introduced a new name *Merulius domesticus* based on finds from Germany.

The American dry rot from Mt Shasta has probably never reached Europe, but it is exciting that *Serpula lacrymans* and *Serpula lacrymans* var. *shastensis* have changed so little that they are able to cross-breed 65 million years after the two continents separated and Europe and America drifted apart. Future DNA analysis may show the relationship more closely.

REFERENCES

Bagchee, K., 1954, *Merulius lacrymans* (Wulf.) Fr. in India, *Sydowia* 8: 80-85.

Bech-Andersen, J., 1985, Basische baustoffe und begrenzte feuchtigkeitsverhältnisse, Antworten auf die frage warum der echte hausschwamm nur in häusern vorkommt, *Material und Organismen* 20(4): 301-309.

Bech-Andersen, J., 1987a, The influence of the dry rot fungus (*Serpula lacrymans*) in vivo on insulation materials, *Material und Organismen* 22(3): 192-202.

Bech-Andersen, J., 1987, Production, function and neutralisation of oxalic acid produced by the dry rot fungus and other brown rot fungi, *IRG/WP* 1330: 2-15.

Bech-Andersen, J., 1992, The dry rot fungus and other fungi in houses, part 1, *IRG/WP* 92-2389: 1-18.

Bech-Andersen, J., Elborne, S.A. & Bech-Andersen, K., 1995, On the biotope of dry rot (*Serpula lacrymans*) in the wild, *IRG/WP* 95-10123: 116-127.

Bech-Andersen, J., Elborne, S.A., Goldie, F., Singh, J., Singh, S. & Walker, B., 1993, Ægte hussvamp (*Serpula lacrymans*) fundet vildtvoksende i Himalayas skove, *Svampe* 27: 17-28.

Coggins, C.R., 1980, *Decay of Timber in Buildings*, Rentokil Ltd., East Grinstead.

Cooke, W.B., 1943, Some basidiomycetes from Mount Shasta, *Mycologia* 35: 277-293.

Cooke, W.B., 1955, Fungi of Mount Shasta (1936-1951), *Sydowia* 9: 94-215.

Cooke, W.B., 1957, The genera *Serpula* and *Meruliporia*, *Mycologia* 49: 197-225.

Falk, R., 1912, Die Merulius-fäule des bauholzes, Neue untersuchungen über unterscheidung, verbreitung, entstehung und bekämpfung des echten hausschwammes, in: *Hausschwammforschungen*, Volume 6 (A. Möller, ed.), pp. 1-405, G. Fischer, Jena.

Harmsen, L., 1960, Taxonomic and cultural studies on brown-spored species of the genus *Merulius*, *Friesia* 6: 233-277.

Patouillard, N., 1897, *Catalogue Raisonné des Plants Cellulaires de la Tunisie, Fungi*, pp. 19-136, Paris.

Persoon, D.C.H., 1801, *Synopsis Methodica Fungorum*, Henricum Dieterich, Gottingae.

Ritschkoff, A. & Viitanen, H., 1989, Preliminary studies of the decay mechanism of some brown-rot fungi, *IRG/WP* 1402.

Rostrup, E. & Weismann, C., 1898, *Hussvampen, En Vejledning for Bygningshaandværkere og til brug i Tekniske Skoler*, Det Nordiske Forlag, Kjøbenhavn.

Schroeter, J., 1889, Hymenomyceten, in: *Kryptog. Fl. Schles.*, Volume 3 (F. Cohn, ed.), p. 466, Breslau.

Tode, H.J., 1783, Beschreibung des verwüstenden adernschwammes, *Merulius vastator*, *Abhandl. Halle Naturf. Ges.* 1: 351.

Wulfen, F.X. von, 1781, in: *Miscellanea Austriaca ad Botanicam, Chemiam, et Historiam Naturalem Spectantia*, Volume 2 (N.J. Jacquin, ed.), pp. 98-113, Vindobonae.

CONTRIBUTORS

K.R. Aneja
Department of Botany
Kurukshetra University
Kurukshetra-136 119
India

A. Archana
Swami Shraddhanand College
Alipur
Delhi-110 036
India

Daljit S. Arora
Department of Microbiology
G.N.D. University
Amritsar-143 005
India

Sapna Arora
Enbee Chemicals Ltd.
Bhopal
India

N.S. Atri
Botany Department
Punjabi University
Patiala-147 002
India

Manju Bansal
Applied Mycology Laboratory
Department of Botany
University of Delhi
Delhi-110 007
India

J. Bech-Andersen
Hussvamp Laboratoriet
Bygstubben 7
DK-2950 Vedbaek
Denmark

B.P. Chamola
Applied Mycology Laboratory
Department of Botany
University of Delhi
Delhi-110 007
India

Myank U. Charaya
195 Sadar Bazar
(Above Ideal Book Depot)
Meerut Cantt.-250 001
U.P. India

Dr. J.L. Faull
Biology Department
Birkbeck College
Malet Street
London WC1E 7HX
UK

B. Giri
Applied Mycology Laboratory
Department of Botany
University of Delhi
Delhi-110 007
India

Rajni Gupta
Applied Mycology Laboratory
Department of Botany
University of Delhi
Delhi-110 007
India

Dr N.S.K. Harsh
Tropical Forest Research Institute
P.O.: R.F.R.C.
Mandla Road
PIN 482 021
Jabalpur
Madhya Pradesh
India

Vandana Joshi
Applied Mycology Laboratory
Department of Botany
University of Delhi
Delhi-110 007
India

Rupam Kapoor
Applied Mycology Laboratory
Department of Botany
University of Delhi
Delhi-110 007
India

N.K. Kapse
National Agricultural Research Project
Sindewahi-441 222
Chandrapur
M.S., India

Ms Mandeep Kaur
Applied Mycology Laboratory
Department of Botany
University of Delhi
Delhi-110 007
India

Dr Sumeet Kaur
Applied Mycology Laboratory
Department of Biology
University of Delhi
Delhi-110 007
India

R.C. Kuhad
Department of Microbiology
University of Delhi South Campus
New Delhi-110 021
India

Raj Kumar
Department of Botany
Kurukshetra University
Kurukshetra-136 119
India

R.S. Mehrotra
Retired Professor
Department of Botany
Kurukshetra University
Kurukshetra-136 119
India

Neelima Mittal
Applied Mycology Laboratory
Department of Botany
University of Delhi
Delhi-110 007
India

Professor K.G. Mukerji
Applied Mycology Laboratory
Department of Botany
University of Delhi
Delhi-110 007
India

Dr J.W. Palfreyman
Dry Rot Research Group
University of Abertay Dundee
Bell Street
Dundee DD1 1HG
UK

Anjula Pandey
Applied Mycology Laboratory
Department of Botany
University of Delhi
Delhi-110 007
India

B.K. Rai
Plant Pathology Department
Jawaharlal Nehru Krishi Vishwa
  Vidyalaya
Jabalpur
Madhyar Pradesh
India

Renuka Rawat
Applied Mycology Laboratory
Department of Botany
University of Delhi
Delhi-110 007
India

Professor S.S. Saini
Department of Botany
Punjabi University
Patiala-147 002
India

Dr T. Satyanarayana
Department of Microbiology
University of Delhi South Campus
New Delhi-110 021
India

Geeta Saxena
Applied Mycology Laboratory
Department of Botany
University of Delhi
Delhi-110 007
India

Mamta Sharma
Applied Mycology Laboratory
Department of Botany
University of Delhi
Delhi-110 007
India

Yash Pal Sharma
Department of Biosciences
University of Jammu
Jammu-180 004
Jammu and Kashmir
India

Brajesh Kumar Singh
Department of Microbiology
University of Delhi South Campus
Benito Jaurez Road
New Delhi-110 021
India

Dr Jagjit Singh
Regional Director
Oscar Faber
Marlborough House
Upper Marlborough Road
St Albans Herts AL1 3UT
UK

V.K. Soni
Forest Pathology Division
Tropical Forest Research Institute
Jabalpur-482 021
Madhyar Pradesh
India

Dr. Geeta Sumbali
Department of Biosciences
University of Jammu
Jammu-180004
Jammu and Kashmir
India

Dr. N. White
Dry Rot Research Group
University of Abertay Dundee
Bell Street
Dundee DD1 1HG
UK

INDEX